Experimental Mass Spectrometry

TOPICS IN MASS SPECTROMETRY

Series Editor: David H. Russell
Texas A&M University

Volume 1 — EXPERIMENTAL MASS SPECTROMETRY
Edited by David H. Russell

A Continuation Order Plan is available for this series. A continuation order will bring delivery of each new volume immediately upon publication. Volumes are billed only upon actual shipment. For further information please contact the publisher.

Experimental Mass Spectrometry

Edited by

David H. Russell

Department of Chemistry
Texas A&M University
College Station, Texas

PLENUM PRESS • NEW YORK AND LONDON

Library of Congress Cataloging in Publication Data

Experimental mass spectrometry / edited by David H. Russell.
    p.   cm. —Topics in mass spectrometry, v. 1
    Includes bibliographical references and index.
    ISBN 0-306-44457-7
    1. Mass spectrometry. I. Russell, David H. (David Harold), 1947–   . II. Series.
QD96.M3E97   1994   93-48868
543′.0873—dc20   CIP

ISBN 0-306-44457-7

©1994 Plenum Press, New York
A Division of Plenum Publishing Corporation
233 Spring Street, New York, N.Y. 10013

Printed in the United States of America

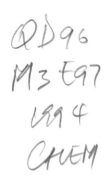
Contributors

Jeanette Adams • Department of Chemistry, Emory University, Atlanta, Georgia 30322

Steven C. Beu • Department of Chemistry and Biochemistry, The University of Texas at Austin, Austin, Texas 78712

John H. Bowie • Department of Chemistry, University of Adelaide, South Australia 5001, Australia

Marc Chevrier • Middle Atlantic Mass Spectrometry Laboratory, Department of Pharmacology and Molecular Sciences, The Johns Hopkins University School of Medicine, Baltimore, Maryland 21205

Tim Cornish • Middle Atlantic Mass Spectrometry Laboratory, Department of Pharmacology and Molecular Sciences, The Johns Hopkins University School of Medicine, Baltimore, Maryland 21205

Robert J. Cotter • Middle Atlantic Mass Spectrometry Laboratory, Department of Pharmacology and Molecular Sciences, The Johns Hopkins University School of Medicine, Baltimore, Maryland 21205

Charles G. Edmonds • Chemical Methods and Separations Group, Chemical Sciences Department, Pacific Northwest Laboratory, Richland, Washington 99352

Jean H. Futrell • Department of Chemistry and Biochemistry, University of Delaware, Newark, Delaware 19716

Tino Gäumann • Institute of Physical Chemistry, École Polytechnique Fédérale de Lausanne, Lausanne, Switzerland

David A. Laude, Jr. • Department of Chemistry and Biochemistry, The University of Texas at Austin, Austin, Texas 78712

Joseph A. Loo • Chemical Methods and Separations Group, Chemical Sciences Department, Pacific Northwest Laboratory, Richland, Washington 99352; *present address*: Parke-Davis Pharmaceutical Research, Ann Arbor, Michigan 48105

Rachel R. Ogorzalek Loo • Chemical Methods and Separations Group, Chemical Sciences Department, Pacific Northwest Laboratory, Richland, Washington 99352; *present address*: Department of Biological Chemistry, University of Michigan, Ann Arbor, Michigan 48109

Kenneth L. Rinehart • School of Chemical Sciences, University of Illinois, Urbana, Illinois 61801

Anil K. Shukla • Departments of Chemistry and Biochemistry, University of Delaware, Newark, Delaware 19716

Richard D. Smith • Chemical Methods and Separations Group, Chemical Sciences Department, Pacific Northwest Laboratory, Richland, Washington 99352

Justin G. Stroh • Pfizer Central Research, Groton, Connecticut 06340

Lynn M. Teesch • Department of Chemistry, Emory University, Atlanta, Georgia 30322

Harold R. Udseth • Chemical Methods and Separations Group, Chemical Sciences Department, Pacific Northwest Laboratory, Richland, Washington 99352

Rong Wang • Middle Atlantic Mass Spectrometry Laboratory, Department of Pharmacology and Molecular Sciences, The Johns Hopkins University School of Medicine, Baltimore, Maryland 21205

Cathy Wolkow • Middle Atlantic Mass Spectrometry Laboratory, Department of Pharmacology and Molecular Sciences, The Johns Hopkins University School of Medicine, Baltimore, Maryland 21205

Amina Woods • Middle Atlantic Mass Spectrometry Laboratory, Department of Pharmacology and Molecular Sciences, The Johns Hopkins University School of Medicine, Baltimore, Maryland 21205

Preface

Mass spectrometry underwent dramatic changes during the decade of the 1980s. Fast atom bombardment (FAB) ionization, developed by Barber and coworkers, made it possible for all mass spectrometry laboratories to analyze polar, highly functionalized organic molecules, and in some cases ionic, inorganic, and organometallic compounds. The emphasis of much of this work was on molecular weight determination. Parallel with the development of ionization methods (molecular weight mass spectrometry) for polar biological molecules, the increased mass range of sector and quadrupole mass spectrometers and the development of new instruments for tandem mass spectrometry fostered a new era in structural mass spectrometry. It was during this same period that new instrument technologies, such as Fourier transform ion cyclotron resonance, radio-frequency quadrupole ion trap, and new types of time-of-flight mass spectrometers, began to emerge as useful analytical instruments. In addition, laser methods useful for both sample ionization and activation became commonplace in almost every analytical mass spectrometry laboratory.

In the last 5 years, there has been explosive growth in the area of biological mass spectrometry. Such ionization methods as electrospray and matrix-assisted laser desorption ionization (MALDI) have opened new frontiers for both molecular weight and structural mass spectrometry, with mass spectrometry being used for analysis at the picomole and even femtomole levels. In ideal cases, subfemtomole sample levels can be successfully analyzed. Sample-handling methods are now the limiting factor in analyzing trace amounts of biological samples. The skeptics once viewed mass spectrometry of peptides as a pipe dream, but it is now common to see papers published on the mass spectrometry of proteins. What distinguishes recent years in mass spectrometry research is that nonspecialists now routinely operate the instruments and acquire their own data. The ESI and MALDI typically are performed on quadrupole and time-of-flight instruments rather than complex, more difficult magnetic sector instruments. In fact, experimental methods at the forefront of analytical mass spectrometry are so radically different from those of 20 years ago that new instruments, and new versions of old instruments, now dominate the field.

This book is a guide for the nonspecialist in the expanding field of experimental mass spectrometry, with chapters contributed by a diverse group. Ion chemistry plays an important role in mass spectrometry, and over the years, excellent reviews and textbooks have been assembled that cover both unimolecular and bimolecular ion chemistry. This text contains chapters on negative ion chemistry and organo-alkali metal ions which are growing areas as new methods of structural information develop. Collision-induced dissociation (CID) is an integral part of tandem/structural mass spectrometry, and the chapter by Shukla and Futrell is the most comprehensive work on the fundamentals of CID since the reference book *Collision Spectroscopy* edited by R. G. Cooks (Plenum, 1978). Two chapters, one by Gaumann and the other by Laude, Jr. and Beu, review fundamental aspects of Fourier transform ion cyclotron resonance (FTICR) mass spectrometry, and Cotter and coworkers review some aspects of protein structure using time-of-flight–based methods. Electrospray ionization and liquid chromotography are covered, respectively, in chapters by Smith and coworkers and Rinehart and Stroh.

Although this book encompasses some of the most important areas of modern biological mass spectrometry, obvious deficiencies are the lack of coverage of matrix-assisted laser desorption ionization and more complete coverage of applications of tandem mass spectrometry. These areas are, however, adequately covered in recent reference text and/or literature reviews. Volume 2 of *Topics in Mass Spectrometry* is now in the planning stage, and will be dedicated exclusively to MALDI of biomolecules. It is hoped that additional volumes will continue to address the rapidly developing field of biological mass spectrometry.

David H. Russell

College Station, Texas

Contents

5.   Experimental Fourier Transform Ion Cyclotron Resonance
      Mass Spectroscopy

      David A. Laude, Jr. and Steven C. Beu

6.   Elucidation of Protein Structure and Processing Using
        Time-of-Flight Mass Spectrometry

*Amina Woods, Rong Wang, Marc Chevrier, Tim Cornish,
Cathy Wolkow, and Robert J. Cotter*

7.   Electrospray Ionization Mass Spectrometry and
     Tandem Mass Spectrometry of Large Biomolecules

     Joseph A. Loo, Charles G. Edmonds, Rachel R.
     Ogorzalek Loo, Harold R. Udseth, and Richard D. Smith

8.   Liquid Chromatography/Fast Atom Bombardment
     Mass Spectrometry

     Justin G. Stroh and Kenneth L. Rinehart

1

The Fragmentations of (M-H)⁻ Ions Derived from Organic Compounds

An Aid to Structure Determination

John H. Bowie

1. INTRODUCTION

The fragmentation behavior of radical anions M⁻· derived by electron capture of organic molecules was studied extensively in the 1970s and early 1980s. The characteristic cleavages of most functional groups have been documented: This area of ion chemistry has been reviewed over the years.[1-4] More recently, the introduction of such techniques as negative ion chemical ionization,[5] secondary ion mass spectrometry,[6] laser-induced mass spectrometry,[7] ²⁵²Cf plasma spectrometry,[8] and fast atom bombardment mass spectrometry[9] has allowed the formation of deprotonated organic molecules (M-H)⁻. While such ions have been used primarily to provide molecular weight information,[4] on occasions, their fragmentations either in the ion source or under collisional activation conditions have been determined and they have yielded useful structural information. The examples shown in Figure 1 illustrate this point: (1) Nucleotides form (M-H)⁻ ions using a variety of techniques,[10-12] and these ions undergo ready P-O bond cleavage, e.g., N.A.D. (Formula *I*);[10] (2) C-O bond cleavage is also facile, e.g., glycoside *II*, undergoes four suc-

John H. Bowie • Department of Chemistry, University of Adelaide, South Australia 5001, Australia.

Experimental Mass Spectrometry, edited by David H. Russell. Plenum Press, New York, 1994.

I will set my burning passions free.

Many of us ignore our passions because we are scared to take a risk. We are afraid that if we express ourselves fully, we will be rejected. However, when we don't follow these drives, we often end up lamenting our lost dreams or resenting others.

At the age of sixteen, I lamented the fact that I was "too old" to learn to play the guitar. Then, when I was in my twenties, I met many great musicians who started playing when they were sixteen. Of course, I again told myself that it was too late to start. Hence, I am not a musician.

If you have a burning desire, set it free. Usually it's something you'll be good at, and you owe it to yourself and the world to share your gifts. Additionally, if you hold a torch for someone, tell him or her. It may feel scary, but without risking it, you'll never know what might have been. A blow to the ego fades a lot more quickly than regret. Life is not a dress rehearsal, so play for real!

cessive losses of glycose from the $(M-H)^-$ ion,[13] while the $(M-H)^-$ ion of avilamycin undergoes the cleavage shown in *III*;[14] (3) collisional activation of peptide $(M-H)^-$ ions yield sequence information, e.g., peptide *IV*,[15] and β-lactam *V* [negative ion cleavages ($-$), positive ion cleavages of the corresponding molecular radical cations ($\sim\!\sim$)];[16] and finally (4) the collision-induced dissociations of carboxylate ion *VI*.[17]

The examples cited in Figure 1 raise the question of whether the noted negative ion cleavages are characteristic of specific compounds, or of particular functionality within those compounds. Within the past 5 years, a start has been made with systematic studies of the fragmentation behavior of even-electron organic negative ions. This *selective* review examines (1) the collision-induced dissociations of $(M-H)^-$ ions of organic compounds, (2) the structures of some of the more interesting product ions, and (3) in conclusion, classifies fragmentations into particular reaction types. The literature has been scanned to the end of 1989.

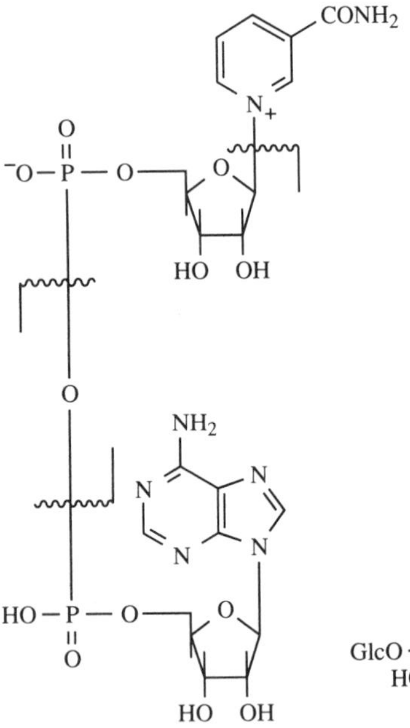

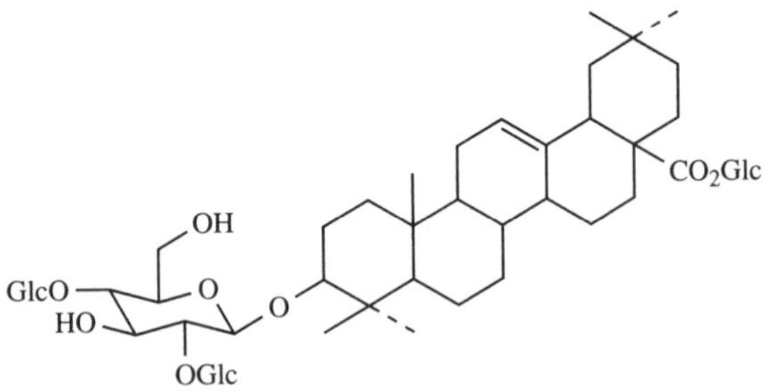

Figure 1. Fragmentations of $(M-H)^-$ ions of some naturally occurring molecules.

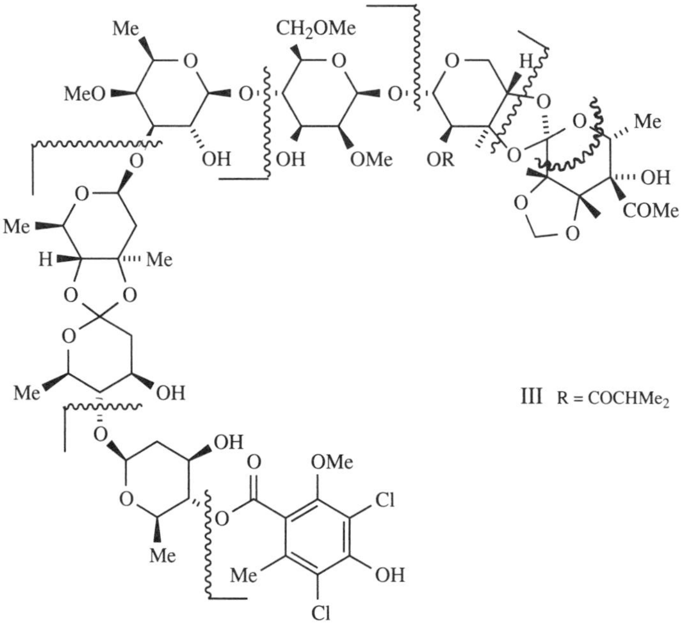

III R = COCHMe₂

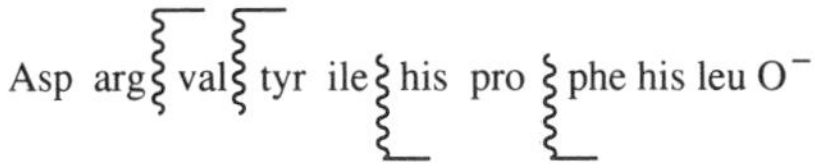

Asp arg⧸ val⧸ tyr ile⧸ his pro ⧸ phe his leu O⁻ IV

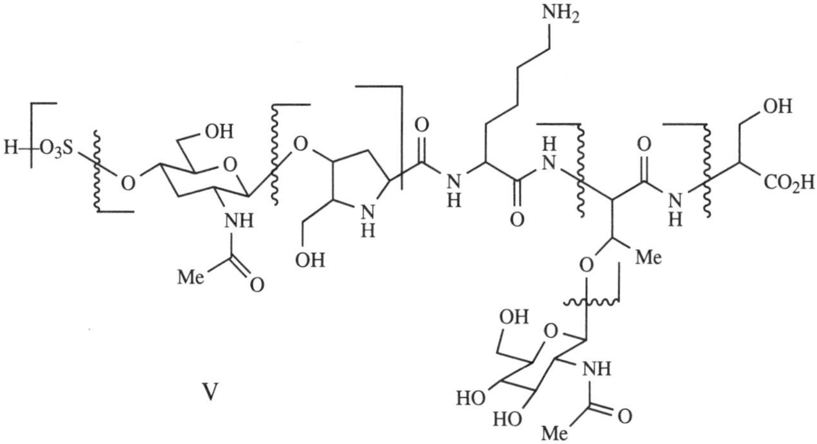

V

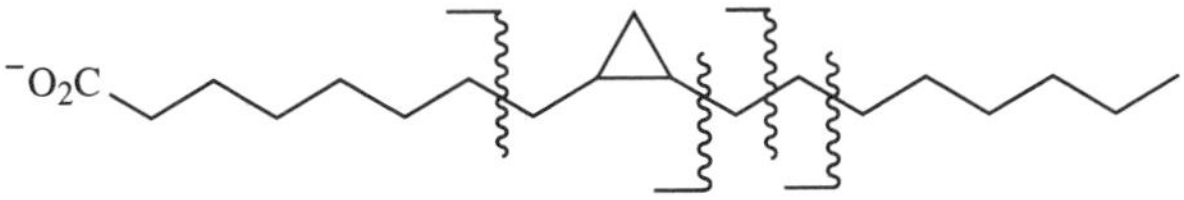

VI

Figure 1. *(Continued)*

2. FRAGMENTATION OF ORGANIC FUNCTIONAL GROUPS

All of the collisional activation mass spectra (MS/MS) recorded as figures in this chapter were determined with the Adelaide VG ZAB 2HF instrument, using either HO^- or NH_2^- negative ion chemical ionization (NICI) or fast atom bombardment mass spectrometry (FABMS). Throughout this chapter, stepwise mechanisms are used whenever possible to rationalize fragmentation processes. Stepwise processes have been shown to operate in a number of cases,[18–22] but a concerted reaction has also been reported.[23] Intermediates in reaction pathways are, for convenience, indicated as anion/neutral complexes, even though radical/radical anion complexes[24–26] are possible in certain cases.*

2.1. Hydrocarbons and Halocarbons

Simple alkanes are so weakly acidic that they are not readily deprotonated by the strongest accessible base NH_2^- (e.g., ΔH_{acid}° CH_4 and NH_3 are 416[27] and 403.5[28] kcal mol^{-1}, respectively). In any case, electron affinities of most simple alkyl radicals are negative, so the anions are unstable with respect to their radicals. Several notable exceptions are the methyl and cyclopropyl anions (electron affinities of Me· and cyclo C_3H_5· are $+1.8$[27, 29] and $+7.9$[30] kcal mol^{-1}, respectively); these ions are best produced indirectly by decarboxylation of RCO_2^- ions.[29]

$$(C_6H_4)^-C(Et) = CH_2 + CH_4 \quad (1)$$

(a)

$$[(PhCH = CH_2)C_2H_3^-] \longrightarrow (C_6H_4)^-CH = CH_2 + C_2H_4 \quad (2)$$

Simple olefins and benzyl derivatives are readily deprotonated. Although the benzyl anion undergoes loss of H· only on collisional activation,[31] substituted benzyl anions show characteristic fragmentations. For

---

* The question is whether AB^- fragments to products through $[A^-(B)]$ or $[A\cdot(B^-\cdot)]$ in a stepwise process. The primary consideration is likely to be the relative electron affinities of A· and B. If the electron affinity of A· is more positive than that of B, then $[A^-(B)]$ should be formed. If in contrast the electron affinity of A· is negative and of B is positive, then there is a possibility that fragmentation may proceed through $[A\cdot(B^-\cdot)]$.

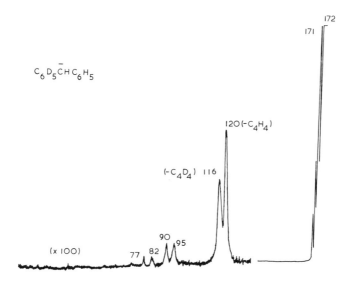

Figure 2. CA HO⁻/NICI spectrum of $C_6D_5\overset{-}{C}HC_6H_5$. VG ZAB 2HF instrument. Electric sector scan. Collision gas helium (10% reduction in main beam). Experimental conditions the same for all figures following.

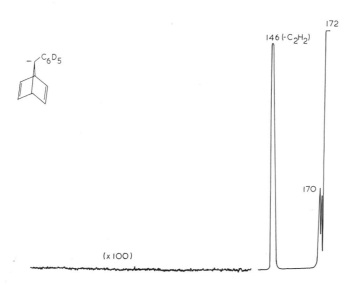

Figure 3. CA HO⁻/NICI spectrum of deprotonated 4-(Phenyl-D_5)-norbornadiene.

example, $Ph\bar{C}Et_2$ loses methane, and isotope effect studies indicate that (1) this loss involves a terminal methyl group and a ring hydrogen and (2) removal of the ring hydrogen is rate determining ($H/D = 3.75$). This evidence suggest that either the reaction is synchronous, or alternatively, it is stepwise (see Equation 1), with the second step being rate determining. In addition, if the reaction is stepwise, then the methyl anion is tightly bound in complex a, otherwise major deprotonation would occur at the more acidic allylic position.[32] Similarly, the characteristic fragmentation of deprotonated 4-phenylbutene is loss of ethene (see Equation 2).[33]

The collisional activation mass spectra of deuteriated derivatives of deprotonated diphenylmethane and phenylnorbornadiene are illustrated in Figures 2 and 3. The diphenylmethane spectrum is characterized by loss of C_4H_4, a reaction that occurs without any scrambling of carbon or hydrogen on the phenyl rings.[32] Such fragmentations are also observed for other PhXR systems ($X = Si^{[34]}$ and $P^{[35]}$) and they are best rationalized by fragmentation through a Dewar benzene (see Equation 3). In contrast, the isomeric norbornadienyl system (Figure 3) shows specific loss of acetylene, presumably by the retro process shown in Equation 4.[36]

$$R-\bar{X}\text{—}\langle\!\langle\rangle\!\rangle \longrightarrow R\bar{X}C \equiv CH + CH_4 \qquad (3)$$

$$\text{[bicyclic structure with Ph]} \longrightarrow \text{[cyclopentadienyl with Ph]} + C_2H_2 \qquad (4)$$

$$CD_3O^- + CH_3CH_2F \begin{array}{l} \xrightarrow{37\%} F^- + CD_3OH + C_2H_4 \qquad (5) \\ \xrightarrow{29\%} [F^-...HOCD_3] + C_2H_4 \qquad (6) \\ \xrightarrow{34\%} {}^-CH_2CH_2F + CD_3OH \qquad (7) \end{array}$$

$$\text{[Newman-type structure with F, H]} \;\rightleftharpoons\; \overset{F^-}{}\; H_2C = CH_2 \qquad (8)$$

Reaction of alkyl halides with strong base in the gas phase generally effects either S_N2 or elimination reactions. Detection of the deprotonated species is not normal, but alkyl fluorides are exceptions to this generalization. The reactions of CD_3O^- with ethyl fluoride are shown in Equations

5–7 with branching ratios shown as percentages. Elimination products are shown in Equations 5 and 6, the β-fluoroethyl anion in Equation 7.[37] The reason for the stability of the β-fluoroethyl beta anion has evoked some controversy[38] but one explanation lies in the negative hyperconjugation argument (see Equation 8).[39] The fragmentation behaviour of such species has not been determined, but it is likely that the β-fluoroethyl anion will form F^- and C_2H_4 on collisional activation.

2.2. Alcohols and Phenols

Alkoxide ion fragmentations have been the subject of a number of studies.[18,22,24,25,40–42] The prototypical case involves the loss of H_2 from EtO^-. A consideration of *ab initio* calculations and deuterium isotopic effect experiments suggest the operation of a 1,2 stepwise elimination (see Equation 9) which proceeds through hydride ion/acetaldehyde complex b. Steps A and B are both rate determining.[18] The analytical applicability of alkoxide negative ion spectra is illustrated for the particular example of $PhCH(Me)O^-$, which undergoes competitive losses of methane, acetaldehyde or benzene.[41] These losses are rationalized in Equations 10–12, *viz*, in Equation 10, the bound methyl anion [one of the strongest bases known $(\Delta H^\circ_{acid} = 416 \text{ kcal mol}^{-1})$[27]] of ion complex c deprotonates the phenyl ring. Complex d may either dissociate to form the phenyl anion (see Equation 11),* or the bound phenyl anion may deprotonate acetaldehyde (see Equation 12) [$\Delta H_{acid}C_6H_6$ and MeCHO are 401[43] and 366[44] kcal mol⁻¹, respectively].

$$MeCH_2O^- \xrightarrow{A} [H^-(MeCHO)] \xrightarrow{B} [H_2(CH_2CHO)^-] \longrightarrow (CH_2CHO)^- + H_2 \quad (9)$$

$$[Me^-(PhCHO)] \longrightarrow (C_6H_4)^-CHO + CH_4 \quad (10)$$
(c)

$$\begin{array}{c} Ph \\ \diagdown \\ CH-O^- \\ \diagup \\ Me \end{array}$$

$$[PH^-(MeCHO)] \longrightarrow (C_6H_5)^-CHO + MeCHO \quad (11)$$
(d)
$$\longrightarrow (CH_2CHO)^- + C_6H_6 \quad (12)$$

(e)   Ph ⁻ (f)   (g)

---

* In contrast, the corresponding methyl anion is not formed from ion complex c: a contributing factor may be the relative electron affinities of Me· and C_6H_5 ($+1.8$[27] and $=23.8$[43] kcal mol⁻¹).

One of the interesting features of RCH_2O^- ions is that the loss of CH_2O often yields R^- when the electron affinity of $R\cdot$ is positive. This has been noted when R is $CH_2=CHCH_2,^{[45]}$ $C_6H_5,^{[41]}$ $PhCH_2,^{[41]}$ and $HSCH_2.^{[46]}$ This is a useful reaction and often produces ions with potentially interesting structures. For example, (1) the loss of CH_2O from $CH_2=CH(CH_2)_3O^-$ gives the base peak of the spectrum;$^{[45]}$ Does the product ion correspond to the β-vinylethyl anion e (whose electron affinity should be small), or to some more stable and rearranged species? (2) The ion $Ph(CH_2)_3O^-$ similarly loses $CH_2O;^{[41]}$ Is the product ion the β-phenylethyl anion f or spiro ion g?

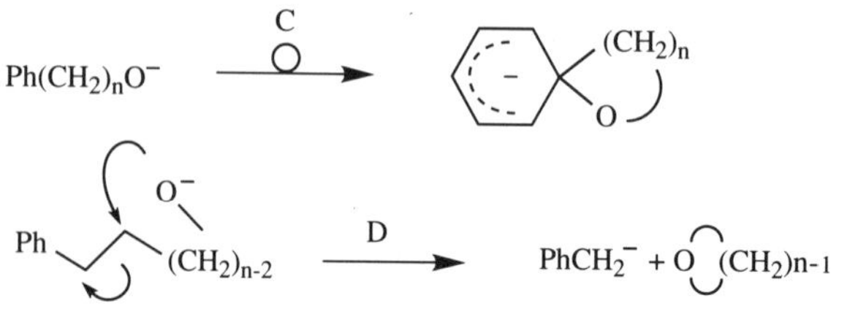

$$\text{Ph}(CH_2)_nO^- \xrightarrow{\quad C \quad}$$

$$\xrightarrow{\quad D \quad} PhCH_2^- + O(CH_2)_{n-1} \qquad (13)$$

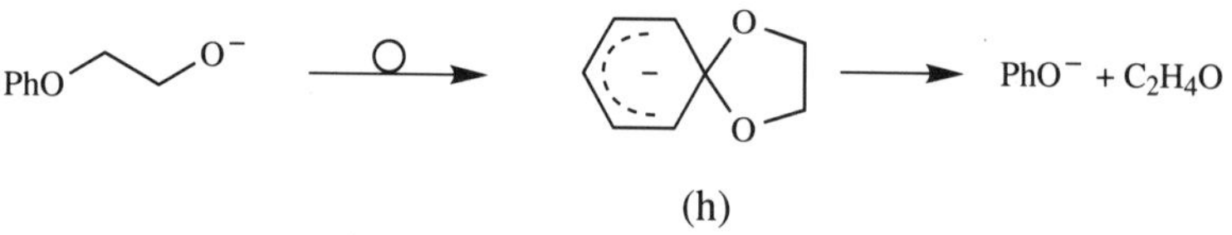

$$PhO \xrightarrow{\quad O \quad} \xrightarrow{\qquad} PhO^- + C_2H_4O \quad (14)$$

(h)

$$\xrightarrow{\qquad} PhO^- + C_3H_6O \qquad (15)$$

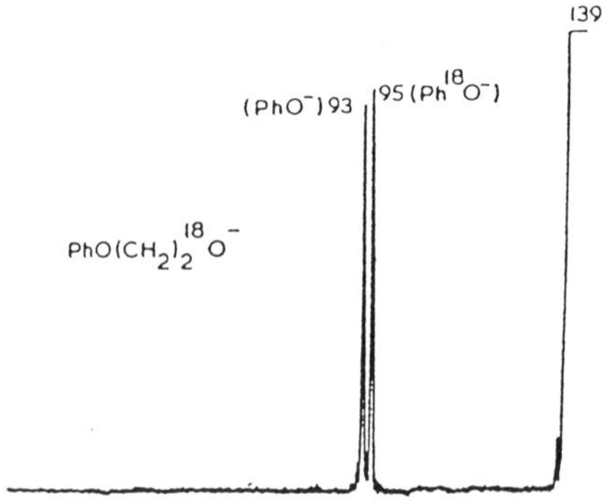

Figure 4. CA NH_2^-/NICI spectrum of $PhO(CH_2)_2{}^{18}O^-$.

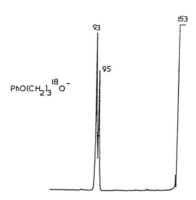

Figure 5. CA NH_2^-/NICI spectrum of $PhO(CH_2)_3{}^{18}O^-$.

Ions $Ph(CH_2)_nO^-$ ($n = 3$–5) fragment to yield $PhCH_2^-$.[41] This could be a fragmentation that occurs remote from, and uninfluenced by the charged site (see Section 2.5). Alternatively, it could occur by nucleophilic aromatic substitution [in this case the Smiles rearrangement[47] (pathway *C*, Equation 13] or by an $S_N i$ displacement reaction (pathway *D*, Equation 13). Since it is difficult to distinguish between these pathways in this case, cognate systems were studied in which a definitive answer may be given. Ions $PhO(CH_2)_nO^-$ ($n = 2$-5) fragment to give the phenoxide ion PhO^- as the sole product ion. ^{18}O and ^{13}C labeling show that when $n = 2$, the reaction proceeds exclusively by the Smiles rearrangement (see Equation 14). Figure 4 shows very nearly equal amounts of $Ph^{16}O$ and $Ph^{18}O^-$ demanding a symmetrical intermediate (e.g., h). When $n = 3$, the Smiles rearrangement and the $S_N i$ reaction (see Equation 15) compete (see Figure 5), and when $n > 3$ only the $S_N i$ reaction occurs. Similar results are obtained for systems $PhS(CH_2)_nO^-$ ($n = 2$-5).[48]

$$C_6H_5O^- \diagdown \diagup \begin{array}{l} C_5H_5^- + CO \qquad\qquad (16) \\[2em] C_5H_4{}^{\bullet-} + CHO^{\bullet} \qquad (17) \end{array}$$

Little work has been reported for simple phenols. The phenoxide ion itself loses CO(C-1) (see Equation 16) and CHO·(C-1 together with an *ortho* H) (see Equation 17).[48] The spectrum of deprotonated juglone (Figure 6) shows successive losses of CHO· and two molecules of CO[49].

2.3. Ethers

Examples II and III cited in the Introduction indicated that cleavage of C-O bonds to form alkoxide ions is both characteristic and facile in the

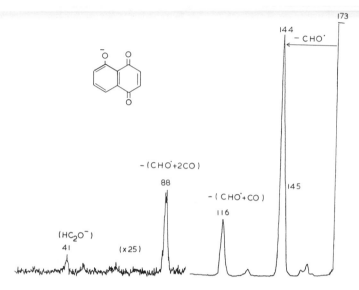

Figure 6. CA NH$_2^-$/NICI spectrum of the (M-H)$^-$ion of juglone.

negative mode. Such product ions could be formed by any one of a number of mechanisms under FAB or NICI conditions in the source of mass spectrometer, e.g., simple cleavage, S$_N$2 displacement, or elimination. The fragmentation behavior of simple deprotonated ethers has now been widely studied and shows a rich ion chemistry: Some reactions are unique to the gas phase, while others show striking analogy to the reactivity of ethers under base-catalyzed conditions in solution.

$$HO^- + EtOEt \longrightarrow EtO^- + C_2H_4 + H_2O \tag{18}$$

$$MeO^- + Me_3SiCH_2CH_2CHO \rightarrow (MeCOCH_2)^- + Me_3SiOMe \tag{21}$$

$$MeO^- + MeCO_2\!-\!\!\triangleleft \longrightarrow (MeCOCH_2)^- + MeCO_2Me \tag{22}$$

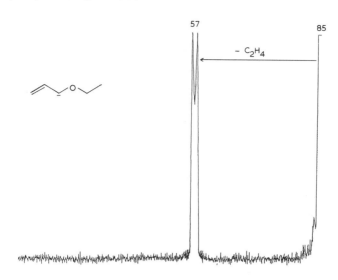

Figure 7. CA NH_2^-/NICI spectrum of CH_2=CHČHOEt.

Simple dialkyl ethers R_2O ($R \geqslant Et$) undergo elimination reactions when allowed to react with strong bases in the gas phase (e.g., see Equation 18) deprotonated species are not observed in these reactions.[50] Unsaturated ethers, however, deprotonate readily at either allylic or benzylic positions, and the carbanions formed undergo a number of collision-induced dissociations. For example, ions CH_2=CH-$\check{C}H(CH_2)_nOEt$(n=1–7) undergo proton transfer to yield the EI_{cb} intermediate CH_2=$CH(CH_2)_{n+1}OCH_2CH_2^-$ which decomposes to form the elimination product CH_2=$CH(CH_2)_{n+1}O^-$. The reaction is most pronounced for n=1; i.e., when the proton transfer proceeds through a six-center state (see Equation 19).[45] In the case of CH_2=CH-$\check{C}HOEt$, the elimination reaction takes a most unusual course. The spectrum is shown in Figure 7. If the mechanism is analogous to that shown in Equation 19, the product ion should correspond to the allyloxy ion CH_2=$CHCH_2O^-$. However, collisional activation and charge reversal MS/MS/MS experiments on the product ion show it to correspond to $(MeCOCH_2)^-$. A mechanism consistent with experimental data is shown in equation 20, i.e., the first formed species is the ion *i*, which rearranges as indicated. This scenario is substantiated by the following experiments: (1) The $S_N2(Si)$ displacement[51] shown in Equation 21, yields $(MeCOCH_2)^-$, and (2) the nucleophilic displacement on cyclopropyl acetate shown in Equation 22, also yields $(MeCOCH_2)$.[45]

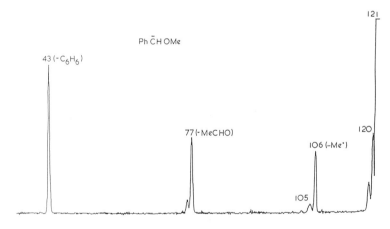

$$\text{(23)}$$

$$\text{(24)}$$

$$PhCHO^{-\bullet} + R^{\bullet} \qquad\qquad\qquad\qquad\qquad (25)$$

$$\text{(26)}$$

Other deprotonated unsaturated ethers show collision-induced dissociations that have direct analogy with solution chemistry. For example, deprotonated diallyl ether undergoes both 1,2 and 1,4 Wittig rearrangements[52] (see Equation 23 and 24), followed by an oxy Cope rearrangement[53] (see Equations 23 and 24). Fragmentations noted in the spectra of all three ions shown in Equation 23 are identical; these are the fragmentations of the final enolate ion.[52] Deprotonated benzyl ethers

Ph $\bar{C}H$ OMe

$43\,(-C_6H_6)$

$77\,(-MeCHO)$

$106\,(-Me^{\bullet})$

105

120

121

Figure 8. CA NH_2^-/NICI spectrum of Ph$\bar{C}$HOMe.

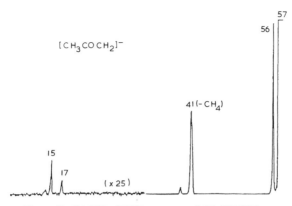

Figure 9. CA NH_2^-/NICI spectrum of $(MeCOCH_2)^-$.

undergo radical cleavage (see Equation 25); apart from that, all decompositions are those of the Wittig rearrangement ion, which could be formed through either an ion/neutral or radical/radical anion intermediate (see Equation 26).[54] An example appears in Figure 8; fragmentations of this Wittig product ion are summarized in Equations 10–12.

2.4. Aldehydes and Ketones

The collision-induced dissociations of enolate ions derived from aldehydes and ketones have been studied extensively.[19,20,55–63] Two examples of alkyl ketones will suffice to illustrate the fragmentation patterns: These are illustrated in Figures 9 and 10.

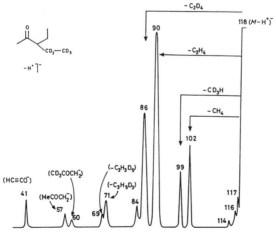

Figure 10. CA NH_2^-/NICI spectrum of the (M-H)⁻ion of $MeCOCH(Et)C_2D_5$.

$$^\bullet CH_2COCH_2^- + H^\bullet \tag{27}$$

$$(MeCOCH_2)^-$$

$$[Me^- (CH_2CO)] \quad \longrightarrow \quad HC\equiv CO^- + CH_4 \tag{28}$$

$$+ C_2H_4 \tag{29}$$

$$+ CH_4 \tag{30}$$

The mass spectrum of the acetone enolate ion is shown in Figure 9. It eliminates a hydrogen atom to yield a stabilized ion radical (see Equation 27) and methane, as shown in Equation 28. The fragmentations of more complex ketone enolates were first reported by Hunt,[55,56] the particular example shown in Figure 10 is 3-ethylpentan-2-one, which has been studied by ^2H and ^{13}C labeling.[20] The major fragmentations shown in Figure 10 are losses of ethene (see Equation 29) and methane (see Equation 30). Fragmentations are shown through the carbanion site for ease of representation. Isotope effect studies show that the deprotonation step (Step E) in Equation 29 is rate determining, whereas for methane elimination (see Equation 30), Steps F and G are both rate determining.

$$+ H_2 \tag{31}$$

$$+ H_2 \tag{32}$$

$$+ C_2H_4 \tag{33}$$

$$MeCO^- + CH_2CO \qquad (34)$$

$$HC \equiv CO^- + MeCHO \qquad (35)$$

$$(MeCOCH_2)^- + CO \qquad (36)$$

$$Me^- + CO + CH_2CO \qquad (37)$$

The cyclohexanone enolate ion fragments by two specific losses of H_2 (see Equations 31 and 32), but the characteristic reaction of the ring system is the retro reaction (see Equation 33), which results in the loss of ethene.[56,60] Substituted cyclohexanes behave similarly.[60] The spectra of deprotonated biacetyl,[64,65] and β-[66] and γ-diketones[67] have been reported. The spectrum (Figure 11) of deprotonated biacetyl is illustrative of such systems. All fragmentations can be rationalized as proceeding through acetyl anion complex *j*. The major fragmentations are direct dissociation to the acetyl anion (see Equation 34), and the deprotonation of ketene [see Equation 35, the acetyl anion is strong base $(\Delta H^{\circ}_{acid} MeCHO = 390 \text{ kcal mol}^{-1})$[68]]. The minor processes involve methyl anion migration (see Equation 36) and methyl anion formation [see Equation 37 (a major collision-induced dissociation of $MeCO^-$ is loss of CO to form Me)[61]].

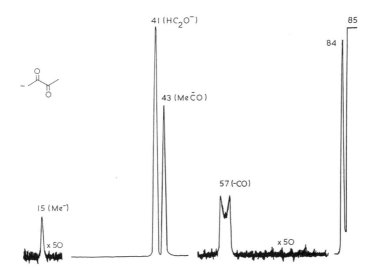

Figure 11. CA NH_2^-/NICI spectrum of $^-CH_2COCOMe$.

2.5. Carboxylic Acids

There have been more reports of negative ion formation from carboxylic acids than from any other functional group.[69] In the prototypical case of acetic acid, deprotonation with a strong base yields mainly $MeCO_2^-$, but the enolate ion $(CH_2CO_2H)^-$ is also formed[70] (ΔH°_{acid} $MeCO_2\underline{H}$ is 348 kcal mol^{-1};[71] $C\underline{H}_3CO_2H$ is 363 kcal mol^{-1}.[72] The isomeric ions have characteristic fragmentations [$MeCO_2^-$ (see Equations 38–40); $(CH_2CO_2H)^-$ (see Equations 41 and 42)], yet they are interconvertible on collisional activation.[72] Deuterium labeling shows that for higher alkyl-carboxylic acids, deprotonation occurs principally at the carboxyl position, and fragmentation may occur through both the carboxylate anion and the enolate ion formed by proton transfer. This feature is illustrated by the spectrum of the labeled 2-methylpropanoate shown in Figure 12.[69] The losses of Me· and H· from *iso*-$PrCO_2^-$ are reactions of the carboxylate ion k; both produce stabilized radical anions (see Equations 43 and 44). Decarboxylation of *iso* $PrCO_2^-$ is not detected, since the electron affinity of the *iso*propyl radical is negative.[29] The formation of HO^- and the loss of CH_4 are reactions of enolate l (see Equations 45 and 46).

$$(CH_2CO_2)^{-\bullet} + H^\bullet \tag{38}$$

$$MeCO_2^-$$

$$Me^- + CO_2 \tag{39}$$

$$[Me^-(CO_2)]$$

$$[O^{-\bullet}(CO)] + Me^\bullet \tag{40}$$

$$(CH_2CO_2)^{-\bullet} + H^\bullet \tag{41}$$

$$(CH_2CO_2H)^-$$

$$[HO^-(CH_2CO)] \longrightarrow HC_2O^- + H_2O \tag{42}$$

Equations (44), (43), (46), (45), (47) — structural chemical equations.

$$\tag{47}$$

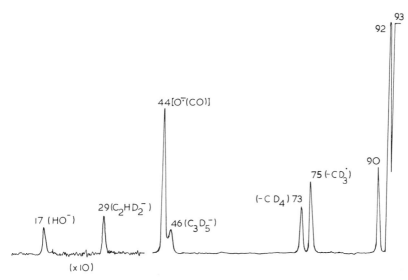

Figure 12. CA NH_2^-/NICI spectrum of $(CD_3)_2CHCO_2^-$.

There have been many reports of long-chain carboxylate anions (and cognate species) undergoing fragmentations remote from the site of the negative charge.[73–75] One example is shown in Formula VI (see Introduction), and a mechanistic proposal is shown in Equation 47 for the loss of C_4H_{10}, from the decanoate anion. In many cases, the fragmenting and

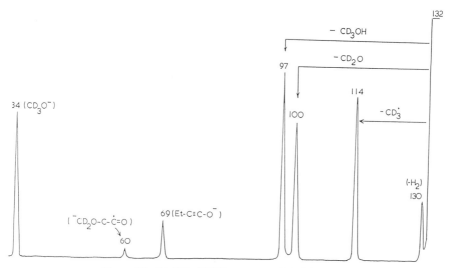

Figure 13. CA NH_2^-/NICI spectrum of $Et_2\overset{\cdot}{C}CO_2CD_3$.

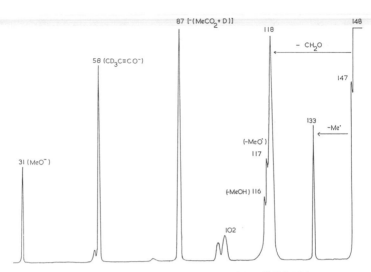

Figure 14. CA NH_2^-/NICI spectrum of $CD_3\bar{C}(CO_2Me)_2$.

charged centers cannot approach closely because of the rigidity of that system.[73–75] Whether such fragmentations conform to that shown in Equation 47, whether they are high-energy processes uninfluenced by the charge center[76] or whether they are induced by proton transfer to, or hydride ion transfer from a charged center is not known at this time.

2.6. Esters

Ester enolate ions undergo a wide variety of fragmentations, which (by now) are generally predictable. Simple alkyl esters,[55,77,78] aceto-acetates,[79] malonates,[80] succinates,[81] adipates,[82,83] and allyl acetates[84] have been studied. Four examples illustrate fragmentation behavior: a simple ester, a malonate, an adipate, and an allyl ester.

$$Et_2\bar{C}CO_2Me \longrightarrow Et_2\bar{C}CO_2^{\bullet} + Me^{\bullet} \qquad (48)$$

$$MeO^- + Et_2C{=}C{=}O \qquad (49)$$

$$[(Et_2C{=}C{=}C{=}O)MeO^-] \qquad Et(Me\bar{C}H)C{=}C{=}O \;\; + MeOH \qquad (50)$$

$$m \qquad Et_2\bar{C}CHO + CH_2O \qquad (51)$$

$$EtC \equiv CO^- + EtOMe \qquad (52)$$

The spectrum of a simple ester enolate ion is shown in Figure 13. Apart from the radical cleavage to form a stabilized radical anion (see Equation 48), all fragmentations in the spectrum can be rationalized as proceeding through the methoxide ion complex *m*. Direct displacement yields MeO⁻ (see Equation 49), deprotonation produces loss of methanol (see Equation 50), hydride transfer induces loss of formaldehyde [(see Equation 51); the methoxide anion is known to be an efficient hydride ion donor[85]]; and S_N2 displacement within the ion complex forms EtC_2O^- (see Equation 52).[78] The malonate spectrum in Figure 14 shows similar fragmentations, but in addition, it shows pronounced loss of methyl formate, best explained by the overall hydride ion sequence shown in Equation 53.[80]

$$+ \ HCO_2Me \qquad (53)$$

$$+ \ MeOH \qquad (54)$$

$$(55)$$

Some ester enolate ions undergo rearrangement reactions of a type observed previously in base catalyzed solution reactions. The best example of these are (1) adipate enolate ions which undergo the Dieckmann condensation[86] in the gas phase (see Equation 54),[82,83] and (2) the anion Claisen ester type rearrangement[87] exhibited by deprotonated allyl phenylacetates (see Equation 55).[84]

2.7. Organonitrogen Compounds

2.7.1. Amines

Alkyl N^- ions undergo much the same type of collision induced fragmentations as do C^- and O^- species.[88] As examples, let us consider the spectra (Figures 15 and 16) of a deprotonated primary amine and a cyclic amine. Deprotonated *iso*propylamine undergoes two specific losses of H_2 (see Equations 56 and 57, and Figure 15); the formation of the alternative methyl anion complex leads to two losses of methane (see Equations 58 and 59, see also Figure 15) together with formation of Me^- (see Equation 60). Saturated cyclic alkylamines lose H_2 (see Equation 61, and Figure 16) in a similar fashion to cyclohexanone enolate ions (see Equations 31, and 32). However, the retro reactions so characteristic of cyclohexanones (see Equation 33), are either minor or absent for cyclic alkylamines. For example, retro reactions do not occur for piperidines, and they are relatively minor processes for piperazines (see Equations 62, 63, and Figure 16).[88] Deprotonated aniline behaves like the phenoxide ion (see Equation 16) in eliminating CNH to form the cyclopentadienyl anion.[88]

$$Me_2CH - \overset{-}{N}H \longrightarrow [H^-(Me_2C=NH)] \underset{\searrow}{\overset{\nearrow}{}} \begin{array}{ll} Me_2C=N^- + H_2 & (56) \\[1em] Me(\overset{-}{C}H_2)C=NH + H_2 & (57) \end{array}$$

$$[Me^-(MeCH=NH)] \underset{\searrow}{\overset{\nearrow}{}} \begin{array}{ll} MeCH=N^- + CH_4 & (58) \\[1em] {}^-CH_2CH=NH + CH_4 & (59) \end{array}$$

$$Me^- + MeCH=NH \qquad\qquad\qquad\qquad\qquad (60)$$

$$\qquad\qquad\qquad\qquad\qquad\qquad\qquad +H_2 \qquad (61)$$

$$\begin{array}{ll} + CH_2NH & (62) \\[1em] CH_2N^- + \quad EtN=CH_2 & (63) \end{array}$$

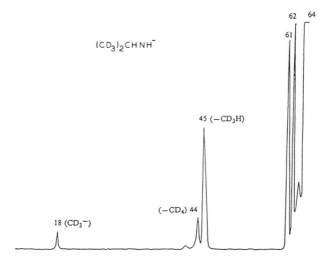

Figure 15. CA NH_2^-/NICI spectrum of $(CD_3)_2CHNH^-$.

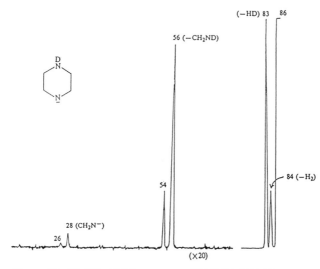

Figure 16. CA NH_2^-/NICI spectrum of $DN(CH_2CH_2)_2N^-$.

2.7.2. Amino Acids

The collisional activation NH_2^- NICI mass spectra[89] and collisional activation FAB spectra[89,90] of $(M-H)^-$ ions derived from a amino acids are similar. Labeling studies indicate that major deprotonation occurs at the carboxylate center; minor deprotonation to produce enolate ions cannot be excluded. The major fragmentations of the carboxylate anions are preceded by specific proton transfer to the carboxylate center. The spectra look complex, but they are readily explicable. Let us choose as examples glycine, serine, and aspartic acid.

$$NH_2\overset{\bullet}{C}HCO_2^- + H^\bullet \tag{64}$$

$$NH_2CH_2CO_2^-$$

$$[O^{-\bullet}(CO)] + NH_2CH_2^\bullet \tag{65}$$

$$^-NHCH_2CO_2H \longrightarrow [H^-(NH=CHCO_2H)] \longrightarrow NH=CHCO_2^- + H_2 \tag{66}$$
n

$$HO-\overset{=}{C}=O + H_2C=NH \tag{67}$$

$$[(NH=CH_2)^- CO_2H] \longrightarrow {}^-CH=NH + HCO_2H \tag{68}$$
o

$$CH_2=N^- + HCO_2H \tag{69}$$

$$^-NHCH_2OH + CO \quad \text{or} \quad {}^-CH_2NHOH + CO \tag{70}$$

The spectrum of deprotonated glycine is shown in Figure 17. Deuterium labeling studies support the following mechanistic proposals.

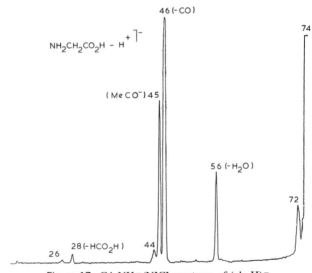

Figure 17. CA NH_2^-/NICI spectrum of (gly-H)$^-$.

Radical cleavages occur as shown in Equations 64 and 65. Loss of CO_2 is not detected since the electron affinity of $NH_2CH_2^{\cdot}$ is $-16.2\ \text{kcal mol}^{-1}$.[46] Collision-induced proton transfer from the amino to the carboxylate position yields *n*, which loses H_2 (see Equation 66). Reaction of *n* through ion complex *o* accounts for all other fragmentations, *viz.*, direct cleavage (see Equation 67), two specific deprotonation reactions (see Equations 68 and 69), and finally decarbonylation (see Equation 70), which produces the base peak of the spectrum. The last reaction is characteristic of complexes containing a bound hydroxycarbonyl anion, since $HOCO^-$, like its analogue the bicarbonate anion $HOCO_2^-$,[91] is an hydroxide anion donor.[92]

$$NH_2\overline{C}HCO_2H + CH_2O \qquad (71)$$

$$NH_2\underset{\overset{|}{{}^-CH_2}}{C}HCO_2H + CO_2 \qquad (72)$$

Deprotonated amino acids containing *a* alkyl side chains generally fragment in a similar fashion to glycine, but in other cases, specific fragmentation through the *a* side chain is noted. This is illustrated for deprotonated serine and aspartic acid. The serine spectrum (Figure 18) is simple, showing characteristic elimination of CH_2O (see Equation 71). The aspartic acid spectrum (Figure 19) is more complex—the base peak is formed by loss of CO_2 (see Equation 72).

2.7.3. Amides and Peptides

The CA NICI spectra of simple amides are quite complex, whereas the CA negative FAB spectra of peptides are deceptively simple at first sight.

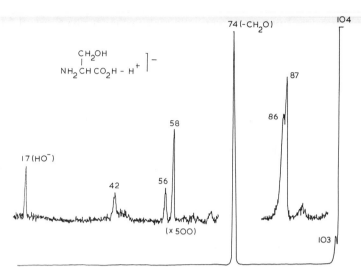

Figure 18. CA NH_2^-/NICI spectrum of (ser-H)$^-$.

The collision-induced dissociations of simple amides have been studied using extensive labeling.[93] In systems like CH_3CONHR, deprotonation occurs on N in accord with the following ΔH°_{acid} values: $MeCONH_2$ (342 kcal mol^{-1}) and CH_3CONMe_2 (375 kcal mol^{-1}).[94] Amides like

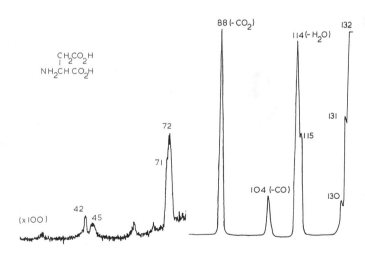

Figure 19. CA NH_2^-/NICI spectrum of (asp-H)$^-$.

CH_3CONR_2 deprotonate to form the enolate ion. Figures 20 and 21 show two examples of amide spectra.

$$MeCO\overline{N}Et + C_2H_4 \qquad (73)$$

$$^-CH_2CONEt_2 \longrightarrow [Et_2N^-(CH_2CO)] \qquad \begin{cases} Et_2N^- + CH_2O & (74) \\ HC_2O^- + Et_2NH & (75) \end{cases}$$

$$\underset{q}{MeCON(Et)\overline{C}HMe} \longrightarrow \underset{r}{[Me\overline{C}O(EtNCHMe)]} \qquad \begin{cases} MeCO^- + EtNCHMe & (76) \\ Me\overline{C}HN=CHMe + MeCHO & (77) \end{cases}$$

$$\longrightarrow [MeCO_2^- (PhCN)]$$

(79) $(C_6H_4)^-CN + MeCO_2H$ $MeCO_2^- + PhCN \qquad (78)$

The spectrum of a labeled diethylacetamide is shown in Figure 20. Deuterium-labeling studies support the following mechanistic proposals. The base peak of the spectrum is formed by loss of ethene. The reaction is summarized in Equation 73, and the first step (the proton transfer) is rate determining.*

This reaction is directly analogous to the elimination reactions of ethers[45] (see Equation 19) and ketone enolates[19,20] (see Equation 29). Other decompositions are as follows: (1) Fragmentation of the initial enolate through ion complex p yields the expected products (see Equations 74 and 75), and (2) specific proton transfer from the 1-position of an N-ethyl substituent to the enolate position yields q, which fragments

---

* The alternative mechanism involving proton transfer to O^- cannot be excluded on the available evidence. The dish-shaped product peak is indicative of a process with a substantial reverse activation energy.[19]

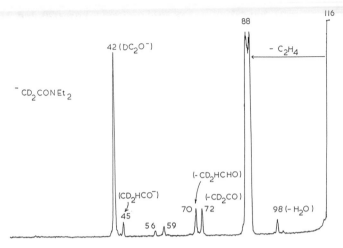

Figure 20. CA NH_2^-/NICI spectrum of $^-CD_2CONEt_2$.

through *r* as indicated in Equations 76 and 77. The second amide example is the bisamide shown in Figure 21. Most of the fragmentations are readily explicable with the exception of the formation of the acetate anion and the elimination of acetic acid. These involve a skeletal rearrangement for which a mechanism is proposed in Equations 78 and 79.

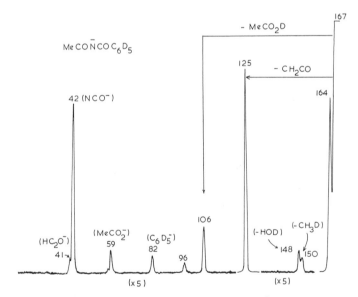

Figure 21. CA NH_2^-/NICI spectrum of $MeCO\bar{N}COC_6D_5$.

It is becoming increasingly clear that MS/MS and MS/MS/MS data from peptide $(M-H)^-$ ions can produce complementary information of the conventional positive ion cleavages[15,16,95,96] although to date negative ion application to peptide analysis has been basically to determine the molecular weight.[95] Only recently have attempts been made to survey the CA FAB spectra of peptide $(M-H)^-$ ions systematically: to date, these studies have been confined to dipeptides and a few tripeptides.[97, 98] There are a number of acidic positions in peptides [$\Delta H^°_{acid}$ ranges are $CO_2\underline{H}$ (340–350), $N\underline{H}CO$ (355–365), and $C\underline{H}CONH$ (360–365 kcal mol^{-1})]. It is not a simple matter to determine the deprotonation site(s) using deuterium labeling, and in any case, both neutral and zwitterion forms of the peptide will be present in the FAB matrix (glycerol with a little water; or in the case of deuterium labeled peptides, glycerol D_3/D_2O). Nevertheless, deuterium labeling indicates that the characteristic decomposition of dipeptides (see Equation 80) occurs following proton transfer to form a decomposing enolate ion. In the case of unlabeled peptides, the FAB ionization process will certainly form *both* carboxylate and enolate ions directly. The analytical applicability of the method is clearly demonstrated in Figures 22 and 23 for the isomeric pair gly ala and ala gly. Tripeptides fragment similarly except that there are now two fragmenting enolate ions to provide sequencing information.[97]

$$NH_2CH(R^1)CONHCH(R^2)CO_2^- \longrightarrow NH_2\overline{C}(R^1)CONHCH(R^2)CO_2H$$

$$NH_2C(R^1)=C=O \ + \ \overline{N}HCH(R^2)CO_2H \qquad (80)$$

$$[(NH_2CHR^1CONHCH=CH_2)HO^-]$$

$$NH_2CH(R^1)CO\overline{N}CH=CH_2 \ + \ H_2O \qquad (81)$$

In Section 2.7.2, we saw that certain amino acids fragment characteristically through the a-side chain. For example, serine loses CH_2O (see Figure 18 and Equation 71). Do dipeptides containing serine fragment exclusively through the side chain, thus masking the normal dipeptide cleavage (see Equation 80)? The answer is shown in Figures 24 and 25. Serine can be readily identified by the loss of CH_2O, and the basic cleavage (see Equation 80) is observed but in reduced abundance. In addition, if a

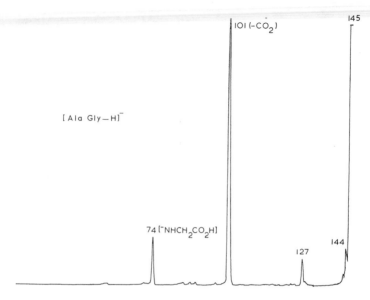

Figure 22. CA FAB spectrum of (ala gly-H)⁻.

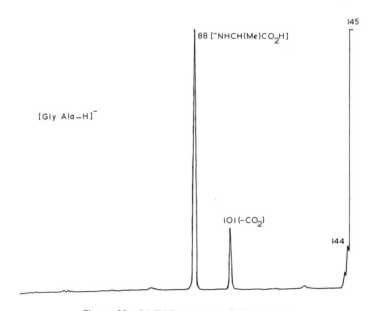

Figure 23. CA FAB spectrum of (gly ala-H)⁻.

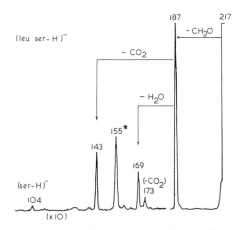

Figure 24. CA FAB spectrum of (leu ser-H)⁻.

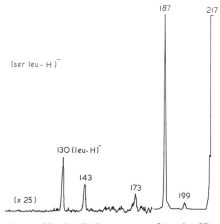

Figure 25. CA FAB spectrum of (ser leu-H)⁻.

peptide contains serine, its position at the C-terminal end of the peptide may be confirmed by the presence of a peak corresponding to loss of $(CO_2 + H_2O)$ [see Equation 81 m/z 155 in Figure 24).[98]

2.7.4. Oximes—The Negative Ion Beckmann Rearrangement

It will be clear to the reader that there are many initially formed anions that cannot fragment directly. These often effect proton transfer or skeletal rearrangement prior to decomposition. The deprotonated oxime is one such group. Either proton transfer to oxygen or some internal rearrangement would be expected, and this expectation is realized.[99] Indeed, the proton transfer shown in Equation 82 is quite facile, since the transfer involves a five-center state, and the acidities at the two positions differ only by some 7 kcal mol^{-1} [e.g., the ΔH°_{acid} values of $Me_2C=NOH$[44] and $(CH_3)_2C=NOMe$[100] are 366 and 373 kcal mol^{-1}, respectively].

$$
\begin{array}{ccc}
\diagdown \\
\diagup C=NO^- \\
-CH_2
\end{array}
\longrightarrow
\begin{array}{c}
\diagdown \\
\diagup C=NOH \\
-CH
\end{array}
\qquad (82)
$$

$$
Pr_2C=NO^- \xrightarrow{-H^{\bullet}} {}^{\bullet}CH_2CH_2CH_2C(Pr)=NO^- \longrightarrow CH_2=C(Pr)N=O^{-\bullet} + C_2H_4 \qquad (83)
$$

$$
Pr_2C=NO^- \longrightarrow
\begin{array}{c}
EtСH \\
\diagdown \\
\diagup C=NOH \\
Pr
\end{array}
\xrightarrow{\quad O \quad} [(EtCH=C=NPr)HO^-]
$$

$$
\downarrow
$$

$$
EtCH=C=\overset{-}{N}CHEt + H_2O \qquad (84)
$$

$$
R_2C=\overset{+}{N}OH_2 \xrightarrow{\quad O \quad} RC=\overset{+}{N}R + H_2O \longrightarrow RCONHR \qquad (85)
$$

$$
Ph_2C=NO^- \xrightarrow{\quad O \quad} Ph-\overset{-}{C}=N-OPh \longrightarrow PhO^- + PhCN \qquad (86)
$$

Figure 26 shows the fragmentations of deprotonated ketoximes for a labeled heptan-4-one ketoxime. Mechanistic interpretation was aided by labeling and product ion studies.[99] There are two major and competing fragmentations shown in Figure 26, *viz.*, the loss of water and the apparent elimination of Et$^{\bullet}$. The latter loss would seem to be a simple radical loss to form a stabilized anion radical. However, Figure 26 shows this process to occur with a pronounced isotope effect (H/D = 3.0); hence the rate-determining step must involve the loss or transfer of a terminal hydrogen. Similar phenomena have been observed for the losses of $C_2H_5^{\bullet}$ from $Et_2CCO_2^{-}$ [69] and $Ph\overset{-}{C}Et_2$,[32] and a suggested mechanism is shown in Equation 83.

Loss of water often constitutes the base peak in the spectra of deprotonated oximes: This is not a usual fragmentation of RO$^-$ species in the gas phase. The mechanism is shown in Equation 84, it involves proton transfer, 1,2 anionic rearrangement of the propyl group to nitrogen, followed by loss of H_2O from the ion complex. This reaction is reminiscent of the Beckmann rearrangement (see Equation 85) a classical rearrange-

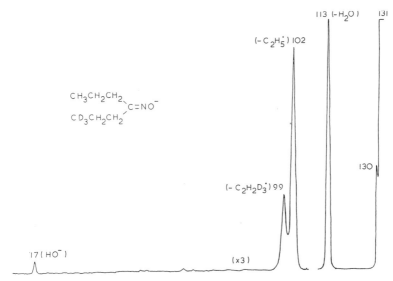

Figure 26. CA NH_2^-/NICI spectrum of $CD_3CH_2CH_2(Pr)C=NO^-$.

ment of positive ions that occurs in both the condensed[101] and gas phases.[102] The negative ion Beckmann reaction occurs only when proton transfer can produce a carbanion site *a* to the double bond. When this is not possible, other rearrangement reactions occur; for example, benzophenone oxime undergoes the phenyl rearrangement shown in Equation 86.[99]

$$PhĊHCH_2CH_2C(Me)=NOH \leftarrow PhCH_2CH_2CH_2C(Me)=NO^- \rightarrow PhCH_2CH_2ĊHC(Me)=NOH$$

$$[(PhCH=CH_2)^-CH_2C(Me)=NOH]$$

$$[PhCH_2^-(CH_2=CHC(Me)=NOH)]$$

$$PhCH=CH_2+{}^-CH_2C(Me)=NOH \quad (89)$$

$$PhCH_2^-+CH_2=CHC(Me)=NOH \quad (87)$$

and

$$PhCH_3+CH_2=CHC(Me)=NO^- \quad (88)$$

Finally, the spectrum (see Figure 27) of $Me(PhCD_2CH_2CH_2)C=NO^-$ is particularly interesting, since it emphasizes the ready proton transfers that may occur for negative ions. In this case, the Beckmann rearrangement is completely suppressed by the more energetically favorable reactions shown in Equations 87–89.[99]

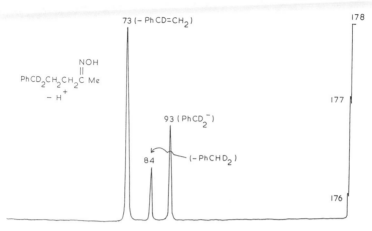

Figure 27. CA NH_2^-/NICI spectrum of $[PhCD_2(CH_2)_2C(Me)=NOH-H]^-$.

3. CLASSIFICATION OF FRAGMENTATIONS OF EVEN-ELECTRON ORGANIC NEGATIVE IONS

The treatment of functional group fragmentation has been selective since (1) only a few examples within each section have been discussed in any detail and (2) a number of systems have not been discussed, including heterocyclics, organoboron, organosulphur, organosilicon, and organophosphorus compounds. The latter work has been omitted because a line had to be drawn somewhere, and in any case, the essential groundwork has already been laid in this chapter to enable classification of fragmentation types.

Most of the negative ion reactions discussed fall into one of four main classification types, together with a few that are at this time classified as miscellaneous. The fragmentation classification is, in brief as follows:

1. Loss of radical to form a stabilized radical anion

2. Direct fragmentation through an intermediate (ion complex) with subsequent competitive fragmentation of that intermediate resulting in the elimination of neutrals

3. Fragmentations preceded by proton transfer to the initial deprotonation site, followed by subsequent elimination of neutral(s) often through an ion complex intermediate

4. Fragmentations preceded by skeletal rearrangement of the first-formed anion.

3.1. Fragmentations Involving Radical Loss

As in Equations 25, 27, 38, 40, 41, 43, 44, 48, 64, 65 and 82. The majority of $(M-H)^-$ ions lose H· to some extent, but in systems that can form stabilized radical anions, such losses are often pronounced, e.g., see Equations 27, 38, 41, 44, and 65. Loss of Me· is often noted for the same reason, e.g., 40, 43, 48. Apparent elimination of higher alkyl radicals ($\geqslant$ Et·) may occur by a two step process (e.g., see Equation 82).

3.2. Fragmentations Proceeding Directly through an Ion Complex

The intermediate may be either an ion/neutral or radical/radical anion complex. Some of the reasons listed here could be synchronous, they are listed together for convenience. The reaction involves formation of the initial intermediate followed by:

1. Direct displacement of the anion (of the ion complex), see Equations 11, 34, 35, 39, 49, 60, 72 and 74

2. The bound anion (of the complex) deprotonates the neutral portion (of the complex), see Equations 1, 2, 7, 9, 10, 12, 28, 31, 32, 42, 50, 56–59, 61, 75, 81

3. The bound anion effects an S_N2 displacement, see Equations 52, 53

4. Hydride transfer from the anion to the neutral, see Equation 51

5. R⁻ donation (from RCO⁻) to the neutral, see Equation 36.

3.3. Fragmentations following Proton Transfer to the Initial Anion

1. Reactions where direct neutral loss (i.e., elimination) follows proton transfer; see Equations 19, 20, 29, 71, and 73. If any of these reactions occurs through an ion complex intermediate, they are best classified under type 2.

2. Ion complex (formed after proton transfer undergoing direct displacement of the anion, see Equations 45, 67, 76, 80, 87, and 89. If any of these reactions are synchronous they are classified under type 1.

3. The bound anion effects S_N2 displacement, see Equation 46.

4. R⁻ donation (from RCO⁻) to the neutral of the ion complex, see Equation 70.

3.4. Fragmentations Occurring following Skeletal Rearrangement

The following types have been described:

1. $S_N i$ reactions (cyclization); see Equations 15, 54 (Dieckmann), 78, and 79
2. 1,2-anion rearrangements; see Wittig (Equations 23, 24, and 26) and Beckmann (Equation 84)
3. Sigmatropic rearrangements; see oxy Cope (Equations 23, and 24) and Claisen ester (Equation 55)
4. Nucleophilic aromatic substitution; see Smiles (Equation 14)
5. Other (Equation 85).

3.5. Unclassified Fragmentations

1. Remote fragmentations, see Equation 47.
2. Retro reactions, see Equations 3, 4, 33, 62, and 63—if these reactions are stepwise (e.g., see Equations 62, and 63), they are best classified as Type 1 in Section 3.2.
3. Losses of CX or CHX· from ArX^- systems (see Equations 16, 17, and loss of CNH for $PhNH^-$).

In conclusion, I hope I have convinced you that (i) negative ion mass spectrometry is a viable analytical technique, (ii) the fragmentations of even-electron anions are explicable within a set of fairly simple rules and (iii), above all, negative ions have a fascinating ion chemistry.

ACKNOWLEDGMENTS

This chapter is dedicated to all of the graduate students and post-doctoral fellows who worked with me in this area. Their names are found in the References. I thank the Australian Research Council for financial support.

REFERENCES

1. von Ardenne, M.; Steinfelder, K.; and Tummler, R. *Electronenanlagerungs Massenspektrographie organischen Substanzen* (Springer-Verlag: Berlin, 1971).
2. Bowie, J. H.; and Williams, B. D., "Negative-Ion Mass Spectrometry of Organic Organometallic and Coordination Compounds," In *International Review of Science,*

Physical Chemistry, series 2, vol. 5, *Mass Spectrometry* (Maccoll., A., ed.), (Butterworths: London and Boston, 1975), pp. 89–127.

3. Bowie, J. H., *Mass Spectrom. Rev.* 1984, **3**, 161.
4. Bowie, J. H., *Mass Spectrometry Specialist Reports* (Chemical Society: London, 1985), **8**, 168 (ed., Johnstone, R. A. W.); 1987, **9**, 177; 1989 **10**, in press (ed Rose, M. E.).
5. Jennings, K. R., Mass Spectrometry Specialist Reports, Chemical Society, London, 1979, **5**, 203; Harrison, A. G. *Chemical Ionization Mass Spectrometry* (CRC Press: Florida, 1983). Budzikiewicz, H. *Mass Spectrom. Rev.* 1986, **5**, 345.
6. Benninghoven, A., *Springer Ser. Chem. Phys.* 1984, **36**, 342, 498, Vickerman, J. C., *Chem. Brit.* 1987, **23**, 969.
7. Bowie, J. H., Mass Spectrometry Specialist Reports. Chemical Society, London, Rose, M. E., ed., 1989, **10**, in the press.
8. Sundquist, B.; Macfarlane, R. D., *Mass Spectrom. Rev.* 1985, **4**, 421.
9. Surman, D. J.; and Vickerman, J. C., *J. Chem. Soc. Chem. Commun.* 1981, 324, Barber, M.; Bordoli, R. S.; Sedgewick, R. D.; and Tyler, A. N., *J. Chem. Soc. Chem. Commun.* 1981, 325; Barber, M.; Bordoli, R. S.; Elliott, G. J.; Sedgewick, R. D.; and Tyler, A. N.; *Anal. Chem.* 1982, **54**, 645A; Devienne, F. M.; Roustan, J.-C. *Org. Mass Spectrom*, 1982, **17**, 173.
10. Fenselau, C.; Vu, V. T.; Cotter R. J.; Hausen, G.; Heller, D.; Chen, T.; and Colvin, O. M. *Spectros. Int. J.* 1982, **1**, 132.
11. Ens, W.; Standing, K. G.; Westmore, J. B.; Ogilvie, K. K.; and Nemer, M. J., *Anal. Chem.* 1982, **54**, 960.
12. McNeal, C. J.; Ogilvie, K. K.; Theriauelt, N. Y.; and Nemer, M. J., *J. Am. Chem. Soc.* 1982, **104**, 976.
13. Domon, B.; and Hostettmann, K., *Helv. Chim. Acta* 1984, **67**, 1310.
14. Mertz, J. L.; Peloso, J. S.; Barber, B. J.; Babbitt, G. E.; Occolowitz, J. L.; Simson, V. L.; and Kline, R. M., *J. Antibiotics* 1986, **39**, 877.
15. Barber, M., Bordoli, R. S.; Sedgewick, R. D.; and Tyler, A. N., *Biomed. Mass Spectrom.* 1982, **9**, 208.
16. Cooper, R.; and Unger, S., *J. Org. Chem.* 1986, **51**, 3942.
17. Tomer, K. B.; Jensen, N. J.; and Gross, M. L., *Anal. Chem.* 1986, **58**, 2429.
18. Hayes, R. N.; Sheldon, J. C.; Bowie, J. H.; and Lewis, D. E., *J. Chem. Soc. Chem. Commun.* 1984, 1431, *Aust. J. Chem.* 1985, **38**, 1197.
19. Stringer, M. B.; Bowie, J. H.; and Holmes, J. L., *J. Am. Chem. Soc.* 1986, **108**, 3888.
20. Stringer, M. B.; Bowie, J. H.; and Currie, G. J., *J. Chem. Soc. Perkin Trans. II* 1986, 1821.
21. Raftery, M. J.; and Bowie, J. H., *Int. J. Mass Spectrom. Ion Processes*, 1987, **79**, 267.
22. Sheldon, J. C.; Bowie, J. H.; and Lewis, D. E., *New. J. Chem.* 1988, **12**, 269.
23. Sheldon, J. C.; Bowie, J. H.; and Eichinger, P. C. H., *J. Chem. Soc. Perkin Trans II* 1988, 1263.
24. Tumas, W.; Foster, R. F.; Pellerite, M. J.; and Brauman, J. I., *J. Am. Chem. Soc.* 1987, **109**, 961.
25. Tumas, W.; Foster, R. F.; and Brauman, J. I., *J. Am. Chem. Soc.* 1988, **110**, 2714.
26. Eichinger, P. C. H.; Bowie, J. H.; and Blumenthal, T. *J. Org. Chem.* 1986, **51**, 5078.
27. Ellison, G. B.; Engelking, P. C.; and Lineberger, W. C., *J. Am. Chem. Soc.* 1978, **100**, 2556.
28. MacKay, G. J.; Hemsworth, R. S.; and Bohme, D. K., *Canad. J. Chem.* 1976, **54**, 1624.
29. Graul, S. T.; and Squires, R. R., *J. Am. Chem. Soc.* 1988, **110**, 607; 1989, **111**, 892.
30. DePuy, C. H.; Bierbaum, V. M.; and Damrauer, R., *J. Am. Chem. Soc.* 1984, **106**, 4051.
31. O'Hair, R. A. J.; and Bowie, J. H., *Rapid Comm. Mass Spectrom* 1989, **3**, 293.

32. Currie, G. J.; Bowie, J. H.; Massy-Westropp, R. A.; and Adams, G. W., *J. Chem. Soc. Perkin Trans. II* 1988, 403.

33. Eichinger, P. C. H.; Bowie, J. H.; and Hayes, R. N., *J. Org. Chem.* 1987, **52**, 5224.

34. O'Hair, R. A. J.; Bowie, J. H.; and Currie, G. J., *Aust. J. Chem.* 1988, **41**, 57.

35. O'Hair, R. A. J.; Hayes, R. N.; and Bowie, J. H., unpublished observations.

36. Stone, D. J. M.; Eichinger, P. C. H.; and Bowie, J. H., *Rapid Comm. Mass Spectrom.* 1989, **3**, 344.

37. Ridge, D. P.; Beauchamp, J. L., *J. Am. Chem. Soc.* 1974, **96**, 3595.

38. Nobes, N. H.; Poppinger, D.; Li, W.-K.; and Radom, L. "Molecular Orbital Theory of Carbanions", in *Comprehensive Carbanion Chemistry*, part C, Buncel, E.; and Durst, T., eds. (Elsevier, Amsterdam, 1987).

39. Schleyer, P. v. R.; and Kos, A. J., *Tetrahedron* 1983, **39**, 1141.

40. Mercer, R. S.; and Harrison, A. G., *Canad. J. Chem.* 1988, **66**, 2947.

41. Raftery, M. J., Bowie, J. H., and Sheldon, J. C., *J. Chem. Soc. Perkin Trans II*, 1988, 563.

42. Eichinger, P. C. H.; Bowie, J. H.; and Hayes, R. N., *Aust. J. Chem.* 1989, **42**, 865.

43. Meot-ner, M.; and Sieck, L. W., *J. Phys. Chem.* 1986, **90**, 6687.

44. Bartmess, J. E.; Scott, J. A.; and McIver, R. T., *J. Am. Chem. Soc.* 1979, **101**, 6047.

45. Waugh, R. J.; Hayes, R. N.; Eichinger, P. C. H.; Downard, K. M., and Bowie, J. H., *J. Am. Chem. Soc.* 1990, **112**, 2537.

46. Downard, K. M.; Sheldon, J. C.; Bowie, J. H.; Lewis, D. E.; and Hayes, R. N., *J. Am. Chem Soc.*

47. Truce, W. E.; Kreider, E. M.; and Brand, W. W., *Organic Reactions* (Wiley: New York, 1970) **18**, 99. Schmidt, D. M.; and Bonvicino, G. E., *J. Org. Chem.* 1984, **49**, 1664.

48. Eichinger, P. C. H.; Bowie, J. H.; and Hayes, R. N., *J. Am. Chem. Soc.* 1989, **111**, 4224.

49. Blumenthal, T.; and Bowie, J. H., unpublished observations.

50. van Doorn, R.; and Jennings, K. R., *Org. Mass Spectrom.* 1981, **16**, 397; DePuy, C. H.; and Bierbaum, V. M., *J. Am. Chem. Soc.* 1981, **103**, 5034; DePuy, C. H.; Beedle, E. C.; and Bierbaum, V. M., *J. Am. Chem. Soc.* 1982, **104**, 6483; Bierbaum, V. M.; Filley, J.; De Puy, C. H.; Jarrold, M. F.; and Bowers, M. T.; *J. Am. Chem. Soc.* 1985, **107**, 2818.

51. DePuy, C. H.; Bierbaum, V. M.; Flippin, L. A.; Grabowski, J. J.; King, G. K.; Schmitt, R. J.; and Sullivan, S. A., *J. Am. Chem. Soc.* 1979, **101**, 6443; Klass G.; Trenerry, V. C.; Sheldon, J. C.; and Bowie, J. H., *Aust. J. Chem.* 1981, **34**, 519.

52. Eichinger, P. C. H.; and Bowie, J. H., *J. Chem. Soc. Perkin Trans II*, 1988, 497.

53. Rozeboom, M. D.; Kiplinger, J. P.; and Bartmess, J. E., *J. Am. Chem. Soc.* 1984, **106**, 1025.

54. Eichinger, P. C. H.; and Bowie, J. H., *J. Org. Chem.* 1986, **51**, 5078, *J. Chem. Soc. Perkin Trans II* 1987, 1499.

55. Hunt, D. F.; Shabanowicz, J.; and Giordani, A. B., *Anal. Chem.* 1980, **82**, 386; *Environ. Health Perspect.* 1980, **36**, 33.

56. Hunt, D. F.; Giordani, A. B.; Shabanowicz, J.; and Rhodes, G., *J. Org. Chem.* 1982, **47**, 738.

57. Moylan, C. R.; Janinski, J. M.; and Brauman, J. I., *Chem. Phys. Lett.* 1983, **1**, 98; Foster, R. F.; Tumas, W.; and Brauman, J. I., *J. Chem. Phys.* 1983, **79**, 4644.

58. Stringer, M. B.; Underwood, D. J.; Bowie, J. H.; Holmes, J. L.; Mommers, A. A.; and Szulejko, J. E., *Canad. J. Chem.* 1986, **64**, 764.

59. Currie, G. J.; Stringer, M. B.; Bowie, J. H.; and Holmes, J. L., *Aust. J. Chem.* 1987, **40**, 1365.

60. Raftery, M. J.; and Bowie, J. H., *Int. J. Mass Spectrom. Ion Processes*, 1987, **79**, 267.

61. Downard, K. M.; Sheldon, J. C.; and Bowie, J. H., *Int. J. Mass Spectrom. Ion Processes*, 1988, **86**, 217.
62. Chowdhury, S.; and Harrison, A. G., *J. Am. Chem. Soc.* 1988, **110**, 7345.
63. Donnelly, A.; Chowdhury, S.; and Harrison, A. G., *Org. Mass Spectrom.* 1989, **29**, 89.
64. Bowie, J. H.; Stringer, M. B.; Hayes, R. N.; Raftery, M. J.; Currie, G. J.; and Eichinger, P. C. H., *Spectros. Int. J.* 1985, **4**, 277.
65. Chowdhury, S.; and Harrison, A. G., *Org. Mass Spectrom.* 1989, **24**, 123.
66. Hayes, R. N.; Bowie, J. H.; and Sheldon, J. C., *Int. J. Mass Spectrom. Ion Processes* 1986, **71**, 233.
67. Raftery, M. J.; and Bowie, J. H., *Aust. J. Chem.* 1987, **40**, 711.
68. DePuy, C. H.; Bierbaum, V. M.; Damrauer, R.; and Soderquist, J. A., *J. Am. Chem. Soc.* 1985, **107**, 3385.
69. Stringer, M. B.; Bowie, J. H.; Eichinger, P. C. H.; and Currie, G. J., *J. Chem. Soc. Perkin Trans. II* 1987, 385 and references cited therein.
70. Grabowski, J. J.; Cheng, X., *J. Am. Chem. Soc.* 1989, **111**, 3106.
71. Fujio, M.; McIver, R. T.; and Taft, R. W., *J. Am. Chem. Soc.* 1981, **103**, 4017.
72. O'Hair, R. A. J.; Gronert, S.; DePuy, C. H., and Bowie, J. H., *J. Am. Chem. Soc.* 1989, **111**, 3105.
73. Adams, J.; Deterding, L. J.; and Gross, M. L., *Spectros. Int. J.* 1987, **329**, 369, Cerny, R. L.; Tomer, K. B.; and Gross, M. L., *Org. Mass Spectrom.* 1986, **21**, 655; Tomer, K. B.; Jensen, N. J.; and Gross, M. L. *Anal. Chem.* 1986, **58**, 2429.
74. Jensen, N. J.; Tomer, K. B.; and Gross, M. L. *J. Am. Chem. Soc.* 1985, **107**, 1863; Tomer, K. B.; Jensen, N. J.; Gross, M. L., and Whitney, *J. Biomed. Envir. Mass Spectrom.* 1986, **13**, 265; Tomer, K. B.; and Gross, M. L. *Biomed. Envir. Mass Spectrom.* 1988, **15**, 89; Adams, J.; and Gross, M. L. *J. Am. Chem. Soc.* 1989, **111**, 435.
75. Adams, J., *Mass Spectrom. Rev.* 1990, **9**, 141.
76. Wysocki, V. H.; Bier, M. E.; Cooks, R. G., *Org. Mass Spectrom.* 1988, **23**, 627.
77. Froelicher, S. W.; Lee, R. E.; Squires, R. R.; and Fraser, B. S. *Org. Mass Spectrom.* 1985, **20**, 4.
78. Hayes, R. N.; and Bowie, J. H. *J. Chem. Soc. Perkin Trans. II* 1986, 1827.
79. Eichinger, P. C. H.; and Bowie, J. H. *Org. Mass Spectrom.* 1987, **22**, 103.
80. Hayes, R. N.; and Bowie, J. H., *Org. Mass Spectrom.* 1986, **21**, 425.
81. Raftery, M. J.; and Bowie, J. H., *Aust. J. Chem.* 1987, **40**, 711.
82. Burinsky, D. J.; and Cooks, R. G., *J. Org. Chem.* 1982, **47**, 4864.
83. Raftery, M. J.; and Bowie, J. H., *Org. Mass Spectrom.* 1988, **23**, 719.
84. Eichinger, P. C. H.; Bowie, J. H.; and Hayes, R. N., *J. Org. Chem.* 1987, **52**, 5524.
85. Ingemann, S.; Kleingeld, T. C.; and Nibbering, N. M. M., *J. Chem. Soc. Chem. Commun.* 1982, 1009; Sheldon, J. C.; Currie, G. J.; Lahnstein, J.; Hayes, R. N.; and Bowie, J. H., *Nouv. J. Chim.* 1985, **9**, 205.
86. Schaefer, J. P.; and Bloomfield, J. J., *Org. React.* 1967, **15**, 1.
87. Brannock, K. C.; Pridgen, H. S.; and Thompson, B., *J. Org. Chem.* 1960, **25**, 1815, Frater, A., *Helv. Chim. Acta.* 1975, **58**, 442; Ireland, R. E.; Mueller, R. H.; and Willard, A. K., *J. Am. Chem. Soc.* 1976, **98**, 2868.
88. Raftery, M. J.; and Bowie, J. H., *Aust. J. Chem.* 1988, **41**, 1477.
89. Eckersley, M.; Bowie, J. H.; and Hayes, R. N., *Int. J. Mass Spectrom. Ion Processes* 1989, **93**, 199.
90. Kulik, W.; and Heerma, W. *Biomed. Envir. Mass Spectrom.* 1988, **15**, 419.
91. Fehsenfeld, F. C.; and Ferguson, E. E., *J. Chem. Phys.* 1974, **61**, 3181; O'Hair, R. A. J.; Bowie, J. H.; and Hayes, R. N., *Rapid Comm. Mass Spectrom.* 1988, **2**, 275.

92. Sheldon, J. C.; and Bowie, J. H., *J. Am. Chem. Soc.* 1990, **112**, 2424.
93. Raftery, M. J.; and Bowie, J. H., *Int. J. Mass Spectrom. Ion Processes* 1988, **85**, 167.
94. Bartmess, J. E.; *The 1987 Gas Phase Acidity Scale.* (Dept. of Chemistry, University of Tennessee, cited as personal communication, Taft, R. W.), July 1987.
95. See reviews listed in Reference 4 and references cited therein.
96. In particular, see Voigt, D.; and Schmidt, J., *Biomed. Mass Spectrom.* 1978, **5**, 44; Bradley, C. V.; Howe, I.; and Beynon, J. H. *J. Chem. Soc. Chem. Commun.* 1980, 562; Bradley, C. V.; Howe, I.; and Beynon, J. H., *Biomed. Mass Spectrom.* 1981, **8**, 85; Buko, A. M.; Phillips, L. R.; and Fraser, B. A., *Biomed. Mass Spectrom.* 1983, **10**, 387; Garner, G. V.; Gordon, D. B.; Tetler, L. W.; and Sedgewick, R. D., *Org. Mass Spectrom.* 1983, **18**, 486; Lyon, P. A.; Stibbings, W. L.; Crow, F. W.; Tomer, K. B.; Lippstreu, D. L.; and Gross, M. L., *Anal. Chem.* 1984, **56**, 8; Bertrand, M. J.; and Thebault, P., *Biomed. Envir. Mass Spectrom.* 1986, **13**, 347; Lignori, A.; Sindona, G.; and Uccella, N., *J. Am. Chem. Soc.* 1986, **108**, 7488; Kulik, W.; and Heerme, W., *Biomed. Envir. Mass Spectrom.* 1988, **15**, 419.
97. Eckersley, M.; Bowie, J. H.; and Hayes, R. N., *Org. Mass Spectrom.* 1989, **24**, 597.
98. Waugh, R. J.; Bowie, J. H.; and Hayes, R. N., *Int. J. Mass Spectrom. Ion Processes* 1991, **107**, 333.
99. Adams, G. W.; Bowie, J. H.; and Hayes, R. N., *J. Chem. Soc. Perkin Trans. II* 1990, 2159.
100. O'Hair, R. A. J.; Gronert, S.; Karrigan, K. E.; Bierbaum, V. M.; DePuy, C. H.; and Bowie, J. H., *Int. J. Mass Spectrom. Ion Processes* 1989, **90**, 295.
101. Beckmann, E., *Chem. Ber.* 1987, **20**, 1507; Donamura, G.; and Heldt, W. Z., *Org. Reactions.* 1960, **11**, 1.
102. Maquestiau, A.; Van Haverbeke, Y.; de Meyer, C.; Duthoit, C.; Meyrant, P.; and Flammang, R., *Nouv. J. Chim.* 1979, **3**, 517; Maquestiau, A.; Van Haverbecke, Y.; Flammang, R.; and Meyrant, I., *Org. Mass Spectrom.* 1980, **15**, 80.

2

Gas-Phase Chemistry of Alkali Adducts of Simple and Complex Molecules

Lynn M. Teesch and Jeanette Adams

1. INTRODUCTION

Gas-phase interactions between alkali metal ions and organic molecules have been of interest since the early 1970's. Initial studies were conducted to determine metal ion affinities of simple monofunctional compounds.[1-6] Thermionic emitters were used as sources of metal ions, and Fourier-transform ion cyclotron resonance (FTICR) mass spectrometers were used for mass analysis. Consequently, alkali metal ions were exploited as chemical ionization reagents.[2d, 6, 7] The role of the metal ion as a reagent has changed as the methods for forming ions and types of mass spectrometers used for analysis have evolved.

Alkali metal ion adducts have been observed in mass spectra of countless molecules. In the majority of cases, the adduct ions have been regarded as impurities and nuisances. Some researchers, however, have used metal ion adducts to determine molecular weight or corroborate molecular weight information obtained from conventional methods.[8] Field desorption and fast atom bombardment (FAB) have been used most extensively to produce metal ion adducts. The development of FAB has particularly simplified the cationization of biomolecules, and *in situ* "derivatization" can be performed easily by mixing a sample with a FAB matrix saturated

Lynn M. Teesch and Jeanette Adams • Department of Chemistry, Emory University, Atlanta, Georgia 30322.

Experimental Mass Spectrometry, edited by David H. Russell. Plenum Press, New York, 1994.

with an alkali salt. More recently, FAB and tandem mass spectrometry (MS) have been combined to study large biomolecules. Interest has increased in the metal ion adducts because the chemistry of alkali metal ion complexes of biologically important molecules has lead to structural information. The chemistry of complexes between alkali metal ions and biomolecules, such as fatty acids, sugars, and peptides, differs from that of $(M + H)^+$ ions. In some cases, fragmentations of metal ion adducts provide more information about structure than fragmentations of $(M + H)^+$ ions.

Alkali metal ions have not been the only metal ions used as reagents. Transition metal ion adducts of biomolecules also have been studied by using chemical ionization and tandem MS,[9] and more recently, laser desorption (LD) with FTICR MS.[10] Complexes between alkaline earth metal ions and peptides[11] and homoconjugated fatty acids[12] have been studied by using FAB and tandem MS. Chapter 2, however, focuses more on the gas-phase chemistry of organo-alkali adducts presenting in part an historical account of early developments and a more detailed discussion of the chemistry of complexes between alkali metal ions and fatty acid derivatives, sugars, and peptides.

2. SIMPLE MOLECULES

In the 1970's, several research groups[1-6] studied the gas-phase "solvation" of alkali metal ions by water, ammonia, dimethylformamide, acetonitrile, and various other simple solvent molecules. Džidić and Kebarle[1a] reported the first gas-phase data on the hydration of alkali metal ions. They used a thermionic emitter to produce the alkali metal ions and a high pressure ion source to study the reactions. The enthalpies and free energies of solvation are largest for Li^+ and decrease as the size of the metal ion increases (see Table 1). The enthalpies and free energies also

Table 1. Thermodynamic Data for the Gas-Phase Equilibrium
$M(H_2O)^+ \rightleftharpoons M^+ + H_2O^a$

M	$\Delta H^{\circ a}$	$\Delta S^{\circ b}$	$\Delta G^{\circ a}$
Li^+	34.0	23.0	25.5
Na^+	24.0	21.5	17.9
K^+	17.9	21.6	11.4
RB^+	15.9	21.2	9.6
Cs^+	13.7	19.4	7.9

Source: Reprinted with permission from reference 1a.
^{a} kcal mol^{-1}.
^{b} Entropy values in eu.

decrease as the number of water molecules in the cluster ions increases (see Table 2), but the entropy increases. This trend is most acute for Li^+ and Na^+ and reflects a crowding of molecules around the central ion. Other researchers observed similar trends for complexes with water,[4] dimethylformamide,[3] acetonitrile,[1c] and ammonia.[5] Some important conclusions were that interactions between these molecules and Li^+ are more covalent in nature and a maximum number of molecules can occupy the metal ions' coordination spheres in the gas phase.

$$Li(H_2O)^+ \rightleftharpoons Li^+ + H_2O$$

Scheme 1

The binding energy or enthalpy of the reaction in Scheme 1, $34.0 \, kcal \, mol^{-1}$, was used as the benchmark for the scale of relative binding energies constructed by Staley and Beauchamp.[2b] They studied 30 small solvent molecules using an FTICR mass spectrometer also equipped with a thermionic emitter. Their purpose was to determine the intrinsic basicities of different molecules toward an acid, which in this case was Li^+. The relative binding energies of various bases to Li^+ were deter-

Table 2. Thermodynamic Data for the Gas-Phase Equilibria
$$M^+(H_2O)_{n-1} + H_2O \rightleftharpoons M^+(H_2O)_n$$

M		1	2	3	4	5	6
					n		
Li^+	$\Delta H^{\circ a}$	34.0	25.8	20.7	16.4	13.9	12.1
	$\Delta G^{\circ b}$	25.5	18.9	13.3	7.5	4.5	2.5
	$\Delta S^{\circ a}$	23.0	21.1	24.9	29.9	31.4	32.0
Na^+	ΔH°	24.0	19.8	15.8	13.8	12.3	10.7
	ΔG°	17.9	13.2	9.3	6.3	3.9	2.9
	ΔS°	21.5	22.2	21.9	25.0	28.1	26.0
K^+	ΔH°	17.9	16.1	13.2	11.8	10.7	10.0
	ΔG°	11.4	8.9	6.3	4.4	3.2	2.3
	ΔS°	21.6	24.2	23.0	24.7	25.2	25.7
Rb^+	ΔH°	15.9	13.6	12.2	11.2	10.5	
	ΔG°	9.6	7.0	5.0	3.8	2.8	
	ΔS°	21.2	22.2	24.0	24.8	25.7	
Cs^+	ΔH°	13.7	12.5	11.2	10.6		
	ΔG°	7.9	5.9	4.2	3.0		
	ΔS°	19.4	22.2	23.7	25.4		

Source: Reprinted with permission from reference 1a.
^{a} $kcal \, mol^{-1}$.
^{b} Entropy values in eu.

mined from the equilibrium reaction shown in Scheme 2. Binding energies increase with methyl substitution ($H_2O < MeOH < Me_2$ and $NH_3 < Me_3N$) and the size of the alkyl substituent ($MeF < EthF < i\text{-}PrF$). A later study[2c] focused on Li^+ binding to a small number of bases that contain oxygen and nitrogen and further reveal the effects of alkyl substituents. The order of increasing binding energies is $H_2O < H_2CO < HCN < C_6H_6 < MeOH < Me_2O < NH_3 < MeNH_2 < Me_3N < Me_2NH$. Interestingly, the ordering of Me_3N and Me_2NH is the reverse of gas-phase proton affinities[13] and theoretical predictions.[14] This important anomaly is believed to be caused by the large electrostatic repulsion between Li^+ and the methyl groups of Me_3N that cause the distance between Li^+ and the N atom to increase enough to reduce the attractive force. Effects of methyl substituents on electrostatic interactions between Li^+ and oxygen-containing bases are not so pronounced as nitrogen-containing bases. Thus, the basicities of the ordering of H_2O, $MeOH$, and Me_2O toward Li^+ is the same as that toward H^+.

$$M_1Li^+ + M_2 \rightleftharpoons M_1 + M_2Li^+$$

Scheme 2

The possible use of alkali metal ions as chemical ionization (CI) reagents was proposed and demonstrated by Hodges and Beauchamp[2d] for determining of molecular weight. Their method involved reacting Li^+, which was produced by thermionic emission, with a reagent gas to form a complex that could then transfer Li^+ to the sample according to the equilibrium in Scheme 2. Ethyl fluoride was tested as a possible reagent gas. More effective reagent gases, however, are hydrocarbons, such as cyclohexane and isobutane, that have a lower binding energy than most other molecules ($24\,\text{kcal mol}^{-1}$ and approximately $20\,\text{kcal mol}^{-1}$ for cyclohexane and isobutane, respectively[2]). Hydrocarbons also tend not to form cluster ions. An example of using this method in mixture analysis is illustrated in Figure 1 for a mixture of six compounds, each approximately 0.1% in isobutane. The mass spectrum in Figure 1 was acquired by using a quadrupole mass spectrometer. Each component forms an abundant Li^+ complex, and a low pressure of 10 mTorr, which is much lower than in conventional CI, reduces cluster ion formation.

Bombick *et al.*[7a] alternatively used K^+ as a CI reagent. They also used a thermionic emitter as a source of alkali metal ions. A variety of compounds that include small solvent molecules, crown ethers, and small peptides were studied using a quadrupole mass spectrometer. All of the compounds studied form abundant $(M + K)^+$ ions. Furthermore, K^+ is

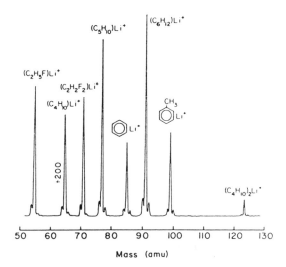

Figure 1. Mass spectrum of a mixture of C_2H_5F, FCHCHF, $CH_3CHC(CH_3)_2$, benzene, 1-hexene, and toluene that have been chemically ionized with Li^+. (Note that the abundance of the Li^+-isobutane peak has been demagnified by a factor of 200.) (Reprinted with permission from Reference 2d.)

the only fragment ion formed from compounds introduced via a batch inlet interface. In contrast, $(M + K)^+$ ions of peptides deposited directly on the emitter fragment to give other ions that contained K^+. The K^+-containing fragments provide some structural information. Bombick and Allison[7b] developed a similar ionization methods called K^+IDS (K^+ ionization of desorbed species) that is based on the K^+CI experiments.

Both of the preceding investigations demonstrate that alkali metal ion CI can be useful for molecular weight determination, little structural information, however, can be obtained. Other groups,[9] however, subsequently began exploiting other metal ions, such as Fe^+, not only for molecular weight information, but also to induce decompositions for structural information. The interesting chemistry of transition metal ion interactions with saturated and unsaturated hydrocarbons and long alkyl chain esters, ketones, and acids is beyond the scope of this chapter.[15]

3. MULTIFUNCTIONAL MOLECULES

There are numerous examples of the use of cationization to determine molecular weights of more complex, multifunctional compounds.[8] With the development of "soft" ionization techniques, applications have been directed more toward polar thermally labile molecules. Furthermore, the novel gas-phase chemistry of the cationized molecules has been exploited by coupling soft ionization techniques to tandem MS. Metastable ion and collision-induced decompositions (CIDs) of $(M + Cat)^+$ complexes are unlike fragmentations of $(M + H)^+$ ions, and they are increasingly being used to structurally determine multifunctional molecules.

Three particular areas of investigations have involved decompositions of alkali metal ion complexes of fatty acid derivatives, sugars and peptides. Studies of these particular biomolecules show the analytical utility of alkali metal ions as tools for structure determination. Many of these studies also have tried to explain the role of the alkali metal ion in the fragmentation processes.

3.1. Fatty Acids and Related Compounds

Levsen et al.[16] were some of the first researchers who coupled soft ionization with tandem MS (MS–MS) to study CID of cationized molecules. They demonstrated that $(M + Li)^+$ ions of alcohols, produced by surface ionization, undergo CID to give product ions that retain the metal ion. The product ions have the general structure $(C_nH_{2n}OH + Li)^+$. The CID spectra also reveal that loss of Li^+ decreases as the size of the alcohol increases. Levsen et al.[10] also specifically noted that Li^+ seems to have an effect on the fragmentation chemistry unlike the effect of H^+.

It was not until 1986, however, when Adams and Gross[17] studied CID of $(M + Cat)^+$ ions of fatty alcohols, that the unusual chemistry of the $(M + Cat)^+$ ions was more fully understood. Jensen et al.[18] had recently reported a new type of fragmentation reaction that occurs away from the site of charge of long-chain fatty acid anions. The charge-remote fragmentations (CRF) involve C-C bond cleavages that occur along the alkyl chain and yield losses of C_nH_{2n+2}. Their proposed mechanism involves concerted 1,4-hydrogen eliminations that yield neutral alkenes, H_2, and terminally unsaturated fatty acid product ions (Scheme 3). Adams and Gross,[17] in their pursuit of an analytical method for structural determination of fatty alcohols, discovered that fatty alcohol $(M + Li)^+$ ions likewise undergo CRF via the mechanism in Scheme 3. This discovery lead to applications in structurally determining fatty alcohols,[17] acids,[19, 20] and esters.[21]

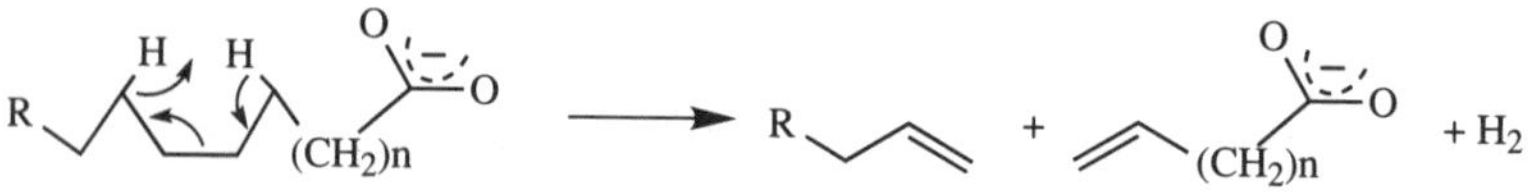

Scheme 3

Furthermore, CRF of $(M+2Li-H)^+$ ions of fatty acids provide more structural information than analogous CRF of $(M-H)^-$ ions.[12, 19b, 20]

Examples of the types of structural information that can be obtained from CID of $(M+Li)^+$ ions of fatty alcohols, esters, and acids are shown in Figures 2 and 3. The CID of $(M+Li)^+$ ions of fatty compounds that contain a saturated hydrocarbon chain results in the general CRF pattern shown in Figure 2A for $(M+Li)^+$ ions of methyl palmitate. The characteristic series of closed-shell product ions 14 u apart begins with the ion of

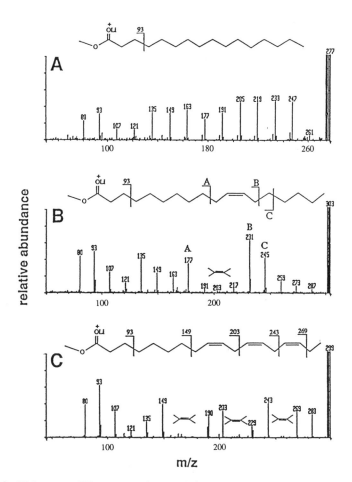

Figure 2. High-energy CID spectra of $(M+Li)^+$ ions of methyl esters of (A) palmitic (hexadecanoic), (B) vaccenic (*cis*-11-octadecenoic), and (C) linolenic (*cis*-9,12,15-octadecatrienoic) acids. Symbols for double bonds are shown over ions that would nominally arise via cleavage through the double bond. (Reprinted with permission from Reference 21.)

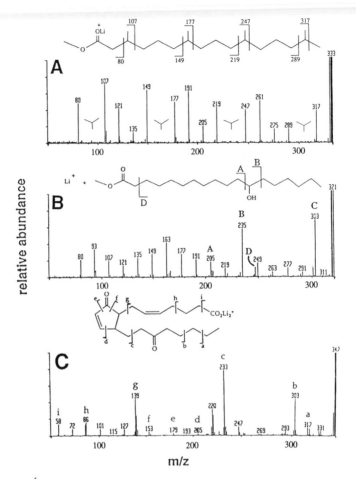

Figure 3. High-energy CID spectra of $(M + Li)^+$ ions of methyl esters of (A) 2,6,10,14-tetra-methyl hexadecanoic and (B) 12-hydroxy-octadecanoic acids, and (C) $(M + 2Li - H)^+$ ions of prostaglandin 13, 14-dihydro-15-keto-PGA$_2$. Symbols for alkyl branch points are shown over ions that would nominally arise via cleavage of C–C bonds on either side of the branch point. (Reprinted with permission from References 21 and 22, respectively.)

m/z 261 (loss of CH$_4$). The presence of a single double bond interrupts the smooth pattern (see Figure 2B), leaving a gap of 54 u. This gap occurs because cleavages vinylic to and through the double bond are energetically unfavorable, whereas cleavages allylic to the double bond are favorable. The fingerprint for a double bond is thus two ions of higher abundances (labeled A and B in Figure 2B) on either side of three ions of much lower abundances. The CID spectrum of linolenic acid methyl ester (see

Figure 2C) shows interruptions in the CRF pattern as the result of multiple double bonds. Here the pattern is somewhat different than the one in Figure 2B, for example, the ion of m/z 203 results from a cleavage vinylic to the 12,13 double bond and from a cleavage allylic to the 9,10 double bond (this is why the ion of m/z 203 is more abundant than normally expected). Other CRF product ions that would result from Li^+ attachment to the double bonds instead of to the carbonyl oxygen do not arise from either $(M + 2Li - H)^+$ ions of fatty acids or $(M + Li)^+$ ions of methyl esters.[19, 20, 21] This phenomenon can, however, be a problem in interpreting CID spectra of $(M + Li)^+$ ions of fatty alcohols because double bonds compete favorably with the hydroxy group for the metal ion.[17]

The analytical utility of this method exists not only for fatty compounds with double bonds, but also for compounds with other functionalities.[19, 22] Examples of CID spectra of $(M + Li)^+$ ions of methyl esters containing alkyl branches and an hydroxy group are shown in Figure 3A and 3B, respectively. Methyl-substituted methyl esters decompose in a manner that causes characteristic gaps of 28 u in the CRF pattern (see Figure 3A). Gaps result from cleavages of $C-C$ bonds on either side of the branch points. The CID spectrum of $(M + Li)^+$ ions of 12-hydroxy-stearic acid methyl ester shows the CRF fingerprint for hydroxy substitution (see Figure 3B). Diagnostic product ions that reveal the hydroxy group are labeled A and B in Figure 3B. They are 30 u apart in mass and result from cleavages of $C-C$ bonds on either side of the hydroxy substituent. The more weakly abundant ion of m/z 219 in Figure 3B is a product of $(M + Li)^+$ ions in which Li^+ is attached to the hydroxy group instead of to the carbonyl oxygen. This ion is always 16 u lower in mass than the ion labeled B. The product ion of m/z 383 (labeled C in Figure 3B) that results from a loss of H_2O is another diagnostic fragmentation of hydroxy substituents. More complex fatty molecules, such as prostaglandins[22] can also be cationized. As $(M + 2Li - H)^+$ ions, these species also undergo CID to give structurally informative CRF (see Figure 3C). As for the simpler esters, substituent-specific alterations in the CRF pattern provide important diagnostics for identifying and locating functional groups, such as double bonds and carbonyl groups of 13,14-dihydro-15-keto-PGA$_2$ (see Figure 3C).

As in earlier study by Levsen *et al.*,[16] the role of the metal ion in the fragmentation chemistry of the $(M + Cat)^+$ and $(M + 2Cat - H)^+$ complexes was questioned. Adams and Gross[17, 19a] noted that relative abundances of CRF decrease with increasing size of the metal ion. As the size of the metal ion increases, loss of Cat^+ effectively competes with the CRF and dominates the CID spectra of Rb^+- and Cs^+-containing species. Thus, a strong metal ion-functional group bond is necessary to localize the charge so that CRF can be observed.[19a] Other possible effects of the metal ion on the fragmentation chemistry is still under investigation.[23]

Other questions raised by studying the CRF of cationized species involve verifying mechanisms for formation of charge-remote product ions, for example, are losses of C_nH_{2n+2} produced by 1,4-hydrogen eliminations or other mechanisms? Are fragmentations involving allylic cleavages analogous to thermal pericyclic reactions, or can they also be explained by radical mechanisms?[24, 25] How "charge-remote" are charge-remote fragmentations?[23] For a more thorough discussion of charge remote fragmentations, the reader is directed to a review by Adams.[26]

3.2. Monosaccharides and Oligosaccharides

Many compounds, including such polyhydroxy compounds as glucose and sucrose, do not yield abundant $(M+H)^+$ ions; they do, however, form stable $(M+Cat)^+$ ions.[27-29] The small amount of fragmentation, if any, that arises from ion source reactions of $(M+Cat)^+$ ions makes molecular weight determination simple. Structure elucidation, however, requires alternative techniques. These alternative techniques have included metastable ion decompositions of cationized monosaccharide cluster ions, high- and low-energy CID of $(M+Cat)^+$ ions of disaccharides, and LD-FTICR, LD-time-of-flight (-TOF), and high-energy CID of cationized oligosaccharides.

3.2.1. Monosaccharides

An important problem in structurally determining monosaccharides has been differentiating stereoisomers. Derivatization has typically been required. Puzo et al.,[30] however, showed that a simple alternative is to study the unimolecular decompositions of (sugars + Cat + matrix)$^+$ cluster ions, where Cat = Li^+ or Na^+ and matrix = glycerol. This technique has been used to differentiate aldohexose stereoisomers. The metastable ion decompositions of the cluster ions yield both (aldohexose + Cat)$^+$ ions and (glycerol + Cat)$^+$ ions. The ratio between the relative abundance of (aldohexose + Cat)$^+$ ions (a in Figure 4) compared to the abundance of (glycerol + Cat)$^+$ ions (b in Figure 4) can differentiate eight aldohexose isomers. The order of ratios in Figure 4 is similar for both Li^+ and Na^+, except for mannose and galactose. For the Li^+ adducts of mannose and galactose, ratios are too similar to differentiate from one another. For the Na^+ adducts, however, ratios are different enough to distinguish the two sugars. Stereoisomers of 2-acetamido-2-deoxyhexose can also be identified by this method.[30c] Furthermore, the best alkali metal ion and matrix for

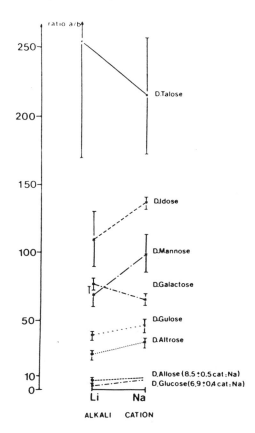

Figure 4. Values of the ratio a/b of the abundances of the ion a (aldohexose + Cat)$^+$ and b (glycerol + Cat)$^+$ resulting from the unimolecular dissociation of the (aldohexose + Cat + glycerol)$^+$ cluster ions generated by FAB for D-aldohexose stereoisomers. (Reprinted with permission from Reference 30b.)

molecular weight determinations of disaccharides seem to be Na$^+$ and either glycerol or diethanolamine, respectively.[30d]

3.2.2. Disaccharides

The cationization of sugars has not only been used for molecular weight information and isomer differentiation, but also for more complete structural elucidation. Both high-[31] and low-energy[32] CID of (M + Cat)$^+$ ions of disaccharides provide more structural information than CID of (M + H)$^+$ ions. The most abundant product ion from high-energy CID of (M + H)$^+$ ions of sucrose results from a loss of H_2O (see Figure 5A). Less abundant ions in Figure 5A arise from the cleavage of one sugar unit to give an ion of m/z 163, and successive losses of H_2O to give the ions of m/z 145 and 127. In contrast, the analogous CID spectrum of the (M + Na)$^+$ ions (see Figure 5B) displays product ions in the high-mass region that

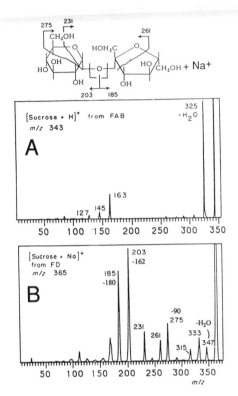

Figure 5. High-energy CID spectra of (A) $(M+H)^+$ and (B) $(M+Na)^+$ ions of sucrose. The $(M+H)^+$ ions were desorbed by FAB, and the $(M+Na)^+$ ions were desorbed by field desorption. The losses shown in B also refer to those listed in Table 3 for low-energy CID. (Reprinted with permission from Reference 31b.)

result from ring-opening reactions to give ions of m/z 275, 261, and 231. The loss of one sugar unit gives the ion of m/z 203, and the subsequent loss of water gives the ion of m/z 185. The $(M+Na)^+$ ions of sucrose in low-energy CID studies, however, decompose to give the two abundant ions of m/z 203 and 185, but the ions that arise from the ring-opening reactions are not observed.[32] Low-energy CID spectra of the Li^+ adducts of disaccharides do, however, contain product ions that arise from ring-cleavage reactions,[33] these are discussed later. For $(M+Li)^+$ ions, the ring-opening cleavages must compete favorably with loss of Li^+, whereas for $(M+Na)^+$ ions, the loss of Na^+ must be preferred.

Leary *et al.*[33] studied low-energy CID of $(M+Li)^+$ and $(M+2Li-H)^+$ ions of disaccharides to determine glycosidic linkages. Product ions formed by low-energy CID of $(M+2Li-H)^+$ complexes of three disaccharide isomers are shown in Figure 6. The presence of product ions resulting from losses of 60, 90, 120 u can differentiate between the glycosidic linkages. Fragmentation patterns of monolithiated $(M+Li)^+$ species also change with the glycosidic linkage. A list of neutral losses from

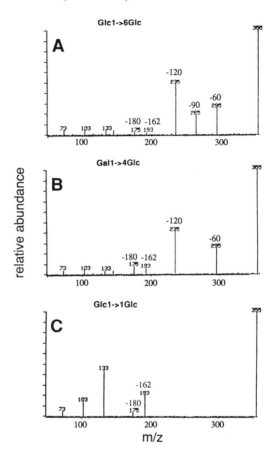

Figure 6. Low-energy CID spectra of $(M+2Li-H)^+$ ions of (A) 1,6-gentiobiose $(Glc(1{\rightarrow}6)Glc)$. (B) 1,4-lactose $(Glc(1{\rightarrow}4)Glc)$, and (C) 1,1-trehalose $(Glc(1{\rightarrow}1)Glc)$. Losses shown are analogous to those in Table 3. (Reprinted with permission from Reference 33a.)

low-energy CID of $(M+Li)^+$ complexes that are useful in differentiating glycosidic linkages is presented in Table 3. Ring-cleavage reactions of disaccharides with a 1,6-linkage result in losses of 60, 90, and 120 u (see Figure 6A and Table 3). For 1,4-, 1,3-, and 1,2-disaccharides, ring-cleavages give only one unique product ion that results from loss of either 60, 90, or 120 u, respectively. Interestingly, $(M+Li)^+$ ions of 1,4-disaccharides do not fragment by low-energy CID to give loss of 120 u, but the $(M+2Li-H)^+$ ions do.[33a] Also, note that high-energy CID[34] of $(M+Na)^+$ ions, and infrared LD[35] of $(M+Na)^+$ and $(M+K)^+$ ions of 1,4-disaccharides, result in loss of 120 u. Disaccharides with a 1,1-linkage can easily be discerned by their total lack of ring-cleavage reactions. The anomeric configuration, a or β of disaccharides can not be differentiated by low-energy CID.[33]

Table 3. Low-Energy CID Neutral Losses from $(M+Li)^+$ Ions of Disaccharides Useful for Differentiating Glycosidic Linkages

	Neutral loss (mass)[a]				
	$-C_2H_4O_2$ (60)	$-C_3H_6O_3$ (90)	$-C_4H_8O_4$ (120)	$-C_6H_{10}O_5$ (162)	$-C_6H_{12}O_6$ (180)
1,6	X	X	X	X	X
1,4	X			X	X
1,3		X		X	X
1,2			X	X	X
1,1				X	X

Source: Adapted with permission from reference 33b.
[a] The X presents the presence of the product ion in the CID spectrum; blanks represent the absence of the product ion. CID spectra of the α and β anomeric configurations are indistinguishable.

Spengler et al.[35] originally proposed that ring-cleavage reactions of polysaccharides occur via retro-aldol reactions. Hofmeister et. al.[33b] further investigated this idea by performing deuterium- and ^{18}O-labeling studies. They proposed, for example, that loss of 60-u ($C_2H_4O_2$) from 1,6-gentiobiose first requires a ring opening via a charge-remote 1,4-migration of H (Scheme 4). The subsequent loss of $C_2H_4O_2$ to give

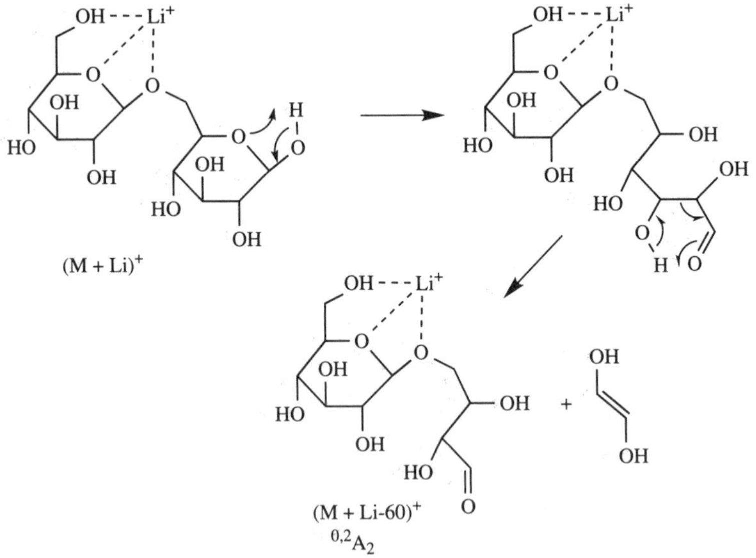

Scheme 4

$(M + Li - 60)^+$ product ions then occurs via a retro-aldol reaction. Successive loss of another 60 u could then occur via another retro-aldol reaction to form the $(M + Li - 120)^+$ fragment ion. Precursor ion scans of the product ion resulting from loss of 120 u indeed reveal that successive losses of 60 u occur. Loss of 120 u occurs primarily from the $(M + Li)^+$ ions, however, and Scheme 5 illustrates the proposed retro-aldol mechanism for their formation from the ring-opened $(M + Li)^+$ ions.[33b]

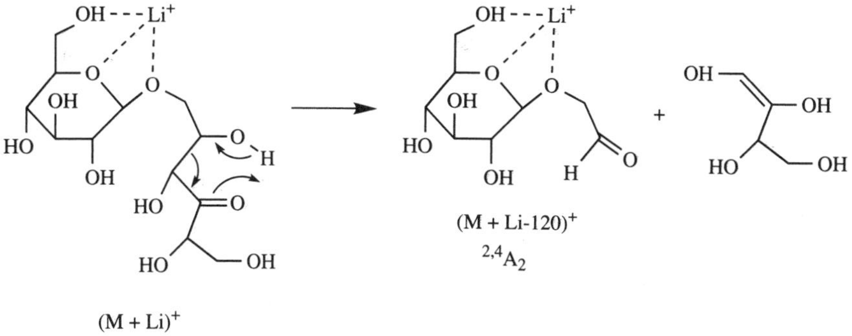

Scheme 5

Hofmeister *et al.*[33b] also addressed possible locations of the alkali metal ion in the cationized sugar complexes by MNDO theoretical calculations, which indicate that stability of the metal ion complex increases with the number of possible Li^+-oxygen bonds. Their calculations also indicate that the most stable complex involves Li^+ in a position bridging the two sugar units. The researchers also suggested that the metal ion strengthens the glycosidic linkage so that ring-cleavage reactions become competitive, although no supporting evidence was given. It is clear, however, that the alkali metal ion has strong effects on fragmentations because the chemistry of the $(M + Cat)^+$ ions is very different from the chemistry of $(M + H)^+$ ions.[31b, 34]

3.2.3. Oligosaccharides

Studies of polysaccharides complexed with alkali metal ions have used two basic approaches. One has involved LD of the cationized species followed by mass analysis via either FTICR or TOF. The other approach has used FAB to produce the ions and high-energy CID to induce fragmentations.

Coates and Wilkins[36] first studied $(M + K)^+$ ions of polysaccharides via infrared LD-FTICR. Several series of fragment ions were formed from masses of $[(162)_n + X + K]^+$, where 162 u is the mass of a single intact hexose ring and X is the mass of a hexose ring fragment remaining after a ring-cleavage reaction. The fragmentation pattern varies with the number of hexose units and with the types of glycosidic linkages in the sugar chain. Because analogous decompositions occur in pyrolysis MS of oligosaccharides, Coates and Wilkins suggested that LD and cleavages of saccharides are thermal or thermal-like processes. They also suggested that cationization does not occur until after the gas-phase thermal-like fragmentation.

Spengler et al.[35] later examined cationized oligosaccharides via infrared LD-TOF. Fragmentations of various sugars with different glycosidic linkages were studied to determine the mechanisms of the ring-opening cleavages and to attempt to differentiate between isomers. The LD-TOF mass spectrum of cellotriose exhibits the characteristic fragment ions formed from oligosaccharides (see Figure 7). Both $(M + Na)^+$ and $(M + K)^+$ ions are observed in addition to fragment ions that contain one of the alkali metal ions. As previously mentioned, Cotter et al. first proposed a retro-aldol reaction for formation of the $^{0,2}A_n$, $^{0,3}A_n$, and $^{2,4}A_n$ fragment ions, with successive retro-aldol reactions accounting for successive cleavages of sugar units along the saccharide chain. Spenger, et al.[35] also agreed with Coates and Wilkins[36] that the ring-opening and fragmentation reactions take place thermally on laser desorption and cationization occurs after degradation. It should be noted, however, that the CID studies by Leary et al.[33] clearly show that cleavage reactions can occur after formation of the $(M + Cat)^+$ species. As with low-energy CID,[32] the α and β anomeric configurations cannot be determined by infrared LD.[35]

Orlando et al.[34] used FAB and MS-MS to study $(M + Na)^+$ complexes of oligosaccharides. They initially studied CID of smaller saccharides to determine the series of product ions produced, and then they studied larger oligosaccharides. The CID spectra of $(M + H)^+$ ions (see Figure 8A) and $(M + Na)^+$ ions (see Figure 8B) of maltodecaose can be compared. The CID spectrum of the $(M + H)^+$ ions contains a series of product ions (labeled ■ in Figure 8A) that are 162 u apart and originate with the ion of m/z 1459 formed by loss of the reducing sugar unit. This series is formed by cleavages of the glycosidic bonds. An analogous series of product ions originate with the ion of m/z 1481 in the CID spectrum of the $(M + Na)^+$ ions (see Figure 8B). In addition, the spectrum of the $(M + Na)^+$ ions shows another series of ions that are 18 u higher in mass than the previous series. Product ions are also formed that result from ring-opening cleavages and retain either the reducing sugar or the non-reducing sugar unit. Particularly, the formation of an abundant series of

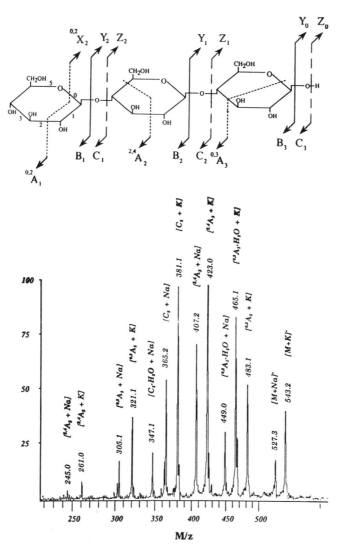

Figure 7. Infrared LD-TOF mass spectrum of a 1,4-linked trisaccharide, cellotriose (β-Glc1→4Glc→4Glc). The structure illustrates the nomenclature from Reference 37 that is indicated in Schemes 4 and 5. (Reprinted with permission from Reference 35.)

ions (labeled □ in Figure 8B) involves ring-cleavage reactions that originate with the opening of the nonreducing sugar. This example illustrates that for larger oligosaccharides, information obtained from CID of $(M + Na)^+$ ions and $(M + H)^+$ ions can be complementary.

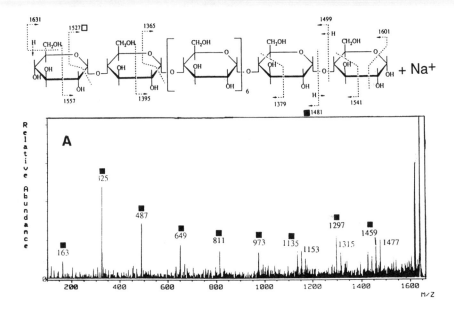

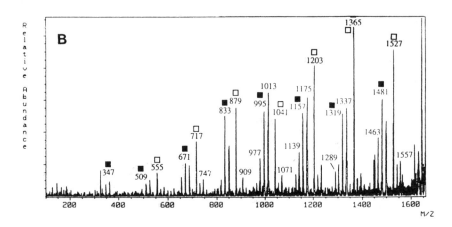

Figure 8. High-energy CID spectra of (A) $(M+H)^+$ and (B) $(M+Na)^+$ ions of a 1,4-linked decasaccharide, maltodecaose $(Glc(1\rightarrow4)[Glc(1\rightarrow4)]_8Glc)$. The peaks marked by ■ correspond to glycosidic cleavages that retain the nonreducing sugar. The peaks labeled □ correspond to ring cleavages to give product ions that contain the reducing sugar. (Reprinted with permission from Reference 34.)

3.2.4. Glycolipids

The CID of $(M + Li)^+$ ions of glycosphingolipids demonstrates a unique combination of the two types of organo-alkali chemistry described thus far.[38] Glycophingolipids are a class of biological compounds that contain a saccharide or polysaccharide attached to a long-chain lipid. The lipid is a ceramide, which is comprised of a sphingoid base with a fatty acyl group substituted at the amino group. The CID spectrum of $(M + Li)^+$ ions of lactosyl-*N*-octadecanoylsphinganine shows not only CRF of the *N*-acyl alkyl chain but also cleavages of the disaccharide (Figure 9). The high-mass end of the spectrum includes the CRF and X_n and Y_n product ions from ring-opening cleavages of the sugar units. The lower mass region of the spectrum includes abundant A_n, B_n, and C_n product ions that result from loss of the ceramide and subsequent cleavages along the sugar chain. The peak labeled E contains the intact sugar but none of the ceramide alkyl chains. Peaks labeled U and T provice molecular weight information about the *N*-acyl chain. All the product ions contain Li^+. Thus, for product ions that contain the sugar,

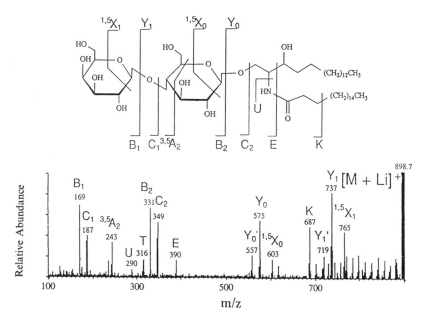

Figure 9. High-energy CID spectrum of $(M + Li)^+$ ions of lactosyl-*N*-octadeca-noylsphinganine. Ions labeled Y_n, B_n, C_n, and $^{3,5}A_2$ are $(Y_n + Li + H)^+$, $(B_n + Li + H)^+$, $(C_n + Li + H)^+$, and $(^{3,5}A_2 + Li)^+$, respectively. (Reprinted with permission from Reference 38.)

Li^+ may be coordinating with the sugar oxygens in positions bridging the glycosidic linkage, as suggested by Hofmeister *et al.*[33b] Alternatively, Li^+ may be bound at the *N*-acyl carbonyl oxygen for those product ions that do not include the sugar chain.

Alkali metal complexes were also used to help elucidate the structure of the lipid A portion of a lipopolysaccharide from gram-negative bacteria.[39] Sugar ring cleavages of the $(M + K)^+$ complexes can be used to establish lengths and locations of fatty acyl substituents.

3.3. Peptides

The gas-phase chemistry of complexes between peptides and alkali metal ions has also been an area of increasing research since the mid-1980's. Cody *et al.*[40] were the first to report that $(M + Cat)^+$ ions of peptides fragment by CID to produce structurally informative ions. They used LD-FTICR to desorb $(M + K)^+$ ions of gramicidin D and S, and CID to fragment them. The CID spectrum of $(M + K)^+$ ions of $Val^{(1)}$-gramicidin A, which is the major component of gramicidin D, contains two series of fragment ions (see Figure 10). One series (labeled ○ in Figure 10) results from cleavages that involve hydrogen rearrangements to give product ions that retain the C-terminus. The other series (labeled ∇ in Figure 10) results from loss of the N-terminal HCO followed by cleavages along the peptide chain to give product ions that retain the N-terminus. The sequence of 12 out of the 15 amino acids can be determined from the spectrum.

Other researchers[41–47] have extended this chemistry to explain the gas-phase reactions of cationized peptides and to use fragmentation patterns to explain metal ion interactions with the different functional groups. Interpretation of the data has been varied, however. Mallis and Russell[41a] who studied CID of $(M + Cat)^+$ complexes of *N*-benzoyl-Gly-His-Leu (Hip-His-Leu), were the first to propose that fragmentations reflect Na^+ binding with more than one functional group. They concluded that the alkali metal ion either is bonded intramolecularly to amide nitrogens and the N-terminal amino group or it is bonded to amide nitrogens and side chain substituents.[41] Renner and Spiteller[42] and Gross *et al.*[43], who later studied both metastable ion and CID of $(M + Li)^+$ complexes, hypothesized instead that the alkali metal ion is bonded to the deprotonated (zwitterionic) carboxylate terminus of the peptide. Westmore *et al.*[44], who studied metastable ion decompositions of $(M + Na)^+$ and $(M + K)^+$ ions, concluded instead that the metal ion is intramolecularly bonded to amide carbonyl oxygens. Teesch and Adams[45], who studied metastable ion and CID of complexes between peptides and all the alkali metal ions, also concluded that the metal ion was intramolecularly bonded

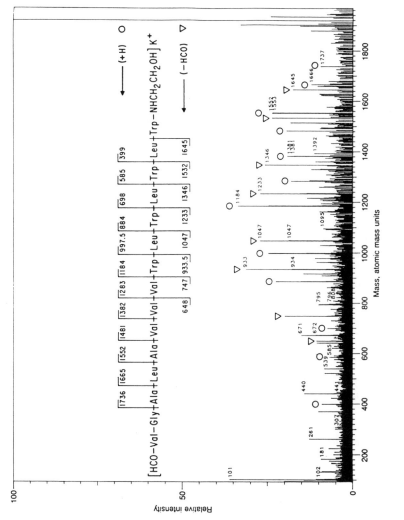

Figure 10. FTICR-CID spectrum of $(M + K)^+$ ions of $Val^{(1)}$-gramicidin A. Peaks labeled $\bigcirc$ and $\triangledown$ correspond to cleavages that include the C- and N-terminus, respectively. (Reprinted with permission from Reference 39.)

to the amide carbonyl oxygens. Leary et al.[46] suggested that fragmentations reflect a combination of the preceding theories, with emphasis on complexation of the metal ion to carbonyl oxygens.

Some of these interpretations of the location of the metal ion were based on studies of the mechanisms of fragmentation, of particular interest has been the mechanism for forming $(b_{n-m} + Cat + OH)^+$ ions.[48] Renner and Spiteller[42] were the first to identify correctly the $(b_{n-m} + Cat + OH)^+$ ions and suggested a mechanism for their formation (Scheme 6). This mechanism involves an intramolecular nucleophilic attack

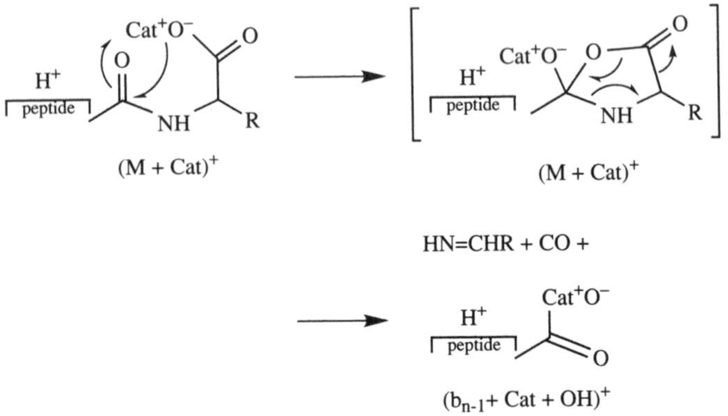

Scheme 6

by a deprotonated and cationized C-terminal carboxylate anion: The alkali metal ion is bonded to a zwitterionic peptide. Westmore et al.,[44] however, suggest an alternative mechanism (Scheme 7) in which the attacking nucleophile is instead the neutral C-terminal carboxylic acid and the metal ion is located on the N-terminal side of the reaction site: The alkali metal ion is bonded intramolecularly to carbonyl oxygens. Grese et al.[43a] offered more evidence in support of the Renner–Spiteller mechanism. They

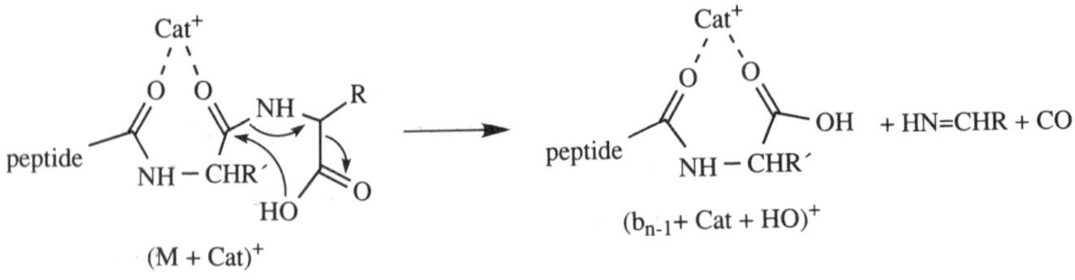

Scheme 7

convincingly argued that the mechanism in Scheme 6 was correct because C-terminal amides and methyl esters do not fragment to give $(b_{n-m} + Cat + NH_2)^+$ and $(b_{n-m} + Cat + OCH_3)^+$ ions, respectively, analogous to the $(b_{n-m} + Cat + OH)^+$ ions that arise from C-terminal acids. That is, they argued that migration of methoxide in C-terminal methyl esters should still occur to yield the analogous $(b_{n-m} + Cat + OCH_3)^+$ ions if the Westmore mechanism (Scheme 7) were instead correct. The importance of these two possible mechanisms involves the location of the metal ion. The Renner–Spiteller mechanism (Scheme 6) involves metal ion interactions that are prevalent in solution-phase chemistry, but the Westmore mechanism (Scheme 7) involves interactions that are prevalent in the gas phase.

Teesch *et al.*[45b] reinvestigated the mechanism for forming $(b_{n-m} + Cat + OH)^+$ ions and offered evidence in support of Westmore's proposal. Their strongest evidence involves results from MS-MS-MS (MS^3) experiments in which the location of the metal ion in the $(b_{n-m} + Cat + OH)^+$ product ions was specifically addressed. For example, CID spectra of $(M + Li)^+$ ions of HLGLAR contain a series of $(b_{n-m} + Li + OH)^+$ ions from b_5 to b_3 (see Figure 11A and 11B). A series of MS^3 experiments, in which the first-generation $(b_{n-m} + Li + OH)^+$ ions were themselves mass and energy selected and collisionally activated, reveal that the $(b_{n-m} + Li + OH)^+$ ions have the structure of the original peptide less the C-terminal amino acid.[43a, 45b] Furthermore, as shown in the MS^3 spectrum in Figure 11C, the first-generation $(b_4 + Li + OH)^+$ product ions from HILGLAR specifically decompose to give a loss of 88 u. Teesch and Adams[45a] presented evidence that the mechanism for a loss of 88 u involves a reaction (Scheme 8) in which the neutral loss that corresponds to 88 u is CO_2 and C_3H_8 from the C-terminal leucine.[49] [This mechanism likewise explains formation of the $(M + Li - 131)^+$ and other $(b_{n-m} + Li + OH - 88)^+$ product ions shown in Figures 11A and 11B.] The most important point is that the reaction in Scheme 8 for loss of 88 u from the $(b_{n-m} + Li + OH)^+$ product ions could not occur if Li^+ were bonded

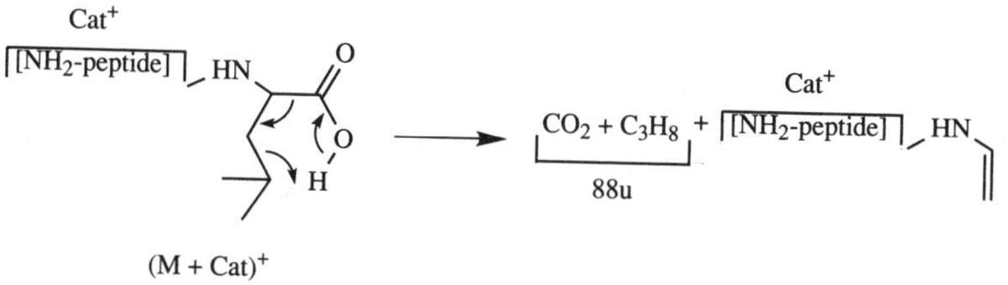

Scheme 8

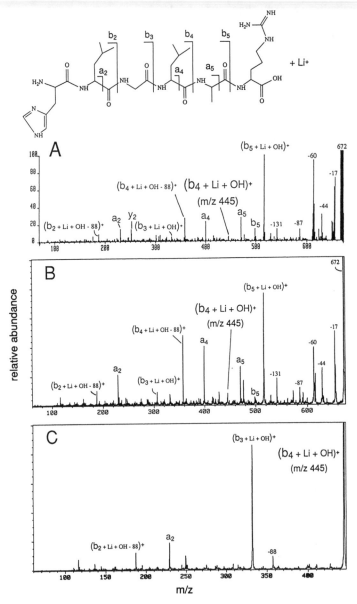

Figure 11. High-energy CID spectra of fragmentations of $(M+Li)^+$ complexes of His-Leu-Gly-Leu-Ala-Arg that occur in (A) the first field-free region between the ion source and ESA of a VG 70-S forward geometry (EB configuration, where E is an electrostatic analyzer and B is a magnetic analyzer) mass spectrometer, and in (B) the third field-free region between MS-I and MS-II of a JEOL HX110/HX110 tandem (E_1B_1-E_2B_2 configuration) mass spectro-

to a C-terminal carboxylate zwitterion: The metal ion *must* be located toward the N-terminus. In addition, analogous $(b_{n-m} + \text{Cat} + \text{OCH}_3)^+$ and $(b_{n-m} + \text{Cat} + \text{NH}_2)^+$ product ions are not observed, because the most nucleophilic site in both esters and amides is the carbonyl oxygen, *not* the methoxy oxygen or the amide nitrogen, respectively.[50] Thus, methyl esters and amides inherently do not undergo intramolecular nucleophilic reactions as in Scheme 7. Teesch and Adams[45a] and Leary *et al.*[46b] also presented evidence that the N-terminal $(a_{n-m} + \text{Li} - \text{H})^+$ ions, such as the $(a_2 + \text{Li} - \text{H})^+$ second-generation ion in Figure 11C, likewise arise from precursor ions that contain the metal ion bonded toward the N-terminus. Thus, the presence of all of these product ions in the MS^3 spectrum of the intermediate $(b_4 + \text{Li} + \text{OH})^+$ ions (see Figure 11C) indicates that the metal ion cannot be bonded to a deprotonated C-terminal carboxylate. Consequently, formation of $(b_{n-m} + \text{Li} + \text{OH})^+$ ions is best described by the Westmore mechanism in Scheme 7. The implication of these results is that gas-phase fragmentations of complexes between alkali metal ions and peptides do not reflect binding interactions that occur in bulk solution, instead, they reflect intrinsic interactions between metal ions and peptide functional groups—interactions that may somewhat reflect binding in less solvated, more hydrophobic interiors of proteins.

As in other areas of organo-alkali gas-phase chemistry discussed in Chapter 2, the role of the metal ion in determining the extent and types of fragmentations that occur from $(M + \text{Cat})^+$ complexes of peptides is still not fully understood. Specifically, the fragmentation patterns of $(M + \text{Cat})^+$ ions of peptides change significantly with increasing size of the metal ion.[45a] For example, abundances of $(b_{n-m} + \text{Cat} + \text{OH})^+$ product ions decrease, but abundances of $(a_{n-m} + \text{Li} - \text{H})^+$ ions remain virtually constant. Teesch and Adams[45b] presented evidence that rates of these different fragmentation reactions are affected by specific binding interactions between the metal ion and peptide side chains. Their results are consistent with formation of $(M + \text{Cat})^+$ complexes in which the metal ions are intramolecularly bonded to several sites in the peptide.

The chemistry of complexes between alkali metal ions and peptides is related to the chemistry of alkaline earth metal ion peptide complexes. For example, Figure 12 shows types of fragmentations that arise from

---

meter. The spectrum in (C) is a MS^3, or second-generation product ion, spectrum of the first-generation $(b_4 + \text{Li} + \text{Oh})^+$ intermediate ions of m/z 445 that were formed in the first field-free region of the tandem MS-I and then collided with He in the third field-free region between MS-I and MS-2. A scan of MS-II at a constant ratio of B/E was then used to obtain the MS^3 mass spectrum. Ions labeled a_{n-m}, b_{n-m} and y_{n-m} are $(a_{n-m} + \text{Li} - \text{H})^+$, $(b_{n-m} + \text{Li} - \text{H})^+$, and $(y_{n-m} + \text{Li} + \text{H})^+$ ions, respectively. (Reprinted with permission from Reference 45b.)

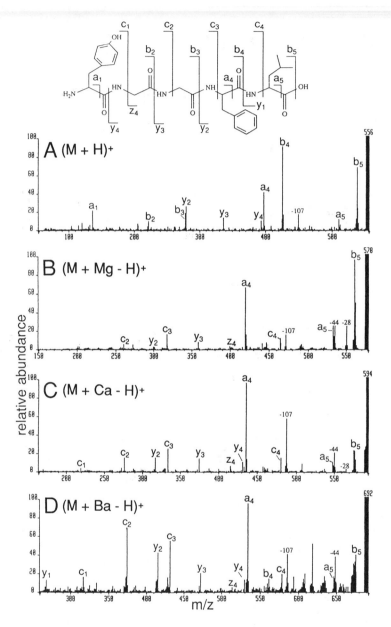

Figure 12. High-energy CID spectra of (A) $(M+H)^+$ and (B–D) $(M+Cat-H)^+$ ions of Leu enkephalin, Tyr-Gly-Gly-Phe-Leu. Ions labeled $a_{n-m'}$, $b_{n-m'}$, $c_{n-m'}$ and y^{n-m} are $(a_{n-m}+Cat-H)^+$, $(b_{n-m}+Cat-H)^+$, $(c_{n-m}+Cat)^+$, and $(y_{n-m}+Cat)^+$, respectively. The z_4 ion in spectra B–D is a $(Z_4+Cat-H)^+$ ion. (Reprinted with permission from Reference 11.)

$(M + H)^+$ and $(M + Cat^{2+} - H)^+$ ions of Leu-enkephalin.[11] Clearly missing in the spectra of the alkaline earth metal ion complexes is the $(b_{n-m} + Cat + OH)^+$ ion that prevails with the alkali metal ion complexes; there are analogous a_{n-m} and y_{n-m} ions, however. Most interesting is the formation of abundant $(c_{n-m} + Cat)^+$ ions from the alkaline earth metal ion complexes: Analogous c_{n-m} ions are not seen in spectra of complexes that contain the alkali metal ions. The complete series of N-terminal $(c_{n-m} + Cat)^+$ ions and an almost complete series of C-terminal $(y_{n-m} + Cat)^+$ ions can help determine the sequence of the peptide. By converting the C-terminal carboxylate into an amide, the relative abundances of the $(y_{n-m} + Cat)^+$ ions can be increased.[11] (This also occurs for alkali metal ion complexes.[45, 47]) Abundances of the lower mass product ions increase with increasing size of the metal ion (see Figures 12B–D). As with alkali metal ions, these trends and proposed mechanisms for forming product ions support the proposal that alkaline earth metal ions are bonded intramolecularly with several sites in the peptide.[11]

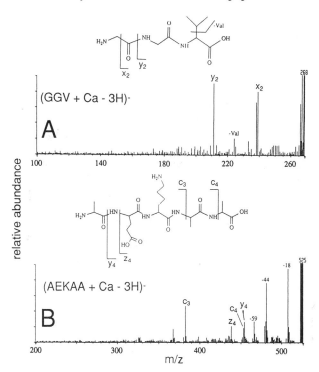

Figure 13. High-energy CID spectra of $(M + Ca - 3H)^-$ ions of (A) Gly-Gly-Val and (B) Ala-Glu-Lys-Ala-Ala. Ions labeled $c_{n-m'}$, $x_{n-m'}$, and z_{n-m} are $(c_{n-m} + Ca - 2H)^-$, $(x_{n-m} + Ca - 2H)^-$, $(y_{n-m} + Ca - 2H)^-$, and $(z_{n-m} + Ca - 3H)^-$, respectively.

Anionic $(M+Ca+3H)^-$ complexes formed between alkaline earth metal ions and peptides have also been studied.[51] Two examples of CID of anionic complexes are shown in Figure 13. The $(M+Ca-3H)^-$ ions of GGV (see Figure 13A) fragment to yield only C-terminal sequence ions, $(x_2+Ca-2H)^-$ and $(y_2+Ca+2H)^-$ ions. In contrast, $(M+Ca-3H)^-$ ions of AEKAA (see Figure 13B) decompose to give both C-terminal $(y_{n-m}+Ca-2H)^-$ and $(z_{n-m}+Ca-3H)^-$ ions and N-terminal $(c_{n-m}+Ca-2H)^-$ ions. The N-terminal $(c_{n-m}+Ca-2H)^-$ product ions are particularly interesting because they are observed only for peptides that contain protic amino acids. Specifically, Adams *et al.*[50] have shown that formation of $(c_{n-m}+Ca-3H)^-$ ions requires either a His, Tyr, Glu, or Asp residue and two adjacent amides that can be deprotonated. The $(c_{c-m}+Ca-2H)^-$ ions are believed to arise from structurally specific complexes containing the metal ion intramolecularly chelated to the deprotonated amino acid side chain and two deprotonated amides.

Gross *et al.*[52b] examined CID of $(M+Sr-3H)^-$ and $(2M+Sr-3H)^-$ ions of peptides that contain only aliphatic (aprotic) amino acid side chains (see Figure 14); again, only C-terminal $(x_{n-m}+Sr-2H)^-$ and $(y_{n-m}+Sr-2H)^-$ sequence ions are formed. For

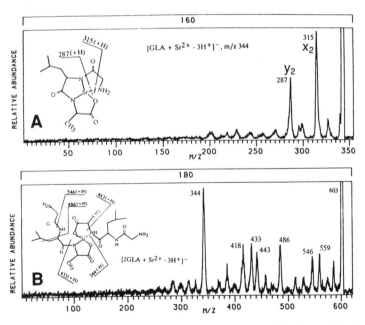

Figure 14. CID spectra of (A) $(M+Sr-3H)^-$ and (B) $(2M+Sr-3H)^-$ ions of Gly-Leu-Ala. (Reprinted with permission from Reference 52b.)

example, the CID spectrum of $(M+Sr-3H)^-$ ions of GLA (see Figure 14A) show the $(x_2+Sr-2H)^-$ ions of m/z 315 and 287, respectively. These fragmentations are believed to arise from complexes in which Sr^{2+} is bonded to the deprotonated C-terminus and the first two adjacent deprotonated amides. Decompositions of the $(2M+Sr-3H)^-$ dimers of GLA (see Figure 14B) are believed to arise from species in which the metal ion is bonded to one singly charged peptide and one doubly charged peptide to gave a $[(M-H)Sr(M-2H)]^-$ complex. The CID of the $(2GLA+Sr-3H)^-$ cluster ions (see Figure 14B) produces several product ions that are thought to arise from cleavages mainly along the singly charged peptide. The dimer also decomposes to give the monomer of m/z 344.

Differences between the chemistry of cationic and anionic complexes between alkaline earth metal ions and peptides is clear from Figures 12–14. From our experiences,[11, 51] positive ion complexes can be produced by FAB at greater abundances and more easily. More sequence information is obtained from fragmentations of the cationic complexes. The anionic complexes, however, provide some interesting fundamental insights into gas-phase binding interactions between multifunctional molecules, such as peptides and metal ions.

4. CONCLUSIONS

Two important conclusions can be drawn from the chemistry discussed in this chapter: (1) From the simplest of molecules to multifunctional peptides, the exact role of the metal ion has yet to be completely understood. We do know, however, that product ion formation is affected by the location of the metal ion and relative stabilities of the resulting $(M+Cat)^+$ complexes.[17, 19a, 33b, 45, 46b] (2) The gas-phase chemistry of cationized complexes yields an abundance of structural information. In some cases, the information is complementary to that obtained from $(M+H)^+$ ions; in other cases, however, decompositions of the $(M+Cat)^+$ complexes produce significantly more information about the molecule than other conventional mass spectrometric techniques. The simplicity of cationizing molecules to determine molecular weights and structure makes it an important analytical tool.

The abundant information available from the gas-phase chemistry of alkali metal ion complexes has influenced the amount of new research that is using metal ion adducts for various purposes. Thus, new applications of metal ion-organo chemistry have been reported. For example, electrospray ionization can be used (1) to quantitate alkali adducts of antibiotics that do not form abundant $(M+H)^+$ ions;[53] (2) to study transition metal ion complexes, such as $[Ru^{II}(bpy)_3]Cl_2$, which undergo reduction during conventional ionization processes;[54] and (3) to perform gas-phase equilibrium experiments with doubly charged metal ions[55] similar to those discussed earlier.[1-5] The LD-FTICR can also be used to form complexes between transition metal ions and porphyrins[10a, b] and peptides.[10c, d]

ACKNOWLEDGEMENTS

The National Science Foundation (grant 9113272), the Emory University Research Fund, and the donors of the Petroleum Research Fund, administered by the American Chemical Society (grant 21755-G5), provided partial support for this research.

REFERENCES

1. (a) Džidić, I.; and Kebarle, P., *J. Phys. Chem.* 1970, **74**, 1466–1474. (b) Kebarle, P., in *Ion-Molecule Reactions*, Vol. 1, Franklin, J. L., ed.; (Plenum Press: New York, 1972), and pp. 315–362. (c) Davidson, W. R.; and Kebarle, P., *J. Am. Chem. Soc.* 1976, **98**, 6125–33. (d) Davidson, W. R.; and Kebarle, P., *J. Am. Chem. Soc.* 1976, **98**, 6133–38.

2. (a) Wieting, R. D.; Staley, R. H.; and Beauchamp, J. L., *J. Am. Chem. Soc.* 1975, **97**, 924–26. (b) Staley, R. H.; and Beauchamp, J. L., *J. Am. Chem. Soc.* 1975, **97**, 5920–21. (c) Woodlin, R. L.; and Beauchamp, J. L., *J. Am. Chem. Soc.* 1975, **100**, 501–8. (d) Hodgers, R. V.; and Beauchamp, J. L., *Anal. Chem.* 1976, 48, 825–29.

3. Rode, B. M., *Chem. Phys. Lett.* 1975, **35**, 517–20.

4. Good, A., *Chem. Rev.* 1975, **75**, 561–83.

5. (a) Castleman, A. W., Jr., *Chem. Phys. Lett.* 1978, **53**, 560–64. (b) Castleman, A. W., Jr.; Holland, P. M.; Lindsay, D. M.; and Peterson, K. I., *J. Am. Chem. Soc.* 1978, **100**, 6039–45.

6. Allison, J.; and Ridge, D. P., *J. Am. Chem. Soc.* 1979, **101**, 4998–5009.

7. (a) Bombick. D.; Pinkston, J. D.; Allison, J., *J. Anal. Chem.* 1984, **56**, 396–402. (b) Bombick, D. D., and Allison, J., *Anal. Chem.* 1987, **59**, 458–66.

8. (a) Rollgen, F. W.; and Schulten, H. R., *Z. Naturforsch* 1975, **30a**, 1685–90. (b) Posthumus, M. A.; Kistemaker, P. G.; Meuzelaar, H. L. C.; and Ten Noever de Brauw, M. C., *Anal. Chem.* 1978, **50**, 985–91. (c) Liu, L. K.; Busch, K. L.; and Cooks, R. G., *Anal. Chem.* 1981, **53**, 109–13. (d) Kambara, H.; and Hishida, S., *Anal. Chem.* 1981, **53**, 2340–44. (e) McNeal, C. J.; Ogilvie, K. K.; Theriault, N. Y., and Nemer, M. S., *J. Am. Chem. Soc.* 1982, **104**, 981–84. (f) Weber, R.; Levsen, K.; Louter, G. J.; Boerboom, A. J. H.; and Haverkamp, J., *Anal. Chem.* 1982, **54**, 1458–66. (g) Siegel, M. M.; McGahren, W. J.; Tomer, K. B.; and Chang, T. T., *Biomed. Environ. Mass Spectrom.* 1987, **14**, 29–38. (h) Evans, C.; Traldi, P.; Bambagiotti-Alberti, M.; Giannellini, V.; Coran, S. A.; and Vincieri, F. F., *Biol. Mass Spectrom.* 1991, **20**, 351–56. [*Also see* Reference 24(b).]

9. (a) Burnier, R. C.; Byrd, G. D.; and Freiser, B. S., *J. Am. Chem. Soc.* 1981, **103**, 4360–67. (b) Freiser, B. S., *Anal. Chim. Acta* 1985, **178**, 137–58. (c) Peake, D. A.; Gross, M. L.; and Ridge, D. P., *J. Am. Chem. Soc.* 1984, **106**, 4307–16. (d) Peake, D. A.; and Gross, M. L., *Anal. Chem.* 1985, **57**, 115–20. (e) Peake, D. A.; and Gross, M. L., *ORGND* 1896, **5**, 1236–43. (f) Peake, D. A.; Huang, S. K.; and Gross, M. L., *Anal. Chem.* 1987, **59**,

1557–63. (g) Schulze, C.; Schwarz, H.; Peake, D. A.; and Gross, M. L., *J. Am. Chem. Soc.* 1987, **109**, 2368–74.

10. (a) Nuwaysir, L. M.; and Wilkins, C. L., *Anal. Chem.* 1989, **61**, 689–94. (b) Irikura, K. K.; and Beauchamp, J. L., *J. Am. Chem. Soc.* 1991, **113**, 2767–68. (c) Speir, J. P.; Gorman, G. S.; and Amster, I. J., *Proc. 39th ASMS Conf. Mass Spectrom. Allied Topics*, 1991, 455–56. (d) Speir, J. P.; Gorman, G. S.; and Amster, I. J., in *Mass Spectrometry in the Biological Sciences: A Tutorial*; Gross, M. L., ed.; (Kluwer: Dordrecht, 1991), Chap. 11.

11. Teesch, L. M.; and Adams, J., *J. Am. Chem. Soc.* 1990, **112**, 4110–20.

12. Crockett, J. S.; Gross, M. L.; Christie, W. W.; and Holman, R. T., *J. Am. Soc. Mass. Spectrom* 1990, **1**, 183–91.

13. Arnett, E. M.; Jones, F. M.; Taagepera, M.; Henderson, W. G.; Beauchamp, J. L.; Holtz, D.; and Taft, R. W., *J. Am. Chem. Soc.* 1972, **94**, 4724–26.

14. Pullman, A.; and Brochen, P., *Chem. Phys. Lett.* 1975, **34**, 7–10.

15. Reviews of gas-phase chemistry of transition metal ion complexes include the following: (a) Allison, J., *Progr. Inorg. Chem.* 1986, **34**, 627–76. (b) Squires, R. R., *Chem. Rev.* 1987, **87**, 623–46. (c) Vairamani, M.; Mirza, U. A.; and Srinivas, R., *Mass Spectrom. Rev.* 1990, **9**, 235–58.

16. Rollgen, F. W.; Borchers, F.; Giessmann, U.; and Levsen, K., *Org. Mass Spectrom.* 1977, **12**, 541–43.

17. Adams, J.; and Gross, M. L., *J. Am. Chem. Soc.* 1986, **108**, 6915–21.

18. Jensen, N. J.; Tomer, K. B.; and Gross, M. L., *J. Am. Chem. Soc.* 1985, **107**, 1863–68.

19. (a) Adams, J.; and Gross, M. L., *Anal. Chem.* 1987, **59**, 1576–82. (b) Adams, J.; and Gross, M. L., *Org. Mass Spectrom.* 1988, **23**, 307–16.

20. Adams, J.; Deterding, L. J.; and Gross, M. L., *Spectrosc. Int. J.* 1987, **5**, 199–228.

21. Contado, M. J.; and Adams, J., *Anal. Chim. Acta* 1991, **246**, 187–97.

22. Contado, J. M.; Adams, J.; and Gross, M. L., *Adv. Mass Spectrom.* 1989, **11B**, 1034–35.

23. Antione, M.; and Adams, J., *J. Am. Soc. Mass. Spectrom* 1992, **3**, 776–8.

24. (a) Adams, J.; and Gross, M. L., *J. Am. Chem. Soc.* 1989, **111**, 435–40. (b) Contado, M. J.; Adams, J.; Jensen, N. J.; and Gross, M. L., *J. Am. Soc. Mass Spectrom.* 1991, **2**, 180–83.

25. Wysocki, V. H.; and Ross, M. M., *Int. J. Mass Spectrom. Ion Phys.* 1991, **104**, 179-211.

26. Adams, J., *Mass Spectrom. Rev.* 1990, **9**, 141–86.

27. (a) Schulten, H. R.; and Games, D. E., *Biomed. Mass Spectrom.* 1974, **1**, 120–123. (b) Rollgen, F. W.; and Schulten, H. R., *Org. Mass Spectrom.* 1975, **10**, 660–68.

28. Stoll, R.; and Rolgen, F. W., *Org. Mass Spectrom* 1979, **14**, 642–45.

29. (a) Prome, J. C.; and Puzo, G., *Org. Mass Spectrom* 1977, **12**, 28–32. (b) Prome, J. C.; and Puzo, G., *Isreal J. Chem.* 1978, **17**, 172–76.

30. (a) Puzo, G.; and Prome, J. C., *Spectros. Int. J.*, 1984, **3**, 155–58. (b) Puzo, G.; Fournie, J. J.; and Prome, J. C., *Anal. Chem.* 1985, **57**, 892–94. (c) Fournie, J. J.; and Puzo, G., *Anal. Chem.* 1985, **57**, 2287–89. (d) Puzo, G.; and Prome, J. C., *Org. Mass Spectrom* 1985, **20**, 288–91.

31. (a) Rollgen, F. W.; Geissmann, U.; Borchers, F.; and Levsen, K., *Org. Mass Spectrom* 1978, **13**, 459–61. (b) Cerny, R. L.; Tomer, K. B.; and Gross, M. L., *Org. Mass Spectrom* 1986, **21**, 655–60.

32. Wright, L. G.; Cooks, R. G.; and Wood, K. V., *Biomed. Mass Spectrom* 1985, **12**, 159–62.

33. (a) Zhou, Z.; Ogden, S.; and Leary, J. A., *Org. Chem.* 1990, **55**, 5444–46. (b) Hofmeister, G. E.; Zhou, Z.; and Leary, J. A., *Am. Chem. Soc.* 1991, **113**, 5964–70.

34. Orlando, R.; Bush, C. A.; and Fenselau, C., *Biomed. Environ. Mass Spectrom.* 1990, **19**, 747–54.

35. Spengler, B.; Dolce, J. W.; and Cotter, R. J., *Anal. Chem.* 1990, **62**, 1731–37.
36. (a) Coates, M. L.; and Wilkins, C. L., *Biomed. Mass Spectrom.* 1985, **12**, 424–28. (b) Coates, M. L.; and Wilkins, C. L., *Anal. Chem.* 1987, **59**, 197–200.
37. Domon, B.; and Costello, C. E., *Glycoconjugate J.* 1988, **5**, 397–409.
38. (a) Ann, Q.; and Adams, J., *J. Am. Soc. Mass. Spectrom* 1992, **3**, 260–63. (b) Ann, Q; and Adams, J., *Anal. Chem.* 1993, **65**, 7–13. (c) Adams, J; and Ann, Q., *Mass Spectrom. Rev.* 1993, **12**, 51–85.
39. (a) Takayama, K.; Qureshi, K.; Hyver, K.; Honovich, J.; Cotter, R. J.; Mascagni, P.; and Schneider, H., *J. Biol. Chem.* 1986, **261**, 10624–631. (b) Cooter, R. J.; Honovich, J.; Qureshi, N.; and Takayama, K., *Biomed. Environ. Mass. Spectrom.* 1987, **14**, 591-98. (c) Qureshi, N.; Honovich, J. P.; Hara, H.; Cotter, R. J.; and Takayama, K., *Biol. Chem.* 1988, **263**, 5502–04.
40. Cody, R. B.; Amster, J. I.; and Mclafferty, F. W., *Proc. Nat. Acad. Sci. USA* 1985, **82**, 6367–70.
41. (a) Mallis, L. M.; and Russel, D. H., *Anal. Chem.* 1986, **58**, 1076–80. (b) Russell, D. H., *Mass Spectrom. Rev.* 1986, **5**, 167–89. (c) Russell, D. H.; McGlohon, E. S.; and Mallis, L. M., *Anal. Chem.* 1988, **60**, 1818–24.
42. Renner, D.; and Spiteller, G., *Biomed. Environ. Mass Spectrom.* 1988, **15**, 75–77.
43. (a) Grese, R. P.; Cerny, R. L.; and Gross, M. L., *J. Am. Chem. Soc.* 1989, **111**, 2835-42. (b) Grese, R. P.; and Gross, M. L., *J. Am. Chem. Soc.* **112**, 5098–5104.
44. Tang, X.; Ens, W.; Standing, K. G.; and Westmore, J. B., *Anal. Chem.*, 1988, **60**, 1791–99.
45. (a) Teesch, L. M.; and Adams, J., *J. Am. Chem. Soc.* 1991, **113**, 812–20. (b) Teesch, L. M.; and Adams, J., *J. Am. Chem. Soc.* 1991, **113**, 3668-75.
46. (a) Leary, J. A.; Williams, T. D.; and Bott, G., *Rapid Comm. Mass Spectrom.* 1989, **3**, 192-96. (b) Leary, J. A.; Zhou, Z.; Ogden, S. A.; and Williams, T. D., *J. Am. Soc. Mass Spectrom.* 1990, **1**, 473–80.
47. Tomer, K. B.; Deterding, L. J.; and Guenat, C., *Biol. Mass Spectrom.* 1991, **20**, 121-29.
48. The terminology used here for the sequence ions is based on Roepstorff, P.; and Fohlman, J., *Biomed. Mass Spectrom.* 1984, **11**, 601.
49. These product ions are also analogous to d_n product ions observed in CID spectra of $(M+H)^+$ ions (Johnson, R. S.; Martin, S. A.; and Biemann, K., *Int. J. Mass Spectrom. Ion Processes.* 1988, **86**, 137, and References 42b and 46).
50. (a) Singh, A.; Andrews, L. J.; and Keefer, R. M., *J. Am. Chem. Soc.* 1962, **84**, 1179-85. (b) Lovins, R. E.; Andrews, L. J.; and Keefer, R. M., *J. Am. Chem. Soc.* 1962, **84**, 3959–62. (c) Andrews, L. J.; and Keefer, R. J., *J. Am. Chem. Soc.* 1959, **81**, 4218–23. (d) Keefer, R. M.; and Andrews, L. J., *J. Am. Chem. Soc.* 1959, **81**, 5329–33. (e) Lawson, W. B.; Gross, E.; Foltz, C. M.; and Witkop, B.; *J. Am. Chem. Soc.* 1962, **84**, 1715–18. (f) Stirling, C. J. M., *J. Am. Chem. Soc.* 1960, 252–62.
51. (a) Zhao, H.; Reiter, A.; Teesch, L. M.; and Adams, J., *J. Am. Chem. Soc.* 1993, **115**, 2854–63. (b) Zhao, H.; Reiter, A.; Teesch, L. M.; and Adams, J., *Int. J. Mass Spectrom. Ion Proc.* 1993, **127**, 17–26. (c) Reiter, A.; Zhao, H.; and Adams, J., *Org. Mass Spectrom* (in press).
52. (a) Hu, P.; and Gross, M. L., *J. Am. Chem. Soc.* 1991, **114**, 9153–60. (b) Gross, M. L.; Cerny, R. L.; Giblin, D. E.; Rempel, D. L.; MacMillan, D. K.; Hu, P.; and Holliman, C. L., *Anal. Chim. Acta* 1991, **250**, 105–30.
53. Schnieder, R. P.; Lynch, M. J.; Ericson, J. F.; and Fouda, H. G., *Anal. Chem.* 1991, **63**, 1789–94.
54. Katta, V.; Chowdbury, S. K.; and Chait, B. T., *J. Am. Chem. Soc.* 1991, **112**, 5348–49.
55. Jayaweera, P.; Blades, A. T.; Ikonomou, M. G.; and Kebarle, P., *J. Am. Chem. Soc.* 1990, **112**, 2452–54.

3

Fundamental Studies of
Collision-Induced Dissociation of Ions

Anil K. Shukla and Jean H. Futrell

1. INTRODUCTION

Collision-induced dissociation (CID) [also called collision-activated dissociation (CAD), tandem mass spectrometry, or MS/MS] of ions in a mass spectrometer plays an increasingly significant role in ion structure determination and analysis of complex mixtures.[1,2] Improvement in MS/MS techniques has been a significant driving force in mass spectrometric instrumental development. The CID has attained even greater importance with the advent of fast atom bombardment,[3] laser desorption,[4] and electrospray[5] techniques for generating ions from large, nonvolatile molecules. These techniques usually produce (quasi- and multicharged) molecular ions with very little fragmentation. Collisional activation of these "molecular" ions and subsequent dissociation to various fragment ions provides structural information about these "molecular ion moities" required for identification and characterization. A few closely related methods, such as angle-resolved[6] and energy-resolved[7] mass spectrometry, surface-induced,[8] and photon-induced[9] dissociations have also been developed and succesfully applied in the past decade. Although it is a well-established method with many important applications to chemi-

Anil K. Shukla and Jean H. Futrell • Departments of Chemistry and Biochemistry, University of Delaware, Newark, Delaware 19716.

Experimental Mass Spectrometry, edited by David H. Russell. Plenum Press, New York, 1994.

cal analysis, CID remains an active research area in mass spectrometry with many nuances being actively explored.

The CID studies of polyatomic ions as presently practiced had their origins in the study of metastable ion dissociation in field-free regions of double-focusing mass spectrometers, first introduced as a technique by Barber and Elliott[10] and by Futrell et al.[11] Collisional activation by increasing flight tube pressure was demonstrated by Jennings[12] and Haddon and McLafferty,[13] a differentially pumped collision cell for CID studies was introduced by Hills and Futrell.[14] These studies showed that CID spectra were qualitatively similar to those obtained by electron impact ionization, and they provided definitive information on fragmentation pathways. It has since been shown that differences between electron ionization and CID spectra can be exploited as a means of mixture analysis to distinguish isomeric structures and reaction mechanism.[15] The development of reversed geometry sector mass spectrometers as well as linked scans for forward geometry mass spectrometers considerably simplified CID experiments.[16] Shortly thereafter, low-energy CID experiments were performed by combining two or more quadrupole mass filters in tandem.[17] Tandem hybrid instruments were logical extensions that exploited advantages of both sector mass spectrometers and quadrupole mass filters.[18]

Basic mechanisms of CID were investigated in early experimental and theoretical studies of simple di- and triatomic ions.[19] Molecular beam-scattering studies of small molecular ions O_2^+, N_2^+, NO^+, N_2^+, and especially H_2^+- have defined the basic mechanisms often invoked in discussing the CID of polyatomic ions at low collision energies.[20] In these small ion systems, particularly at a low energy, the dominant dynamics feature is a pronounced peak of product ion intensity that forward scattered with essentially the velocity of the primary ion beam. This follows predictions of the classic spectator-stripping model, where one atom (or molecular fragment) is essentially stripped away and the remainder of the ion (or molecule) proceeds in flight with little or no change in momentum.

For the di- and triatomic ion cases cited, it is postulated that lengthening the bond of the struck atom or an electronic state change to a weakly repulsive or nonbonding state causes the bond energy to decrease during the collision to a negligible value so that momentum exchange with the non interacting ion or fragment is minimal.[21] Also noted in these experiments were velocity and angular broadening of ion intensity about the spectator-stripping peak, suggesting that the mechanism was more complicated than the simple stripping model implies. Additional mechanisms were implied by the presence of diffuse product intensity at large scattering angles and at higher and lower velocities than the spectator-stripping peak.[21] The detailed mechanisms were specific for each of the systems studied, however, all had the common feature of a prominent product ion peak in the spectator-stripping position.

Because its properties are so wellknown both theoretically and experimentally, the CID of H_2^+ has been studied in greatest detail. The energy (velocity) dependence of the product velocity spectra of H^+ from H_2^+ CID provides the primary information for CID schemes classification by Durup[18] that is often cited in articles on CID mechanisms. The well-known behavior of this system at high-collision energies and at low energies where conventional ion molecule reactions compete effectively with CID provides very useful information about plausible mechanisms in CID despite the fact that no one realistically expects polyatomic ions to follow the same mechanistic pathways. CID of H_2^+ has been studied in crossed beam experiments by Haveman *et al.*[22] and in single-beam state-selected experiments by Chupka and Russell,[23] Anderson *et al.*,[24] and by Guyon *et al.*[25, 26] Recent theoretical calculations used to interpret these experiments include the work of Whitton and Kuntz,[27] Kumar and Sathyamurthy,[28] Vogler and Seibt,[29] and Gislason and Guyon.[30a] These recent results, which have direct parallels in our molecular beam dynamics studies of the CID of polyatomic ions are reviewed in Section 5.1.

These fundamental studies of small ions have provided invaluable insights into basic CID mechanisms and in particular, changes in mechanisms as a function of collision energy. However, it can hardly be expected that these principles will carry over directly to polyatomic ions. For this reason, the pioneering dynamics study carried out by Herman *et al.*[31] on the CID of CH_4^+ and $C_3H_8^+$ provides a superior starting point for interpreting the CID of polyatomic systems. For CH_4^+, the reaction

$$CH_4^+ + He \rightarrow CH_3^+ + H + He \tag{1}$$

was investigated and a number of important characteristics of polyatomic ion CID were established. A point of particular interest was to determine whether the stripping mechanism that figures prominently in diatomic and triatomic ion CID is equally important for this pentatomic ion.

Figure 1, taken from Reference 31, demonstrates that the stripping reaction found to be such a prominent feature for small ions is actually quite unimportant for CH_4^+. Two circles in Figure 1 circumscribe the regions where the CH_3^+ product is found for the stripping reaction (smaller circle) and the alternative model where activation of the entire ion is involved (larger circle). [Details of constructing and interpreting center-of-mass (CM) scattering diagrams are presented in Section 3.] The two circles follow directly from the different CM velocities of the He−H and

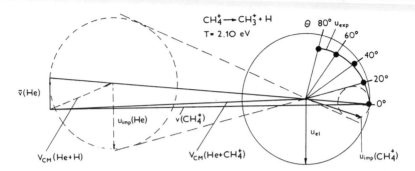

Figure 1. Schematic Newton diagram showing the most probable CM velocities of collisionally activated CH_4^+ at 2.10-eV collision energy. The solid circle corresponds to the elastic scattering of CH_4^+ by He, and the small broken circle shows the locus of the CH_4^+ CM velocities expected from the impulsive spectator-stripping model, where only 1 H atom takes part in the collision process. The construction follows directly from the CM velocities of the systems $(He+CH_4^+)$ and $(He+H)$, respectively. The large dashed circle locates the region where He and H are scattered, and this circle is projected through the system $(CH_4^+ +He)$ CM to construct the dashed circle where CH_3^+ is found (as required by conservation of momentum). The solid circle is constructed using the principles described in detail in the text. The solid points are the maximum in the experimental contour diagram (Ref. 31), and the fact that they fall within the solid circle (and outside the dashed circle) is evidence that the scattering collision particles are CH_4^+ and He.

$He-CH_4^+$ collision frames. The plotted experimental points show conclusively that the second model—collisional excitation of the entire CH_4^+ moiety—is the operative mechanism.

In addition, the role of vibrational energy in CID of CH_4^+ generated by 70-eV electron impact was clearly demonstrated. At zero deflection angle, the observed endothermicity for Reaction 1 was 0.08 eV rather than the 1.54-eV minimum energy required to dissociate ground state CH_4^+. This demonstrates both that internal energy present as vibrational excitation is efficiently used in CID and that highly vibrationally excited ions are preferentially dissociated at a zero-scattering angle. With an increasing angle, ions with less vibrational energy are progressively excited into the dissociation continnum. Over the collision energy range from 1.2–4 eV CM, an approximately linear dependence of energy transfer on a CM-scattering angle was observed.

These scattering characteristics are consistent with a correlation between energy deposition and impact parameter. Long-range interactions deposit less energy, while small-impact parameter, impulsive collisions result in large angular scattering accompanying larger energy transfers. The limit of zero impact parameter results in the largest transfer of energy and back scattering from the CM. These concepts have been demonstrated elsewhere in molecular beams study of energy transfer in inelastic collisions by Cheng *et al.*[21] and Qian *et al.*[32] These studies also supported the view that the CID of polyatomic ions can be adequately rationalized using the concepts of the quasi-equilibrium theory[33] (QET) of mass spectra. All of the energy content of the molecular internal energy deposited in the ionization step—plus that acquired in the collision—is available potential energy for driving dissociation into fragment ions.

Although there are a great number of studies on the effects of ion kinetic energy and collision gas on ion fragmentation patterns and kinetic energy release on dissociation, very little quantitative information is available about actual mechanisms of energy transfer and dissociation of collisionally activated polyatomic ions. With the exception of the single low-energy study previously described,[31] no detailed experiments investigating the dynamics of polyatomic ion CID were carried out prior to the research initiated in our laboratory in 1988. In Chapter 3, we describe these recent studies of the dynamics of low- and medium-collision energy CID carried out using an angle- and energy-resolved crossed beam tandem mass spectrometer developed especially for investigating this problem.

The apparatus used for these mesurements is shown schematically in Figure 2 and described in detail elsewhere.[34] The primary beam is mass- and energy-selected by a double-focusing mass spectrometer, decelerated and focused on a perpendicularly moving supersonic beam of neutral

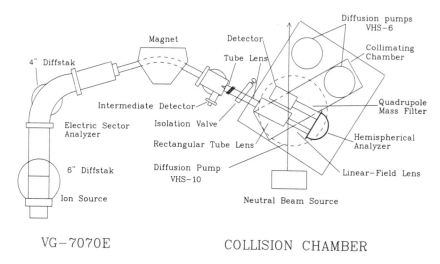

Figure 2. Schematic diagram of the tandem hybrid mass spectrometer.

atoms. The mass, energy, and scattering angle of detected products are measured by the second-stage spectrometer, which pivots around the molecular beam axis. All these parameters must be measured for complete analysis of reaction kinematics. The first studies using this apparatus have revealed many surprising features in the activation and dissociation steps in polyatomic ion CID, especially at low-collision energies.

2. COLLISION-INDUCED DISSOCIATION: A TWO-STEP PROCESS

It is useful to consider CID as proceeding via two steps that are separable in time, *viz.*, collisional activation where a fraction of an ion's kinetic energy is transferred into the internal energy of the ion, followed by unimolecular dissociation of the internally excited ion. The formal requirement for applying the two-step mechanism assumed by Equations 2 and 3

$$M_1^+ + M_2 \rightarrow M_1^{+*} + M_2 \tag{2}$$

$$M_1^{+*} \rightarrow M_3^+ + M_4 \tag{3}$$

is that the neutral M_2 not be present in the force field of the excited ion M_1^{+*} during the dissociation step. This interpretation is readily rationalized in terms of typical collision times for CID. For example, if we assume an interaction length of 2 Å, the collision time for an 8-keV ion of m/z 44 with a low-velocity neutral is 10^{-15} sec, far too short for nuclear motion to occur. At 5 eV, collision time increases to 4×10^{-14} sec, still much shorter than the time scale typical for ion dissociation (10^{-12}–10^{-5} sec) in mass spectrometry. The extreme mismatch of characteristic excitation and dissociation times allows us to consider the two processes as proceeding sequentially.

A weaker assumption often made in interpreting CID experiments is that the relative kinetic energy of ion and neutral separation after fragmentation is small. Consequently, the two fragments move collinearly in the laboratory (LAB) frame with nearly the same velocity—i.e., there is negligible kinetic energy in the reaction coordinate as the fragment ion and neutral separate. [For those cases where kinetic energy release cannot be neglected, the maximum in the velocity distribution remains interpretable as the average velocity of the excited parent ion, while the distribution is broadened by energy release (and other factors).] Based on these assumptions, measured fragment ion velocity vectors are used to map velocity vectors for collisionally excited parent ions. Transforming these velocity vectors into the CM reference frame provides a simple and quantitative means of evaluating energy transfer in CID. These relationships have been used to construct ΔT circles in scattering diagrams discussed later in this chapter. As noted, kinetic energy release on dissociation broadens only the energy distribution and does not affect the measurement of average energy-transferred in CID.

The activation step has been the most elusive goal in understanding the CID. Knowledge of the internal energy distribution of activated ions is the essential first step in applying unimolecular decay theories (e.g., QET/Rice–Rampsperger–Kassel–Marcus (RRKM)[35] theories) to describe the fragmentation step. The second characteristic feature of the excitation step—which has received even less attention than energy deposition—is angular scattering of the collisionally activated ion. These features—energy deposition and angular scattering—are the key parameters in reaction kinematics. The first systematic measurements of these quantities for polyatomic ion CID have been carried out in our laboratory using the apparatus shown in Figure 2.

3. REACTION KINEMATICS

Before discussing experimental results, we review some general features of scattering experiments, with emphasis on the merits of presenting data in CM coordinates. All experimental measurements in a collision process are obviously made in the laboratory frame (LAB) of reference. However, mathematical analysis of the collision process is greatly simplified in the CM reference frame. In this coordinate system, the parent ion and neutral collide collinearly at 180°, and the excited ion and recoiling neutral retreat from the CM after the collision. The CM moves with respect to the LAB system, so that its origin is always coincident with the CM of the colliding particles. The kinetic energy and momentum of the CM of the collision pair is unchanged in the collision. It follows that the motion of the CM should be subtracted, since only the relative motion of the two particles is significant. Motion of the two particles can be further reduced to that of a single particle of reduced mass μ. In this frame, the kinetic energy of the system is

$$1/2\,\mu v_r^2 \qquad \text{where } \mu = M_1 \cdot M_2/(M_1 + M_2),$$

M_1 and M_2 are the masses of the colliding particles, and v_r is the relative velocity.

Since the origin of the coordinates is the CM of the colliding reactants, collision takes place at the origin in the CM frame. Both before and after collision, momentum vectors are symmetric with respect to the relative velocity vector. Expressing experimental results in this reference frame displays the conservation of energy and momentum in the collision in an easily interpretable form. In particular, it provides a direct means of deducing kinetic energy transformed into internal energy of the ion. It also provides a way of testing various models for the collision process that predict different angular-scattering characteristics.

Figure 3 shows the geometric transformation from LAB to CM coordinates when an ion and neutral collide at right angles; V_1 and V_2 are LAB velocity vectors, and U_1 and U_2 are CM velocity vectors for the ion and neutral beams of masses M_1 and M_2, respectively. The intersection of the two beams is taken as the origin of the LAB reference frame. In the CM reference frame, the ion and neutral move collinearly and collide at the CM determined from conversation of momentum

$$\frac{M_1}{M_2}=\frac{U_2}{U_1}.$$

The velocity of the CM is given by the relation

$$\vec{V}_{CM}=\frac{M_1}{M_1+M_2}\cdot\vec{V}_1+\frac{M_2}{M_1+M_2}\cdot\vec{V}_2 \tag{4}$$

and the CM velocities are given by

$$\vec{U}_1=\vec{V}_1-\vec{V}_{CM} \tag{5}$$

and

$$\vec{U}_2=\vec{V}_2-\vec{V}_{CM} \tag{6}$$

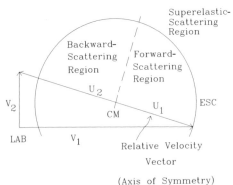

Figure 3. Schematic illustration of the geometric transformation from laboratory to center-of-mass reference frame.

Since the kinetic energy of the CM is conserved in a collision process, only the relative kinetic energy of the collision partners is available for conversion into internal energy of the colliding ion. The initial energy relative to the CM

$$T = \frac{1}{2} M_1 U_1^2 \left(1 + \frac{M_1}{M_2}\right) \tag{7}$$

defines the upper limit for energy transfer in a collision process.

In Figure 3 the ion and neutral beams have been represented as single well-defined vectors; experimentally, both vectors have a finite distribution of velocity, which can be mapped into a distribution of CM velocity vectors. This problem is greatly aggravated for low-energy experiments performed using a collision cell where the random Boltzmann velocity distribution of the neutrals introduces very large uncertainties at low-collision energies. The energy and angular band-width of the second-stage analyzer also contribute to experimental uncertainty. The combined effects of these uncertainties can be determined from a standard propagation of errors treatment and used to ascertain whether structural features detected are real or simply an enhancement of noise.

During the collision, a fraction of the ion's relative kinetic energy is converted into internal energy, and particles are scattered from their initial trajectories. The excited ion and neutral recoil from the CM with their relative postcollision velocities following the inverse mass relationship,

$$\frac{M_1}{M_2} = \frac{U_2'}{U_1'}.$$

Consequently, the full dynamic consequences of reactive scattering are completely determined by measuring any three of the velocity vectors involved. Conventionally, the reactant and product ion and reactant neutral are defined in our experiments.

The relative kinetic energy after the collision is given by

$$T' = \frac{1}{2} M_1 U_1'^2 \left(1 + \frac{M_1}{M_2}\right) \tag{8}$$

where U_1' is the relative velocity of the excited ion. The difference $\Delta T = T' - T$ the translational exoergicity of the collision process, is the magnitude of the energy transfer into internal modes. The magnitude of the energy transfer reaches a maximum when the final relative kinetic energy of the system becomes zero in the CM, i.e., when all translational energy is converted into internal energy. In this case, the velocity distribution of product ions is centered at the CM origin.

4. DATA PRESENTATION

Experimental data are obtained for the energy and intensity distributions of fragment ions as a function of a LAB-scattering angle. These energy distributions are transformed into velocity distribution by the relationship

$$T = \frac{1}{2} mv^2.$$

Data are finally presented as Cartesian velocity probability contour diagrams, which display relative probabilities for product ions to be scattered at a particular CM angle and velocity, with the primary ion velocity vector defining zero deflection angle. Details of data transformation have been described in detail elsewhere,[36] and key relationships are given in Table 1.

Points of equal intensity in the transformed velocity distributions are joined together as contour lines to define peaks and valleys in the scattering diagram, which is a plane containing the CM-relative velocity vector previously defined. The distribution is necessarily symmetric about this

Table 1. Relationships for Velocity and Intensity Transformations

---

Laboratory-measured quantity: intensity as a function of mass, kinetic energy, and scattering angle

$$I(m, E, \Theta, \phi)$$

Laboratory velocity:

$$\left(\frac{2eV}{m}\right)^{1/2}$$

Laboratory velocity from laboratory energy:

$$I(v, \Theta, \phi) = m \, V . I(m, E, \Theta, \phi)$$

Laboratory Cartesian velocity from laboratory polar velocity:

$$I(V_x, V_y, V_z) = \frac{m}{V} I(m, E, \Theta, \phi)$$

CM Cartesian velocity from laboratory Cartesian velocity:

$$I(U_x, U_y, U_z) = I(V_x, V_y, V_z)$$

In Cartesian space, intensity is independent of the coordinate system, since the volume elements in the two systems are the same.

---

vector, and the total relative cross section is obtained by integrating this distribution—e.g., converting it into a spherical distribution having the relative velocity vector as its polar axis. From this geometric consideration, it is evident that the relative probability distribution displayed in these planar diagrams must be multiplied by $\sin\phi$, where ϕ is the CM-scattering angle, to obtain the true relative intensity of scattering features.

This transformation Jacobian is particularly important in evaluating energy deposition in CID. The kinetic energy distribution of product ions is evaluated using the expression[37]

$$P(T) \, \alpha \, U \int_0^\pi P_c(U_1, U_2, U_3) \sin\phi \, d\phi \qquad (9)$$

where U is the CM velocity of the product ion; P_c is the relative intensity corresponding to a particular set of collision coordinates U_1, U_2, and U_3 and ϕ is the CM-scattering angle. The average energy transferred in CID is evaluated by subtracting the mean of this distribution from the initial distribution, both expressed in the CM system.

Before discussing experimental data, it is instructive to discuss a hypothetical CM contour diagram and the information that can be derived using a graphical construction showing the LAB→CM transformation. For this, we return to construction of laboratory velocity vectors of reactant ion and neutral beams and the relative velocity vector in Figure 3. The measured beam energies, laboratory intersection angle, and masses of the two particles determine CM velocities shown in Figure 3. Collision takes place at the CM, and velocity vectors of the ion and neutral after collision are marked U_1 and U_2, respectively.

In an elastic collision, energy is not exchanged between kinetic and internal modes; hence, relative velocity vectors remain the same. Accordingly, we define the elastic scattering circle (ESC) as the circle drawn from the CM having as a radius the ion's relative velocity vector. Since it inscribes all velocity vectors for a collision in which no interconversion of translational and internal energy occurs, the ESC is a useful boundary line in discussing CID. Since CID is actually the result of an inelastic collision in which kinetic energy is converted into internal energy, excited parent ions (hence, fragment ions from the subsequent dissociation step) generally fall inside the ESC. In the exceptional case of superelastic collisions where internal energy is converted into kinetic energy, CID fragment ions are located outside the ESC. The difference in the relative energies before and after collision provides quantitative information on collisional energy transfer.

A line drawn at the CM perpendicular to the relative velocity vector

divides the scattering diagram into forward and backward regions, even
though ions continue to move forward in the LAB frame. This separation
into forward and backward scattering in the CM frame provides information on the role of the impact parameter in these collisions. Ions in the
forward-scattered region correspond to the recoil of an ion and neutral
particles from the CM in their initial direction, and the reverse is true for
backward-scattered ions. Forward scattering is generally associated with
large-impact-parameter (glancing) collisions, and back scattering with
small-impact-parameter (rebound) collisions.

Figure 4 describes scattering patterns for two hypothetical forward-
and backward-scattered CID processes. Since no additional information is
obtained by including the neutral product velocities, only the ion distribution is presented in these diagrams. We also point out that the relative
velocity vector in Figure 4 is a plane of symmetry in the planar depiction
of scattering events and an axis of symmetry for the three-dimensional-
scattering event we are actually describing by these reduced coordinates. It
is noteworthy that CID product ions found in forward- and backward-
scattering regions have very different kinetic energy distributions in the
LAB frame for these two examples; nonetheless, the amount of energy
transferred into internal modes is identical. This is obvious in data presented in the CM frame but obscured in the LAB frame, underscoring the
importance of analyzing CID experiments using CM coordinates.

The CM frame for collisional excitation is constructed using the
masses and velocities of the primary ion and neutral, as we have just
described. Further, the collisionally excited ion dissociates in its own CM

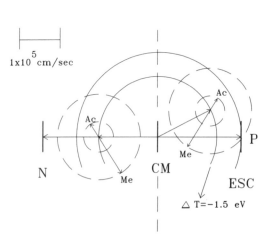

Figure 4. Hypothetical CM-scattering contour diagram illustrating
forward and backward scattering
of acetone ions in an inelastic CID
process at 3-eV collision energy.
The circle marked ESC is the
elastic-scattering circle representing no kinetic energy transferred
into the ion's internal energy. The
inner circle marked $\Delta T = -1.5 \, \text{eV}$
corresponds to a transfer of 1.5 eV
of translational energy into internal modes of acetone ions on
collision. The dashed circles show
velocity vectors for the isotropic
dissociation of acetone ions into
acetyl ions and methyl radicals
with the release of 0.1-eV energy
on dissociation.

frame. Since no orientation forces are operative, the randomized velocity vectors generate a spherical distribution of fragment ions with the energized primary ion vector at the origin. Any kinetic energy release as the fragment ion and neutral dissociate broadens the experimental distribution.[38] This spherical distribution is displayed as a circle in Figure 4, which depicts the sampling plane defined by the reactant ion and neutral vectors.

In this hypothetical example, we have displayed results that would be observed for two uniquely defined velocity vectors for the reactants and products, each depositing 1.5 eV in the collisional excitation step and dissociating with 0.1 eV released into translational energy in the fragmentation step. Similarly, unique angles are depicted, corresponding to precise impact parameter trajectories. In real cases, a distribution of reactant velocity vectors, impact parameters, scattering angles, and kinetic energy release values inevitably broaden the distribution. Indeed, it was our initial expectation that these broadening effects would smear out most of the information content in CID-scattering experiments. We consider it rather remarkable that our experimental results have exhibited significant structure, permitting us to reach general conclusions about specific mechanisms, as discussed later in this chapter.

It is interesting to compare collision cell CID experiments with crossed beam experiments using these same concepts, so that the dynamics implications of both kinds of experiments can readily be appreciated. The general Newton diagram previously discussed is somewhat simplified in collision chamber experiments, provided the thermal velocity of the collider gas is assumed to be zero. When the neutral has such low velocity that it can be neglected compared to ion beam velocity (e.g., for high-energy CID), the neutral vector becomes a point. Since the velocity vector V_2 has zero length, the laboratory velocity vector for the ion beam becomes the CM relative velocity vector once the CM velocity of the parent ion is calculated as follows:

$$\vec{U}_1 = \frac{M_2}{M_1 + M_2} \, \vec{V}_1 \tag{10}$$

The CM energy in this case is given by the simple relationship

$$T = \frac{1}{2} \frac{M_1 M_2 V_1^2}{M_1 + M_2} \tag{11}$$

It can also be expressed in terms of laboratory kinetic energy (E_{lab}) as

$$T = \frac{M_2}{M_1 + M_2} E_{lab} \tag{12}$$

The energy transfer in the collision activation step can be determined in the same way as described earlier. For high-energy (keV) collisions, the kinetic energy transferred into internal modes becomes approximately equal to the loss in the CM kinetic energy of the projectile ion. This would be strictly true if zero angle forward-scattering were the operative mechanism, as is certainly the case for some CID reactions. However, we show that most CID activation collisions result in significant scattering of ions, which can distort such experimental parameters as kinetic energy loss, energy release, and fragmentation patterns observed in collision chamber experments. This distortion is especially severe for heavier neutrals and low-energy CID experiments in general.

5. SELECTED RESULTS

5.1. The Hydrogen Molecular Ion: A Comparison of Theory and Experiments

During the long period of investigating this fundamental CID dynamics system results have been interpreted as proceeding via non-adiabatic transitions from vibrationally excited levels of the ground electronic state to specific electronic states that then dissociate. This is clearly the correct limiting case mechanism at high energy where vertical Frank–Condon transitions constitute the collisional excitation mechanism. Within this limit, atoms in H_2^+ are essentially frozen in position during the several femtosecond time scale of the collision, and optical selection rules for H_2^+ excitation apply.

At low-collision energies, properties of the HeH_2^+ surface govern transition probabilities. At low and intermediate energies, CID reaction probability is a strong function of vibrational quantum number, proceeding very efficiently for highly excited states. Champion et al.[39] suggested that this occurred because the colliding atom (Ar or D_2 in their experiments, conducted at about 10 eV) promoted transitions from the $^2\Sigma_g^+$ ground state to the $^2\Sigma_u^+$ repulsive state of H_2^+. This perturbation by the third atom is expected to be much stronger at lower collision energies and to become most important when CID competes with the formation of bound HeH^+ (or ArH^+) products.

The most detailed dynamics of H_2^+/He CID experiments in this low-energy range are those by Havemann et al.[22] who investigated the dissociation of H_2^+ at 3- and 5-eV CM collision energies. These experiments were interpreted using results obtained from detailed three-dimensional trajectory calculations on a modified Whitton and Kuntz[27] surface. The excellent agreement between the calculated and experimental results that compared experimental data with calculations for trajectories °on the ground-state surface led to the conclusion that the reaction occurs adiabatically in this energy range. This is a very satisfying result, suggesting that the CID mechanism occurs adiabatically at low energy, nonadiabatically at high energy, and with mixed mechanisms at low-to-moderate collision energies.

However, these experiments and calculations have recently been reinterpreted by Gislason *et al.*[30] in light of the conflicting results of Guyon *et al.*[26] that vibrational dissociation from the ground-state surface cannot account for the observed velocity spectra of H^+ from the CID of vibrationally selected H_2^+. The latter experimental results and others were rationalized on the basis of a mechanism specific to the H_2^+/rare gas system, which involves Demkov-type, nonadiabatic transitions occurring between two nearly parallel potential surfaces.[30] The upper excited state of the HeH_2^+ system has H_2^+* ($^2\Sigma_u^+$) + He and $HeH + H^+$ as assymptotic limits, while the ground-state surface has H_2^+ ($^2\Sigma_g^+$) + He and $HeH^+ + H$ as its limits. The curves are nearly parallel over a broad range of internuclear distances, and Demkov-type transitions are a prevalent curve-crossing mechanism. As Gislason and Guyon pointed out, the degeneracy of the overall CID mechanism

$$H_2^+ + He \rightarrow H^+ + H + He \qquad (13)$$

is broken by the rare gas atom at *all* finite internuclear distances. Consequently, the proximity of H^+ and He when the dissociation occurs defines whether the product originated on the ground or excited state potential energy surface, as discussed in the following paragraphs.

This approach provides a unified theory for CID reactions. At a high energy, H_2^+, initially in its ground electronic $^2\Sigma_g^+$ state, is excited to $^2\Sigma_u^+$ in a vertical transition and promptly dissociates on this repulsive curve into $H^+ + H$. At a low energy, collision of H_2^+ ($^2\Sigma_g^+$, v) with He leads to both H^+ (CID) and HeH^+ products as the reactants approach on the ground-state HeH_2^+ surface. The high efficiency of generating HeH^+ and H^+ as competing processes just above threshold for CID is readily interpreted by this theory. As He approaches H_2^+ on the ground-state potential energy surface, the positive charge migrates to the H closest to the He nucleus; simultaneously, the H_2^+ bond elongates and weakens. The smooth continuation of this process leads to the formation of HeH^+ as one of the observed products. Over some range of R, there is high probability that electron exchange occurs and the system transfers to the excited-state potential energy surface. Here, the repulsive force between H^+ and H leads to the CID product, producing an H^+ that recedes from the (weakly bound or dissociated) HeH molecule. This nonadiabatic transition explains why CID has such a large cross section near its thermochemical treshold.

This analysis leads to the interesting conclusion that CID originating from the HeH_2^+ ground electronic state always results in similar postcollision velocities of H^+ and He, while CID occurring on the upper surface leads to similarly correlated velocities of H and He. Thus, the CM recoil velocities of He and H^+ are directly correlated with the occurrence of curve crossing and the hypersurface on which CID actually occurs. Therefore, the slow peak in beam experiments is made up exclusively of H^+ ions originating from the ground electronic state, the situation involving the fast peak is ambiguous. Very rapid particles are produced on the excited state, but others could be in either electronic state, depending on the actual final velocity of the H atom product.

This result may be compared with predictions of the well-known, high-energy model[40] for CID, as extended by Gislason and Guyon.[30] The initial step in this mechanism is a vertical excitation of H_2^+ by Ar to the repulsive $^2\Sigma_u^+$ state. Thus, all CID orginates on the excited potential energy surface of the triatomic system. Although the Ar atom is far away as the H_2^+ dissociates, its presence splits the two potential energy surfaces by a small amount, ΔE. In the limit of very small ΔE, the Demkov model predicts that 50% of the dissociating molecules will migrate to the lower surface, and the resulting product H^+ ions will be observed as the slow experimental peak. The remaining 50% of the molecules will remain on the excited surface, and the resulting H^+ ions will be seen in the fast peak. This model correctly rationalizes the observed equal intensities of fast and slow protons.

The low-energy predictions of this theoretical model are consistent with experimental results of Havemann et al.[22] Who reported differential cross sections for CID of H_2^+, by He at $E_{cm} = 3.03$ eV. The broad range of velocities observed reflects the uncertainty in the H_2^+ internal energy. This uncertainty, which equals the dissociation energy of the molecule, 2.65 eV, is nearly equal to the relative energy of collision. Consequently, very few collisions should give products on the ground state with velocity close to the maximum allowed for the ground-state mechanism, as is observed experimentally. Thus, it is certain that some, perhaps one-third, of the H^+ ions produced in the experiment originate from the excited electronic state potential energy surface (PES). However, both calculation and analysis of experimental results suggest that the majority of H^+ ions are formed from vibrationally excited ions propagated on the ground electronic state hypersurface.

Havemann et al.[22, 41] have also carried out trajectory calculations for this process using the Whitton–Kuntz potential energy surfaces.[27] While their computations were restricted to the ground electronic state of HeH_2^+, in their work, they assigned the possitive charge randomly to either of the two protons at the end of the collisional dissociation step. (This is inconsistent with the assumption that the system is in the ground electronic state; for in this case, the positive charge must always be on the proton nearest the He atom.) This is tantamount to producing half of the products from the excited-state surface. There are two problems with this procedure, and the close agreement between calculation and experiment that they reported must be regarded as fortuitous. First, the velocities of H^+ ions on the excited surface will not be correct because trajectories were not integrated on the excited PES after reaching the Demkov crossing point. Secondly, this procedure overestimates the fraction of products formed from the excited state.

5.2. Propane Cation CID

It is appropriate to begin our discussion of polyatomic ion CID dynamics with a summary of our results for the propane molecular ion. This ion is the archetypical example of molecular ions whose unimolecular decay is well described by the quasi-equilibrium theory of mass spectra.[42–44] Decomposition mechanisms and energetics of the principal reaction steps are well known. Regardless of how energy is deposited in the ion, it rapidly interconverts into vibrational excitation of the ground electronic state. Dissociation of the molecular ion appears to be nearly entirely statistical, and we may infer that its CID fragmentation would be accurately described by these well-established concepts. Thus, the principal new information obtained by studying its scattering dynamics are details of energy deposition and angular scattering.

We have carried out such a study[45] for the principal primary fragmentation processes

$$C_3H_8^{+\cdot} \rightarrow C_3H_7^+ + H^\cdot \tag{14}$$

$$C_2H_4^{+\cdot} + CH_4 \tag{15}$$

$$C_2H_5^+ + CH_3^\cdot \tag{16}$$

and summarize our principal findings in the following sections. We have also conducted an extensive baseline CID study using a conventional tandem mass spectrometer.[46] These results are used to deduce how parameters measured in conventional instruments can be perturbed by dynamics effects.

5.2.1. Ethylene Ion Formation

We have investigated the reactive scattering of propane molecular ions to form $C_2H_4^{+\cdot}$, Reaction 15, at CM collision energies of 1.7, 8.6, and 49.0 eV using He and Ar collision gases. The lowest energy results were in excellent agreement with our previously published 2.1-eV scattering diagram,[31] within the probable errors of two quite different crossed beam instruments. All of the contour diagrams exhibit similar scattering dynamics characteristics—namely, their intensity maxima are located at CM-scattering angles greater than zero degrees. As discussed earlier, non-zero scattering is consistent with our expectations for this more endothermic channel. Nonzero scattering angle further implies small impact parameter and that impulsive collisions with angular momentum exchange dominate this CID reaction at all energies.

Figure 5 shows a typical velocity vector contour diagram for collision energy of 8.6 eV. The relative intensities of product ions are labeled as points of the same intensity as a function of angle and relative velocity to present three-dimensional data in two dimensions. The relatively close spacing of the contour lines shows that this product is surprisingly sharply peaked. We also remark that positive and negative scattering angles are indistinguishable in scattering experiments and the relative velocity vector line defines a symmetry plane for velocity contour diagrams. In three dimensions, the $C_2H_4^{+\cdot}$ ion product having the characteristics shown in Figure 5 would be observed as a core symmetrically distributed around the primary ion beam, forward scattered with a most probable deflection angle of 13°.

Contour diagrams for $C_2H_4^{+\cdot}$ do not show distinctive change in dynamics when moving from energy close to threshold to high CM energy even for kilovolt laboratory energies. These results clearly demonstrate that there is no change in the observed dynamics with collision energy, as expected for ions whose decomposition is well described by statistical (RRKM/QET) models. Since only a single peak of intensity is observed, it

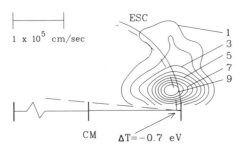

Figure 5. The CM-scattering contour diagram for the CID of propane molecular ion to ethylene ion in collision with helium at 8.6-eV collision energy: $C_3H_8^+ + He + 8.6\,eV \rightarrow C_2H_4^+ + CH_4 + He$. The numbers marked on contours represent the relative intensity of each contour.

cannot be concluded from these data whether loss of CH_4 proceeds via two or more competing pathways. Most likely, the dynamical behavior for different channels is not distinguishable for the same reason they are not clearly separated in the breakdown graph for this cation[47] shown in Figure 6.

Energy deposition functions, or P(T) diagrams, for CID Reaction (15) at various collision energies have been constructed using Equation (8); an example is shown in Figure 7, also obtained at 8.6-eV collision energy. As collision energy increases from 1.7–49 eV, most probable energy deposited into the ion does not change significantly. The fundamental reason for this behavior is implicit in the breakdown graph shown in Figure 6. The relatively narrow peaking of the $C_2H_4^{+\cdot}$ curve in the breakdown graph strongly limits the energy transferred. If more than several tenths of eV energy is transferred, the mechanism changes to Reaction (16), which we discuss in the following sections.

5.2.2. Ethyl Ion Formation

Figure 8 shows the analogous velocity contour plots for the CID of propane ion to $C_2H_5^+$ at 8.6-eV collision energy. This contour map is qualitatively similar to Figure 5, obtained at the same CM collision energy. We have investigated this reaction over the collision energy range 1.7–449 eV. Just as we found for $C_2H_4^{+\cdot}$, our data do not show a significant change in dynamics with increasing collision energy. The scattering

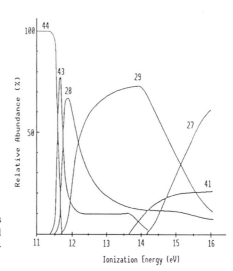

Figure 6. Experimental breakdown graphs for the propane ion determined by threshold photoelectron-photoion coincidence technique. (Reproduced from Reference 47.)

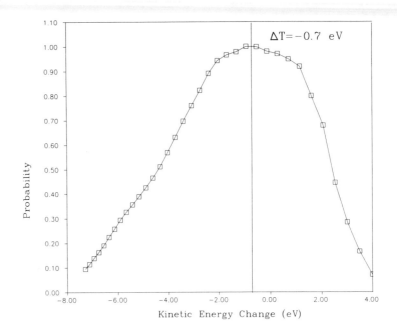

Figure 7. Translational energy distribution of $C_2H_4^{+\cdot}$ ion expressed in terms of translational exoergicity. Negative ΔT represents conversion of kinetic energy into internal energy from the collision of $C_3H_8^{+\cdot}$ with He at 8.6-eV collision energy.

angle where the maximum intensity of $C_2H_5^+$ is observed is always larger than that for $C_2H_4^{+\cdot}$ at the same CM collision energy, and the energy transferred is correspondingly larger. It follows that the mechanism for this reaction is essentially the same as for Reaction 15 but with quantitative differences. A simple bond scission as in Reaction 16 is usually described as proceeding via a "loose" transition state[48] where the dissociation rate constant increases very rapidly as soon as the reaction threshold is reached.

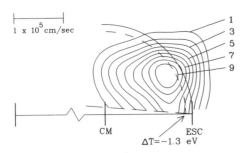

Figure 8. The CM-scattering contour diagram for the CID of propane molecular ion to ethyl ion on collision with helium at 8.6-eV collision energy: $C_3H_8^{+\cdot} + He + 8.6\ eV \rightarrow C_2H_5^+ + CH_3^{\cdot} + He$. The numbers marked on contours represent relative intensity of each contour.

This is particularly the case with the propane ion, which has a large number of low-lying energy states. As collision energy is increased, the internal energy distribution slowly shifts to the higher energy, and the distibution is broadened. The relative intensity of $C_2H_5^+$ also increases gradually with increasing collision energy. Since $C_2H_5^+$ formation from propane molecular ion is more endothermic than $C_2H_4^{+\cdot}$ formation, more violent, small-impact-parameter collisions are necessary to generate ethyl ions by CID. A broader distribution reflects both a broader range of energy deposition and greater kinetic energy release into the fragments.

5.2.3. H-Atom Elimination

The lowest energy dissociation of the propane molecular ion is H-atom elimination to form propyl ions, Reaction 14. At threshold, a tight transition-state-structure usually represented as a cyclized propane structure,[42] is invoked theoretically: at higher energy, the s-propyl ion may become the dominant structure. This tight transition-state-structure is required to explain the very intense metastable ion that is observed over an extremely broad range of observation times.[49] This long-lived metastable ion is also an extremely strong interference in our experiments and thwarted our attempts to characterize the detailed CID dynamics of Reaction 14.

Although propyl ions formed by CID were detected in this experiment, the signal from metastable dissociations also occurring in this zone were too intense for us to obtain statistically significant data for CID even after many hours of signal averaging. (We note parenthetically that much less than 1% of the incident propane ions are collisionally excited in our beam–beam experiment and the cross section for formation of propyl ions by CID is lower than that for Reactions 15 and 16.) Although total counts in our (signal plus noise) channel registers consistently exceeded (noise) channel registers, the high level of "chemical noise" in these experiments precluded our construction of a meaningful scattering diagram. Our observation based on this fact is that the CID product distribution is indistinguishable from that of the corresponding metastable ion in either energy or angular coordinates. From this, we conclude that the propyl ion CID product has essentially the same dynamics characteristics as the metastable ion itself—namely, scattering with maximum intensity at zero-scattering angle and with no detectable energy shift other than 1-amu mass loss.

The significance of this result is that these are precisely the dynamics features predicted for the propyl ion in earlier study of methane and propane ion CID.[31] In this work, we demonstrated a correlation between the scattering angle and energy transfer that was approximately linear over the low-energy range. This predicts zero (or very small) energy deposition for ions scattered at zero angle and correspond to large-impact-parameter collisions with very modest perturbation of ion trajectories. We also demonstrated that the internal energy remaining in ions once they were formed by electron impact is fully available to drive the decomposition process. This implies that only those ions promoted from the highest vibrational levels of the ground state can be collisionally induced to dissociate at zero-scattering angle; consequently, the least endoergic process, Reaction 14, is the only one that will be observed at zero-scattering angle. Finally, the sharply peaked propyl ion distribution in the breakdown graph (see Figure 6) implies that only a very narrow window of excitation energies for propane ion excitation are observable as propyl ion. Taken together, these considerations lead to the prediction that these ions will be observed at zero-scattering angle with a very small kinetic energy shift and will fall off sharply with an increasing scattering angle. This is entirely consistent with our inability to resolve the CID product from metastable ions decaying at the same time, with the same energy, and at zero-scattering angle.

As previously discussed, propane ions generated by electron impact in our experiments have a distribution of internal energy ranging from zero to the maximum allowed by the breakdown graph—e.g., limited by the threshold energy for ions to dissociate to propyl ions with observable decay rates. The metastable ion decay rate sampled in our beam experiments correponds to a unimolecular decay rate of 3×10^4 sec^{-1}, and the internal energy is limited to about 0.7 eV. We make the further assumption that energy deposition by ion–neutral collision simply moves that energy distribution along the x-axis of the breakdown graph and this energy content in turn determines the product ion that will be observed as a CID product. For the reasons just discussed, propyl ions are restricted to near-zero scattering angles. Deposition of slightly larger amounts of energy is associated with nonzero-scattering angles and decomposition into the more endothermic Reactions (15) and (16) previously discussed.

5.2.4. Application of the Massey Criterion to Propane Ion CID

We have modeled our propane ion CID results using the Massey criterion energy deposition function. Massey[50] suggested that the cross section for endothermic charge transfer should be at its maximum value when collision time is of the same order as the natural period of the oscillator; this is a consequence of the uncertainty principle $\Delta E \cdot \Delta t \approx h$, applied to collision processes. It follows that the probability for inelastic energy transfer is maximized when

$$a/v = h/\Delta E \tag{17}$$

where a is the distance over which the transition can take place, v is the relative velocity of the colliding particles, h is Planck's constant, and ΔE is the energy transferred. The terms a/v and $h/\Delta E$ are described as the collision time and time for energy transfer, respectively, and a is called the adiabatic parameter. This parameter is a characteristic radius of action of the intermolecular forces governing the transition, and it is of the order of molecular dimensions. The cross section for the energy transfer is therefore maximum when $v = a\Delta E/h$ and falls off exponentially at lower and higher velocities. In more modern language, the strength of nonadiabatic coupling between two states is given by[51, 52]

$$\xi = \frac{\Gamma \Delta E}{\hbar v} \tag{18}$$

where Γ is the width of the region of nonadiabatic interaction, ΔE is the smallest energy gap between two adiabatic states, $\hbar$ is the $h/2\pi$, and the other parameters have already been defined. For small ξ ($\xi \ll 1$), the system will follow the diabatic potential surfaces in a collision. For large ξ ($\xi \gg 1$), the system follows the adiabatic potential surface throughout the collision. For ξ of the order of unity, transitions (energy transfer ΔE) are likely, with the exponential dependance given by the traditional Massey formalism.

This concept has been widely invoked to describe energy transfer in collisions, including CID.[31, 53, 54] It is consistent with the often-quoted concept that electronic excitation is a high-energy collision excitation mechanism (high-frequency transitions, short collision times) and vibrational or rovibrational excitation is the dominant excitation mechanism at low and intermediate energy (low-frequency oscillations and long collision times). For example, taking the interaction length, or adiabaticity parameter $a = 5$ Å, a value suggested by Massey for endothermic charge transfer reactions, the optimum collision time is 21 femtosec for a rovibrational transition requiring 0.2 eV, and the corresponding collision energy of propane ions with Ar is 63 eV. For 2-eV transitions, the optimum collision occurs in 2.1 femtosec and the collision energy is 6.3 keV.

The probability of energy transfer for a two-state system is expressed as[55]

$$P \propto e^{-a|\Delta E - \Delta E_0|/hv} \tag{19}$$

The appropriate formalism we suggest for a multistate, high-level density (e.g., polyatomic ion) system involves solving Equation 17 for the most probable energy transfer ΔE_0, for the experimental relative velocity under consideration. Equation 19 is then used to compute the probability of depositing any other amount of internal energy ΔE. The importance of velocity as the key parameter in energy deposition is evident in this equation.

If this model properly describes energy transfer in collisions, the QET description of fragmentation as a function of internal energy provides the formalism for calculating both fragmentation patterns and energy deposition for individual ions produced by CID. Taking *a* equal to 1 Å (a value shown in our work to be appropriate for impulsive collisions.[45]), we can calculate probability for energy transfer as a function of collision energy. This is illustrated in Figure 9 for the collision energies investigated. The product of this function with the Figure 6 breakdown graphs gives the most probable energy deposition values for each fragment ion. As shown quantitatively in Table 2, the Massey Criterion/RRKM model provides a ready explanation for the general observation that collisional activation is ineffective in depositing large amounts of energy in gaseous ions. However, this model cannot describe the non-RRKM behavior of acetone and nitromethane molecular ions discussed later. The value deduced by this method are in excellent agreement with our experimentally determined

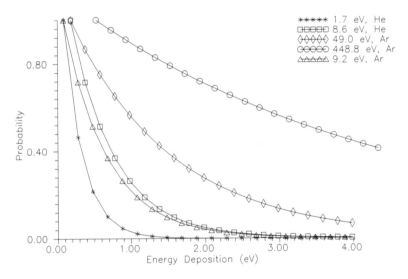

Figure 9. Probability of energy transfer into the propane ion at different collision energies calculated using the Massey criterion (see Equations 17 and 19).

Table 2. Experimental and Predicted Most Probable Energies Deposited in the Propane Ion

Fragment ion	Average energy deposited (eV) at designated collision energy											
	1.7		8.6		9.2		49.0		448.8		5360	
$C_2H_4^+$	0.6	0.6	0.7	0.7	0.6	0.7	0.6	0.7				
$C_2H_5^+$	1.0	0.9	1.3	1.3	1.4	1.3	1.3	1.4	1.6	1.6		
$C_3H_7^+$		0.2		0.2		0.2	0.2	0.22	0.2	0.25	0.2	0.25

values, as demonstrated in Table 2. The essentially quantitative agreement of theoretical results with our scattering experiments strongly supports this model for energy deposition in propane ion CID.

This model has also found direct application in interpreting results obtained in an extensive study of propane ion CID conducted at kilovolt ion energies using a Kratos MS50TA tandem mass spectrometer at the University of Nebraska.[46] These experiments determined fragmentation patterns, energy transfer, and kinetic energy release on fragmentations for seven collision gases and primary beam attenuation at collision energies from 500–8000 V (LAB). A striking feature of this study was the relatively weak dependance of all the parameters measured on these variables.

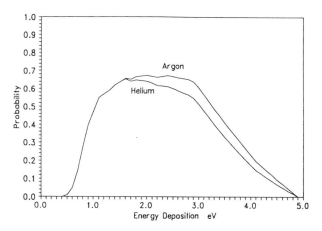

Figure 10. Energy distributions of the propane ions dissociating into ethyl ions in collision with helium and argon neutrals at 8-keV collision energy. These curves were determined by the product of the breakdown graph and energy deposition function calculated from the Massey criterion.

These results and those previously discussed are readily understood in terms of the exponential decay of the Massey function (Equation 19 and Figure 10) and the width of product ion peaks in the breakdown graph of propane in Figure 6. As already discussed, ethyl ion energy deposition depends somewhat more strongly on CM collision energy than ethylene ion, reflecting their respective widths in Figure 6. The quantitative prediction of our model is obtained by multiplying the energy deposition function in Figure 10 with the breakdown graph in Figure 7. The result is the combined probability for depositing a range of energy in the propane ion, which results in decomposition into $C_2H_4^{+\cdot}$, $C_2H_5^+$ (or any other fragment ion), respectively.

This variation and intensitivity to CM collision energy is illustrated quantitatively in Figure 10 and Table 2. Figure 10 presents the calculated energy deposition into $C_2H_5^+$ product ions when 8-keV propane ions are impacted on He and Ar collision gases. Although the CM collision energy is ten times greater for Ar collisions, the distribution is shifted only slightly to a higher energy. A corollary of this prediction is that the kinetic energy spread of $C_2H_5^+$ ions, connected to internal energy via the QET, is nearly the same for both collision gases, as observed. The energy deposited in propane ion for fragmentation to $C_2H_4^{+\cdot}$ is even less dependent on collision gas (or CM energy) because its breakdown graph distribution function (see Figure 6) is narrower than that for $C_2H_5^+$. For the propyl ion, $C_3H_7^+$,

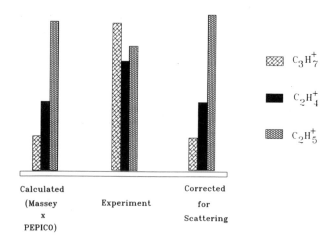

Figure 11. Comparison of calculated and experimental relative abundances of primary fragmentation processes in the CID of propane ion on collision with helium at 666-eV CM collision energy: $C_3H_8^+$/He at 667 eV. Experimental data were also corrected for scattering losses and provide an excellent agreement with calculated abundances.

energy deposition is entirely controlled by the breakdown graph rather than the energy deposition function in Figure 9; this is shown quantitatively in Table 2.

The application of this model was further tested by comparing the relative intensities of the three products of Reactions (14–16) with that calculated by forming the product of the Massey criterion distribution with the breakdown graph. In making this comparison, experimental results must be corrected for scattering. Scattering correction factors are estimated by computing the fraction of integrated total intensities of $C_2H_4^{+\cdot}$ and $C_2H_5^+$ bounded by the sampling core at $0 \pm 1°$. Correction factors for He CID at 8 keV are rather large: 3.4 for ethylene ions and 5.7 for ethyl ions. When this correction is made, corrected experimental data are in nearly perfect agreement with experimental results; this is demonstrated in Figure 11. It is also evident from the correction factors that scattering severely distorts the relative intensities of CID product ions in high-energy collision chamber experiments even for such low-mass collision gases as helium.

5.3. Non-RRKM CID Processes

A fundamental tenet of QET/RRKM theories of unimolecular dissociation is that dissociation takes place from the ground electronic hypersurface following rapid energy redistribution from upper vibronic hypersurfaces no matter how the energy is deposited initially. Detailed mechanisms and rates of intramolecular vibrational relaxation of neutral and ionized systems continue to be intensively studied both experimentally and theoretically as one of the most important questions in reaction dynamics. For our purpose, propane ion CID has defined the dynamics characteristics of systems following the statistical models. Explicity, there is no distinctive change in mechanism as collision energy is varied by orders of magnitude, and the energy transfer and fragmentation processes observed can be readily rationalized within the framework of these theoretical models.

It is equally important to define CID dynamics for molecular ions whose decay characteristics are not readily described within this theoretical framework and to search explicitly for abrupt changes in relative abundances, angular or energy distributions as a function of collision energy. For the first example investigated from this perspective, we chose to investigate acetone cation CID. A series of papers have described our results in detail.[32, 56–59] We summarize our principal findings, in the following sections.

5.3.1. Acetone

Several experiments have suggested that coupling upper electronic states to the ground states in ketones is impeded by some kind of bottleneck to energy exchange. Nystrom et al.[60] observed a second onset for the increase in the abundance of fragment ions in the CID of 2-pentanone ions when monitored as a function of laboratory kinetic energy. This onset corresponded to the energy required to excite ground-state 2-pentanone ions to their first excited state; they suggested that above 2-eV CM collision energy, the excitation mechanism changes from vibrational to electronic excitation. The most convincing evidence for nonstatistical behavior was obtained in photoelectron-photoion coincidence (PEPICO) studies[61-63] of acetone ion dissociation, in which parent acetone ions were detected 2.3 eV above the ionization potential. Further, kinetic energy relese in these experiments could not be fully accounted for from QET predictions. Bombach et al.[63] pointed out that the excited A state of acetone ions is long lived and suggested that the dissociation reaction is slowed by the requirement of isomerization of the ion from the keto form to the enol structure prior to dissociation. Stace and Shukla[54] performed high-energy CID of acetone ions, and based on the analysis of measured and theoretically calculated CID peak profiles, they suggested that dissociation of acetone ions does not follow QET behavior. It was shown that QET-based kinetic energy release distribution superimposed over a broader distribution reproduced experimental results, suggesting that two different dissociation mechanisms may be operative simultaneously. Lorquet and Takeuchi[64] analyzed in some detail the mechanisms responsible for very inefficient coupling between electronically excited formaldehyde cations and the ground state using ab initio potential energy surfaces to describe the relevant intersections of these surfaces.

Our recent dynamics studies of the CID of acetone ions in the low-energy range from near threshold to > 100-eV collision energy support the isolated state hypothesis. Specifically, these studies supported earlier observations of long-lived excited state and non-QET behavior for ketonic cations. For example, the amount of kinetic energy transferred into internal energy of dissociating acetone ions reaches a maximum of about 6 eV at high-energy collision, exceeding by an order of magnitude the thermochemical threshold. This contradicts QET predictions that an ion should dissociate as soon as it acquires the critical energy for the process, as was observed for propane cations. We concluded that collisional activation proceeds via electronic (or vibronic) excitation rather than vibrational excitation mechanisms. These observations were supported in a detailed study by Martinez and Ganguli[65] of relative abundances of all possible dissociation channels as a function of ion kinetic energy using a triple-quadrupole mass spectrometer. Conclusive proof of the proposed excited state mechanism was obtained in our study of threshold energy CID.

5.3.1.1. Dissociation of Long-Lived Excited States in Acetone Ion

CID. The QET framework asserts that ions generated by electron impact rapidly relax to their ground electronic state with a distribution of vibrational energy at the time of mass analysis. Since the time between ion formation and collision in our collision experiments is of the order of tens of microseconds, any electronically excited state(s) present at the time of collision are by definition long-lived excited states, sometimes described as isolated states. A major finding in our study of the acetone cation is that low-energy CID is dominated by fragmentation characteristics of a significant population of long-lived electronically excited states present in our primary ion beam.

Figure 12 shows the contour map for the CID of acetone molecular ions in collision with He at 2.4-eV collision energy. In sharp contrast to propane ion CID, where only a single peak was observed in the scattering diagram, three distinct peaks are observed for this cation. One peak is forward scattered, and two of the peaks are located in the backward-scattering region. The peak in the forward-scattering region corresponds to the (endoergic) energy transfer of 2.2-eV translational energy into internal modes. This energy transfer is complemented by a 2.2-eV (exoergic) energy transfer from internal into translational mode, and it is found as backward scattered in the CM references frame. The third peak located in the backward region corresponds to about a 1.3-eV (endoergic) energy transfer.

The energetics for the exothermic CID process can be accounted for only by excess internal energy of the ion being converted into relative translational energy following collision with the neutral and preceding its dissociation to acetyl ion. If this energy were released after dissociation, it would broaden only the peak rather than shift the mean value outside the ESC. In the present case, the superelastic backward-scattered peak is

Figure 12. The CM-scattering contour diagram for the CID of acetone molecular ion to acetyl ion on collision with argon at 2.4-eV collision energy: $CH_3COCH_3^+$ + Ar + 2.4 eV → CH_3CO^+ + CH_3 + Ar. Circles marked $\Delta T = 2.2$ eV and $\Delta T = -2.2$ eV represent transfer of 2.2 eV from internal to translational and translational to internal modes, respectively.

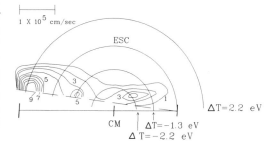

located about 2.2 eV away from the ESC; i.e., dissociating acetone ions have gained 2.2-eV translational energy prior to dissociating into acetyl ion and methyl radical. It follows that the translationally exothermic CID phenomenon can be explained only if 70-eV electron ionization of acetone results in the formation of acetone ions in highly excited state(s) that survive long enough to reach the collision center without redistributing its internal energy and dissociating on the ground-state hypersurface. This superelastic peak was the dominant feature in all experiments performed below 5 eV (even at and below the thermochemical threshold—e.g., at 0.45-eV collision energy) when acetone molecular ions were produced by 70-eV electron ionization.[57] Reducing electron energy to 12.5 eV eliminated this peak in the 0.45 eV collision energy experiment, demonstrating unequivocally that a long-lived excited state is responsible for this unusual behavior.

The unique behavior observed in acetone ion CID (compared to propane ion CID in the same energy range) can be partially understood by considering their respective photoelectron spectra. The propane spectrum (see Figure 13) consists of overlapped electronic states because of the strong vibrational coupling between states. It follows that rapid intersystem crossings occur, leading to complete energy randomization prior to decomposition from the ground electronic hypersurface. In contrast, the acetone spectrum in Figure 14 shows a large "window" between the ground and first electronic bands where no energy deposition is possible by photon absorption. Coupling between these levels is weak, and energy randomization is not complete within the time frame of our experiments. Hence, dissociation proceeds directly from the excited state surface, resulting in the unique dynamic features we have described. Moreover, the adiabatic energy difference between the ground and first excited states is 2.2 eV, corresponding within our experimental error to the superelasticity measured in our low-energy CID.

An independent confirmation of this mechanism and further characterization of this long-lived excited state were obtained in a triple-cell ion cyclotron resonance (ICR) study by Fenistein *et al.*[66] The He-induced

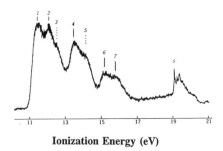

Ionization Energy (eV)

Figure 13. Photoelectron spectrum of propane. (Reproduced from Reference 80.)

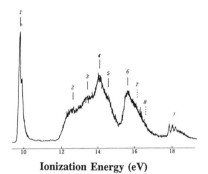

Figure 14. Photoelectron spectrum of acetone. (Reproduced from Reference 63.)

Ionization Energy (eV)

CID of excited acetone ions (generated by 30-eV electron energy) were investigated as a function of CM collision energy and a function of time delay between ion formation and detection of CID-induced acetyl ion formation. Charge transfer with O_2, which can occur exothermically from the A-state of acetone cation but is endothermic from the X-state, was used as a second chemical reaction monitor of the fraction of excited states present in this ion-trapping experiment. The radiative lifetime of excited acetone ions was determined to be nearly 5 milliseconds in these experiments. Further, the CM collision energy threshold for inducing this dissociation of excited acetone ions was determined as 0.15 ± 0.03 eV after correcting the observed 0.08-eV threshold for thermal motion of particles. The lifetime is so much longer than the transit time from ion source to collision region (38 μsec in our experiments) that the fraction of excited ions present in our ion beam is essentially unchanged from the nascent distribution formed by electron impact.

Since the lifetime of excited acetone ions is essentially infinite compared with the experimental time frame of a CID experiment, it was possible to demonstrate microscopic reversibility in collision activation/deactivation. In particular, we examined the reverse process of collisionally exciting acetone ions at the ground state to the first electronic state.[59] Figure 15 shows the inelastic-scattering diagram for this process. Here, acetone ions formed in the ground state using low-energy electrons are collided with He at 2.2-eV collision energy. The peak at the CM results from an inelastic process in which all available kinetic energy has been converted into the ion's internal energy. This experiment conclusively demonstrates the efficiency with which translational energy is converted into electronic excitation energy in acetone.

We have suggested the curve-crossing mechanism for this particular acetone ion CID pathway, illustrated schematically in Figure 16. This three-dimensional pseudodiatomic view shows collision of the acetone

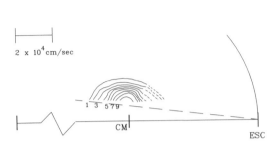

2 x 10⁴cm/sec

CM   ESC

Figure 15. The CM-scattering contour diagram for the inelastic collisions of acetone molecular ions with helium at 2.2-eV collision energy: $CH_3COCH_3^+(X) + He + 2.2\ eV \rightarrow CH_3COCH_3^+(A) + He$. The peak centered at the CM suggests that all available kinetic energy is transferred into internal energy and excited acetone ions did not dissociate between collisional activation and mass analysis.

molecular ion with a neutral particle and dissociation of acetone ions into acetyl ions and methyl radicals from both the ground- and the first, excited electronic-state surfaces. Both excitation and deexcitation of acetone ions are shown in Figure 16 with double-headed arrows, consistent with our experimentally demonstrated microreversibility shown in Figures 12 and 15. Collisions lead to deexcitation of acetone ions to the ground state, releasing excess energy (the adiabatic energy difference between the two states) to translational recoil of the acetone ion and neutral collision particles. Vibrational excitation present in the ion is preserved as internal energy in the ground state and provides the energy required to promote the dissociation process. Conversely, starting with ground-state ions containing little vibrational energy, collisions can induce bound–bound transitions, observed as inelastic absorption of 2.2 eV. Ions with more vibrational

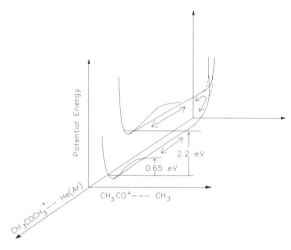

Figure 16. Schematic three-dimensional representation of the curve-crossing mechanism for the CID of acetone molecular ions.

energy absorb 2.2 eV and dissociate on the upper electronic level into acetyl ions and methyl radicals. This mechanism is the forward-scattered endoergic process in Figure 12.

The interesting superelastic backward scattering in acetone ion CID was observed only at low-collision energies. When collision energy was increased above 5 eV, endothermic backward scattering was observed that slowly turned into endothermic forward scattering at higher energies. It turns out that this dynamical "signature" is an important characteristic of reaction pathways that involve curve-crossing mechanisms (crossing seams in polyatomics).

It is interesting to note that the analogous fragmentation of the formaldehyde cation into the formyl ion and hydrogen atom is also an example of isolated state decomposition.[64] This simpler case has been treated theoretically by Lorquet and Takeuchi,[64] who draw attention to the fact that the CO bond stretch has such different equilibrium positions in the ground and first excited states of the ion. This requires discounting the degree of freedom corresponding to CO vibration in the theoretical description of interconversion between these states and introduces a Franck-Condon factor into the intramolecular energy relaxation rate for electronically excited ions. Thus, the nonadiabatic transition between these states has a transmission coefficient much less than unity at the point of lowest energy along the nonadiabatic crossing seam connecting the two hypersurfaces. (A formally similar factoring of degrees of freedom is also required to describe the methyl nitrite cation dissociation).[67] The appearance of Frank-Condon factors in the formal calculation of the dissociation rate implies that impulsive collisions, which distort the molecule, can markedly affect the coupling between isolated states. A qualitatively similar description was proposed in our paper reporting the existence of a long-lived electronically excited state in the acetone cation.[32]

5.3.2. Isomerization of Ions: NO⁺ from Nitromethane Cation CID

The last example involves two CID reaction channels for this ion, as follows:

$$CH_3NO_2^{+\cdot} \rightarrow NO^+ + CH_3O^\cdot \tag{20}$$

$$\rightarrow NO_2^+ + CH_3^\cdot \tag{21}$$

These reactions have been extensively investigated in several CID studies,[68–70] and the general features of kinetic energy dependance can be correlated with internal energy effects. The direct bond breakage, Reaction (21), exhibits interesting, but not extraordinary, dynamics features discussed elsewhere,[71, 72] and it is not described in detail here. The second path, Reaction (20), provided new dynamics features relating to both isomerization and non-QET behavior, which we now describe. Figure 17 is a simplified schematic view of the relevant potential surface for describing nitromethane ion CID. In particular, the barrier heights for rearrangement to form methyl nitrite and relative energy levels for nitromethane and methyl nitrite cations and their decomposition products are shown. Since fragmentation of nitromethane ion to NO_2^+ requires 1 eV and isomerization, 0.64 eV, isomerization and subsequent dissociation into NO^+ compete very effectively with the formation of NO_2^+ at low-collision energy. The dissociation of nitromethane ions into NO_2^+ and NO^+ follow two distinct pathways—NO_2^+ via direct $C-N$ bond cleavage and NO^+ via isomerization into $CH_3ONO^{+\cdot}$. This accounts for the observation of a strong metastable ion for NO^+, while NO_2^+ is formed only by CID.[73, 73] It is also plausible that significant differences in the dynamics of these two reaction channels may be observable.

The formation of NO^+ from $CH_3NO_2^{+\cdot}$ involves the simultaneous cleavage of $C-N$ and $N-O$ bonds and the formation of a new $C-O$ bond. This can take place either as a two-step mechanism—isomerization

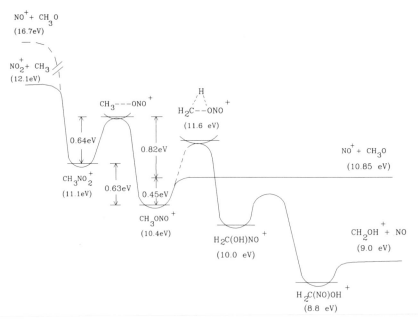

Figure 17. Schematic potential energy diagram for the dissociation of nitromethane ion to NO_2^+ and NO^+. (Reproduced from Reference 72.)

into methyl nitrite ion followed by dissociation—or via a "tight" transition state (rearrangement) as the dissociation proceeds. Both mechanisms involve energy barriers not present in single-bond dissociation. It has been shown by both metastable ion and CID studies that the nitromethane ion can isomerize to the energetically favored (more stable) methyl nitrite ion, which later dissociates via simple $O-N$ bond cleavage.[73, 74] The barrier to isomerization has been estimated to be 0.64 eV, and ions interconverting to the methyl nitrite ion must have an excess energy of 1.27 eV.[75, 76] Fragmentation of $CH_3ONO^{+\cdot}$ into NO^+ requires 0.45 eV, and all ions crossing the barrier must have sufficient internal energy for dissociation. These energy relationships are summarized in Figure 17.

Figure 18 is a contour map for NO^+ formation at 3.1-eV collision energy. A remarkable feature of this diagram is the nearly symmetric forward- and backward-scattered peaks with scattering angles of 0° and >170°. Forward-backward symmetry is traditionally interpreted as evidence of formation of an orbiting complex between the ion and neutral that survives several rotational periods before fragmentation. However, in this case, kinetic energy is much larger than the binding energy for an ion neutral $CH_3NO_2^{+\cdot}--$ He complex that an alternative explanation must be sought. Evidently, NO^+ ions in the forward- and backward-scattered sectors are formed by two dynamically different mechanisms with similar reaction endoergicities.

Close inspection of Figure 18 and detailed analysis of the energy deposition by applying Equation 9 to the forward- and back-scattered peak (e.g., integrating from $0-\pi$ and $\pi/2-\pi$) shows that the latter NO^+ mechanism is more endothermic by about 0.3 eV. As we have already discussed, backward scattering is associated with impulsive CID (occurring via small-impact-parameter collisions). This likely involves a population of vibrational excited $CH_3NO_2^{+\cdot}$ ions located low in the potential well of Figure 17, which is described as vibrationally excited nitromethane cations. These ions are activated by hard collisions that move them energetically to

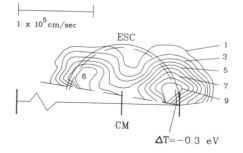

Figure 18. The CM-scattering contour diagram for the CID of nitromethane molecular ion to NO^+ on collision with helium at 3.1-eV collision energy: $CH_3NO_2^+ + He + 3.1\text{-eV} \rightarrow NO^+ = CH_3O + He$.

the top of the barrier and induce the nuclear motion required for rearrangement as the methyl nitrite structure.

The less endothermic (nearly thermoneutral) forward-scattered peak is readily interpreted by further considering the schematic potential surface in Figure 17. Following electron impact ionization, the "nitromethane" ions include a population near the top of the barrier to isomerization into methyl nitrite cation, the lowest energy decomposition path. This will include surviving metastable ions with decay rates equal to the reciprocal of our sampling time for CID experiments (*ca.* 40 μsec). For ions near the barrier height, adding a very small amount of additional energy in a large-impact-parameter glancing collision is sufficient to induce the decomposition. Moreover, ions near the barrier height approach the transition state structure, which we can describe qualitatively as an ion structure with such high-amplitude oscillations that it can no longer be described as either a methyl nitrite or a nitromethane cation. Thus, it is not necessary to perturb nuclear motion to a significant degree to induce decomposition; these ions have already been rearranged appropriately for dissociation into NO^+ and $CH_3O^.$.

Once the methyl nitrite ion is formed with 0.8 eV more than the minimum energy for dissociation following either of these dynamic excitation mechanisms (see Figure 17), it proceeds adiabatically along the potential surface to form NO^+. Leyh-Nihant and Lorquet[67] have thoroughly discussed this particular example of unimolecular reactions following a nonadiabatic pathway. The relative probability of retaining constant electronic configuration and dissociating into NO^+ on the diabatic surface was estimated to be 10^4 greater than following the adiabatic surface in a single passage through the critical region. This explains many of the puzzling features about methyl nitrite cation decompsositions, and also explains the absence of $CH_2OH^{+.}$ as a significant CID product and the prompt decomposition of $CH_3NO_2^{+.*}$ collisionally energized by the two low-energy mechanisms we have described.

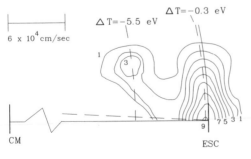

Figure 19. The CM-scattering contour diagram for the CID of nitromethane molecular ion to NO^+ on collision with argon at 18.6-eV collision energy: $CH_3NO_2^+ + Ar + 18.6$-eV $\rightarrow NO^+ + CH_3O + Ar$.

Figure 19 displays a velocity contour diagram for NO^+ formation at 18.6-eV CM collision energy; no backward-scattered peak is observed at this energy, consistent with our earlier discussion suggesting that higher collision energies favor larger impact parameter CID. However, two distinct peaks in the forward-scatterred region are still observed. The two peaks are clearly separated in both energy shift and angle, with one peak located near the relative velocity vector at $0°$ and the other at a scattering angle of about $8°$ degrees. The two peaks are characterized by energy transfers of 0.3 and 5.5 eV, respectively, at this collision energy. This new high-energy mechanism, which we have shown dominates high-energy CID, readily accounts for the conclusion by Zwinselman *et al.*[70] that collisions deposit anomalously large amounts of energy in nitromethane cations.

The small energy transfer, forward-scattered CID peak in Figure 19 originates from nitromethane ions that have traversed the methyl nitrite cation isomerization path, as previously described. The unexpectedly large energy transfer of about 5.5 eV clearly arises from a different mechanism. We therefore suggest that dissociation into NO^+ proceeds in parallel on two distinct potential energy surfaces at this energy. This necessarily implies that the high-energy peak is formed on an electronically excited hypersurface. Exciting nitromethane ions by adding 5.5-eV internal energy causes them to decompose faster than they can randomize energy among all internal degrees of freedom, and internal conversion into the ground states does not occur. This distinct excitation mechanism is responsible for the discrete structure obtained in the scattering contour diagram in Figure 19.

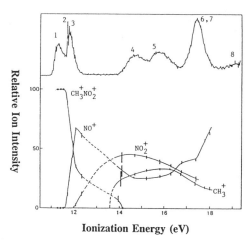

Figure 20. Photoelectron spectrum and breakdown graphs of nitromethane. (Reproduced from Reference 76.)

The PEPICO experimental results[75, 76] provide useful information supporting this interpretation of high-energy loss CID mechanism. As shown in Figure 20, the sixth ionization band of nitromethane is located 5.7 eV above the ground state. The fact that our energy transfer measurement closely matches the energy gap between the ground state and the sixth ionization band ($^2A_1$ state) of nitromethane ion[77] suggests that electronic excitation from the ground state is the operative activation mechanism. A relatively well-separated and strong peak for the sixth band in the photoelectron spectrum demonstrates that this transition has a favorable Frank-Condon factor. It is therefore highly probable that ions are excited to this state when sufficient energy is available to do so. If these ions randomize energy among all internal states, nitromethane ions would have a very broad distribution of energy in the ground state, resulting in a broad fragment ion peak distribution. This is not observed experimentally. The PEPICO experiments have also shown that nitromethane ions excited to this energy state preferentially dissociate into NO^+; this appears as the second rise in NO^+ intensity in the breakdown graph (see Figure 20). Indeed, it is the sharp rise in NO^+ formation at 5.7 eV above the ionization potential in the breakdown graph that enables us to resolve this dynamics feature clearly in the contour diagrams.

The hypothesis that NO^+ may dissociate directly from the sixth electronic state of nitromethane is further supported by kinetic energy release distribution measurements.[78] At a high internal energy range (5–6 eV), the mean kinetic energy release is significantly lower than that predicted by RRKM theory if internal energy is completely randomized in both electronic and vibrational modes. From these results, we can propose that the sixth electronic state of nitromethane ion is weakly coupled to the ground-state potential surface, and energy randomization at this level of excitation of nitromethane cation is much slower than dissociation.

Our global conclusion about the CID mechanism for nitromethane cations is that certain distinctive dynamics features are similar to propane and others are more like acetone. In particular, low-energy collisions are typical of statistical (QET, RRKM) behavior. Using the language suggested by Figure 20, the first three electronic states are closely coupled, and decomposition apparently proceeds via accessible low-energy pathways. (The rapid diabatic passage through a highly excited methyl nitrite structure does not affect the general conclusion.) However, at high-collision energy, a new reaction channel opens, and the reaction proceeds on an electronically excited hypersurface. Dissociation on this surface is faster than intramolecular vibrational relaxation to the ground-state hypersurface, as was observed in the acetone system; above 20 eV, this becomes the dominant CID reaction mechanism.

6. CONCLUSIONS AND PROSPECTS

Our dynamics studies of the CID of polyatomic ions have demonstrated interesting complexities involved in the collisional activation and dissociation of the activated ion. The general formulations of CID mechanisms proposed very early by Durup,[19] Loss,[20] and Singh *et al*,[79] which were mainly based on high-energy studies, are clearly inadequate to describe these processes. We infer from our quantitative success in modeling propane ion CID that polyatomic ions whose unimolecular fragmentation is adequately described by RRKM/QET theories are satisfactorily understood. The mechanisms for acetone and nitromethane cation are markedly different; they exhibit characteristic dynamics features resulting from small-impact-parameter collisions. Impulsive collisions that promote nonadiabatic, curve-crossing excitation are important mechanisms for these ions; very likely, they are important in many heteroatom systems with large energy gaps separating electronic bands. In all three cases, extensive fragment ions scattering has been observed, suggesting significant losses of fragment ions in conventional MS/MS experiments. This effect in combination with multiple collisions may be responsible for discrepancies between CID results for the same primary ions performed in different laboratories.

Collision-induced dissociation of ions continues to provide an invaluable method for analytical and structural characterization of complex mixtures and molecules in parallel with basic mass spectrometric techniques. As mass spectrometric applications have expanded to cover high-molecular-weight polymers and biological substances, the role of CID for characterization of structures of quasimolecular ions has become increasingly important. Limitations imposed by CM collision energy in the case of high-molecular-weight ions and light neutrals imposed by the Massey criterion formulation discussed in this chapter predict that single-collision CID with He-collision gas is ineffective for these ions. Nevertheless, CID is observed with high efficiency. Evidently, multiple collisions are operative, excitation mechanisms are of the curve-crossing type we observed for nitromethane and acetone, or both. It is highly desirable to extend these dynamics studies to high-molecular-weight and multiply charged ions, so that ions generated by fast atom bombardment, laser desorption, and electrospray techniques can be properly characterized.

ACKNOWLEDGMENTS

We gratefully acknowledge many helpful discussions on various aspects of CID we have had with Professors Graham Cooks, Tom Baer,

Chava Lifshitz, Doug Ridge, and Tom Lehman and Drs. Richard Martinez, Robert Boyd, and Alan Rockwood over the last 5 years. We would like to extend our sincere thanks to former graduate students Stephen Howard, Stephen Anderson, Karl Sohlberg, Runzhi Zhao, and especially, Kuangnan Qian—who carried out a good portion of the work described in Chapter 3. We also thank Betty Painter for her help in typing and revising this manuscript numerous times. This work was supported by the National Science Foundation through grants CHE 87962191 and CHE 8312069.

REFERENCES

1. Busch, K. L.; Glish, G. L.; and McLuckey, S. A., *Mass Spectrometry/Mass Spectrometry: Techniques and Applications of Tandem Mass Spectrometry* (VCH Publishers: New York, 1988).
2. McLafferty, F. W., ed., *Tandem Mass Spectrometry* (Wiley: New York, 1983).
3. Barber, M.; Bordoli, R. J.; Elliott, G. J.; Sedgwick, R. D., and Tyler, A. N., *Anal. Chem.* 1982, **54**, 645A.
4. Karas, M.; Bahr, U.; and Giebmann, U., *Mass Spectrom. Rev.* 1991, **10**, 335.
5. Smith, R. D.; Loo, J. A.; Ogorzalek Loo, R. R.; Busman, M.; and Udseth, H. R., *Mass Spectrom. Rev.* 1991, **10**, 359.
6. Laramee, J. A.; Carmody, J. J.; and Cooks, R. G., *Int. J. Mass Spectrom. Ion and Phys.* 1979, **31**, 333.
7. Nacson, S.; and Harrison, A. G., *Int. J. Mass Spectrom. Ion Processes* 1985, **63**, 325.
8. Mabud, M. A.; Dekrey, M. J.; and Cooks, R. G., *Int. J. Mass Spectrom. Ion Processes* 1985, **67**, 285.
9. Mukhtar, E. S.; Griffiths, I. W.; Harris, F. M.; and Beynon, J. H., *Org. Mass. Spectrom.* 1980, **15**, 51.
10. Barber, M.; and Elliott, R. M., *Twelfth Annual Conference on Mass Spectrometry and Allied Topics* (1964), Montreal, Canada, June, P. 150.
11. Futrell, J. H., Ryan, K. R., and Sieck, L. W., *J. Chem. Phys.* 1965, **43**, 1832.
12. Jennings, K. R., *Int. J. Mass Spectrom. and Ion Phys.* 1968, **1**, 227.
13. Haddon, W. F., and McLafferty, F. W., *J. Am. Chem. Soc.* 1968, **90**, 4745.
14. Hills, L. P., and Futrell, J. H., *Org. Mass Spectrom* 1971, **5**, 1019.
15. Holmes, J. L., *Org. Mass Spectrom.* 1985, **20**, 169.
16. Beynon, J. H., and Cooks, R. G., *Res. Dev.* 1971, **22**, 26.
17. Yost, Y. A., and Enke, C. G., *J. Am. Chem. Soc.* 1978, **100**, 2274.
18. Glish, G. L.; McLuckey, S. A.; Ridley, T. Y., and Cooks, R. G., *Int. J. Mass Spectrom. Ion Phys.* 1982, **41**, 157.
19. Durup, J., *Recent Developments in Mass Spectrometry*, K. Ogata and T. Hayakawa, eds., (University Park Press: Baltimore, 1970).
20. Loss, J., *Ber. Bunsenges. Phys. Chem.* 1973, **77**, 640.
21. Cheng, M. H.; Chiang, M.; Gislason, E. A.; Mahan, B. H.; Tsao, C. W.; and Werner, A. S., *J. Chem. Phys.* 1970, **52**, 5518.
22. Havemann, U.; Pacak, V.; Herman, Z.; Schneider, F.; Zuhrt, Ch.; and Zulicke, L., *Chem. Phys.* 1978, **28**, 147.
23. Chupka, W. A.; and Russell, M. E., *J. Chem. Phys.* 1968, **49**, 5426.
24. Anderson, S. L.; Houle, F. A.; Gerlich, D.; and Lee, Y. T., *J. Chem. Phys.* 1981, **75**, 2153.

25. Govers, T. R.; and Guyon, P. M., *Chem. Phys.* 1987, **113**, 425.
26. Guyon, P. M.; Baer, T.; Cole, S. K.; and Govers, T. R., *Chem. Phys.* 1988, **119**, 145.
27. Whitton, W. N.; and Kuntz, P. J., *J. Chem. Phys.* 1976, **64**, 3624.
28. Kumar, S.; and Sathyamurthy, N., *Chem. Phys.* 1989, **137**, 25.
29. Vogler, M.; and Seibt, W., *Z. Phys. A* 1968, **21**, 337.
30. (a) Gislason, E. A.; and Guyon, P. M., *J. Chem. Phys.* 1987, **86**, 677 (b) Sizun, M.; Parlant, G.; and Gislason, E. A., *J. Chem. Phys.* 1988, **88**, 4294.
31. Herman, Z.; Futrell, J. H.; and Friedrich, B., *Int. J. Mass Spectrom. Ion Processes* 1984, **58**, 181.
32. Qian, K.; Shukla, A.; and Futrell, J., *J. Chem. Phys.* 1990, **92**, 5988.
33. Rosenstock, H. M.; Wallenstein, M. B.; Wahrhaftig, A. L.; and Eyring, H., *Proc. Nat. Acad. Sci. USA* 1952, **38**, 667.
34. Shukla, A. K.; Anderson, S. G.; Howard, S. L.; Sohlberg, K. W.; and Futrell, J. H., *Int. J. Mass Spectrom. Ion Processes* 1988, **86**, 61.
35. Marcus, R. A.; and Rice, O. K., *J. Phys. & Colloid Chem.* 1951, **55**, 894.
36. Friedrich, B.; and Herman, Z., *Collect. Czech. Chem. Commun.* 1984, **49**, 570.
37. Wolfgang, R.; and Cross, Jr., R. J., *J. Phys. Chem.* 1969, **73**, 743.
38. Cooks, R. G.; Beynon, J. H.; Caprioli, R. M., and Lester, G. R., *Metastable Ions* (Elsevier: New York, 1973).
39. Champion, R. L.; Doverspike, L. D.; and Bailey, T. L., *J. Chem. Phys.* 1966, **45**, 4377.
40. Los, J.; and Govers, T. R., *Collision Spectroscopy*, R. G. Cooks, ed. (Plenum: New York, 1978).
41. Zuhrt, Ch.; Schneider, F.; Havemann, U.; Zulicke, L.; and Herman, Z., *Chem. Phys.* 1979, **38**, 205.
42. Vestal, M.; and Futrell, J. H., *J. Chem. Phys.* 1970, **52**, 978.
43. Gilman, J. P.; Hsieh, T.; and Meisels, G. C., *J. Chem. Phys.* 1982, **76**, 3497.
44. Brehm, B.; Eland, J. H. D.; Frey, R.; and Schuffe, H., *Int. J. Mass Spectrom. Ion and Phys.* 1976, **21**, 373.
45. Shukla, A. K.; Qian, K.; Anderson, S. G.; and Futrell, J. H., *Int. J. Mass Spectrom. Ion Processes* 1991, **109**, 227.
46. Anderson, S. G., Ph. D. dissertation, University of Utah, 1992.
47. Stockbauer, R.; and Inghram, M. G., *J. Chem. Phys.* 1976, **65**, 4081.
48. Williams, D. H.; and Cooks, R. G., *Chem. Commun.* 1968, 663.
49. Smith, R. D.; and Futrell, J. H., *Int. J. Mass Spectrom. Ion and Phys.* 1975, **17**, 233.
50. Massey, H. S. W., *Rep. Prog. Phys.* 1949, **12**, 249.
51. Nikitin, E. E., *Theory of Elementary Atomic and Molecular Processes in Gases*, translated by M. J. Kearsley (Clarendon: Oxford, 1974).
52. de Froidmont, Y.; Lorquet, A. J.; and Lorquet, J. C., *J. Phys. Chem.* 1991, **95**, 4220.
53. Kim, M. S.; and McLafferty, F. W., *J. Am. Chem. Soc.* 1978, **100**, 3279.
54. Stace, A. J.; and Shukla, A. K., *Int. J. Mass Spectrom. Ion and Phys.* 1981, **37**, 35.
55. Hasted, J. B., *Physics of Atomic Collisions* (Butterworths: London, 1964).
56. Shukla, A. K.; Qian, K.; Howard, S. L.; Anderson, S. G.; Sohlberg, K. W.; and Futrell, J. H., *Int. J. Mass Spectrom. Ion Processes* 1989, **92**, 147.
57. Qian, K.; Shukla, A. K.; Howard, S. L.; Anderson, S. G.; and Futrell, J. H., *J. Phys. Chem.* 1989, **93**, 3889.
58. Shukla, A. K.; Qian, K.; Anderson, S.; and Futrell, J. H., *J. Am. Soc. Mass Spectrom.* 1990, **1**, 6.
59. Qian, K.; Shukla, A.; and Futrell, J., *Chem. Phys. Lett.* 1990, **175**, 51.
60. Nystrom, J. A.; Bursey, M. M.; and Hass, J. R., *Int. J. Mass Spectrom. Ion Processes* 1983/84, **55**, 263.

61. Johnson, K.; Powis, I.; and Danby, C. J. *Chem. Phys.* 1981, **63**, 1.
62. Mintz, D. M.; and Baer, T., *Int. J. Mass Spectrom. Ion and Phys.* 1977, **25**, 39.
63. Bombach, R.; Stadelmann, J. P.; and Vogt, J., *Chem. Phys.* 1982, **72**, 259.
64. Lorquet, J. C.; and Takeuchi, T., *J. Phys. Chem.* 1942, **94**, 2279.
65. Martinez, R. I.; and Ganguli, B., *J. Am. Soc. Mass Spectrom.* 1992, **3**, 427.
66. Fenistein, S.; Futrell, J. H.; Heninger, M.; Marx, R.; Mauclaire, G.; and Yang, Y., *Chem. Phys. Lett.* 1991, **179**, 125.
67. Leyh-Nihant, B.; and Lorquet, J. C., *J. Chem. Phys.* 1988, **88**, 5606.
68. Hubick, A. R.; Hemberger, P. H.; Laramee, J. A.; and Cooks, R. G., *J. Am. Chem. Soc.* 1980, **102**, 3997.
69. Todd, P. J.; Warmack, R. J.; and McBay, E. H., *Int. J. Mass Spectrom. and Ion Phys.* 1983, **50**, 299.
70. Zwinselman, J. H.; Nacson, S.; and Harrison, A. G., *Int. J. Mass Spectrom. Ion Processes* 1985, **67**, 93.
71. Qian, K.; Shukla, A.; and Futrell, J., *Rapid Comm. Mass Spectrom.* 1990, **4**, 222.
72. Qian, K.; Shukla, A.; and Futrell, J., *J. Am. Chem. Soc.* 1991, **113**, 7121.
73. Lifshitz, C.; Rejwan, M.; Levin, I.; and Peres, T., *Int. J. Mass Spectrom. Ion Processes* 1988, **84**, 271.
74. Egsgaard, H.; Carlsen, L.; and Elbel, S., *Ber. Bunsen. Ges.* 1986, **90**, 369.
75. Gilman, J. P.; Hsieh, T.; and Meisels, G. G., *J. Chem. Phys.* 1983, **78**, 1174.
76. Niwa, Y.; Tajima, S.; and Tsuchinya, T., *Int. J. Mass Spectrom. Ion and Phys.* 1981, **40**, 287.
77. Rabalais, J. W., *J. Chem. Phys.* 1972, **57**, 960.
78. Ogden, I. K.; Shaw, N.; Danby, C. J.; and Powis, I., *Int. J. Mass Spectrom. Ion and Phys.* 1983, **54**, 41.
79. Singh, S.; Harris, F. M.; Boyd, R. K.; and Beynon, J. H., *Int. J. Mass Spectrom. Ion Processes* 1985, **66**, 131.
80. Kimura, K.; Katsumata, S.; Achiba, Y.; Yamazaki, T.; and Iwata, S., *Handbook of HeI Photoelectron Spectra of Fundamental Organic Molecules* (Halsted Press: New York, 1981).

4

Multiple Pulses and Dimensions in FTICR

Tino Gäumann

1. INTRODUCTION

This chapter introduces two-dimensional FTICR. Since two-dimensional spectroscopy depends on many details that are not directly connected to it but are nevertheless important for the success of a two-dimensional experiment, a few points are discussed at the beginning for those who wish to work in this field. A more detailed review of new achievements in FTICR has been presented elsewhere.[1]

Fourier-transform ion cyclotron resonance is a pulsed method operating in the kHz-to-MHz frequency range. Its basic pulse sequence is shown in Figure 1. A quench pulse empties the ion source of the ions from a preceding experiment; usually, an electron beam of a few milliseconds duration produces the ions. The peak emission current of the filament is low (typically a few mA) in spite of the short duty cycle, since the total number of ions per pulse has to be kept low (below 10^5 ions/pulse) to avoid space charge effects. The ions are initially homogeneously distributed across the source in a cylinder of 1 to several millimeters in diameter along the axis of the magnetic field. Because of the trapping field, the ions oscillate with the trapping frequency and tend to be concentrated at the center of the ion source. This process may be mass dependent, and it is not

Tino Gäumann • Institute of Physical Chemistry, École Polytechnique Fédérale de Lausanne, Lausanne, Switzerland.

Experimental Mass Spectrometry, edited by David H. Russell. Plenum Press, New York, 1994.

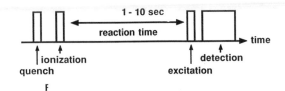

Figure 1. Basic pulse sequence in an FTICR instrument.

always very rapid.[2, 3] Usually a few milliseconds after ionization, the ions are excited by an rf-pulse that contains their resonance frequency. This time is long enough so that any fragmentation or isomerization has come to an end (see, e.g., References 4 and 5). Compared with a mass spectrum obtained on a magnetic instrument, the metastable decompositions are also summed up in the fragmentation spectra obtained with a FTICR instrument. Although the difference with a classical spectrum is not great, it is often noticeable.[6] Due to the open construction of the ion source, the translational temperature of the ions corresponds roughly to room temperature. Since the time constant for infrared emission of excess internal energy is of the order of ten to several hundred milliseconds long and the collision frequency is small because of the low pressure (10^{-8}–10^{-9} Torr) used for high resolution, the ions may possess quite an amount of internal vibrational energy.[4-13] The excitation pulse is usually a swept frequency pulse (chirp) over the frequency (mass) range of interest.

After excitation by an rf-pulse of the appropriate frequencies, the ions start to radiate with their resonance frequency. This emission transient is detected and amplified. Since signal averaging is nearly always used to improve the signal-to-noise ratio, timing between the excitation pulse and the detection transient is critical and must have a very low jitter. The decay time of the transient is mainly determined by the ion packets dephasing due to collisions. This varies from 10 msec to 10 sec in the pressure range used (10^{-6}–10^{-9} Torr). At the low pressure end, FTICR is inherently capable of a very high-mass resolution that decreases with increasing m/z ratio for a given transient length. It should be recalled that in FTICR, the frequency resolution depends only on the the number of points measured, i.e., the length of the transient. For high masses, the resonance frequency is low, and thus long transients and correspondingly low pressures are needed; an increase in the magnetic field allows shorter transients.

The time required for a transient for high resolution may last several seconds. If a large mass range is measured, the acquisition frequency is given by the lowest mass, and it is high. To have a sufficient number of data points and thus sufficient resolution at high masses (low frequencies), computer memory capabilities in the order of several Mbytes are necessary.

This is in itself feasible, but at the same time, signal averaging demands very rapid updating that has not been commercially achieved at present. Two solutions to this problem are used: (1) The whole mass range is detected in one short transient at a corresponding lower mass resolution (unity resolution is still easily obtained) in what is known as broad band detection; and (2) only part of the frequency range is detected by mixing the transient with a frequency close to the resonance frequency of the ions to be measured, i.e., a heterodyne or single-line detection. The improvement in resolution is coupled with a considerable gain in the signal-to-noise ratio because of the much smaller bandwidth needed for amplification. Its disadvantage (only a multiplet of 1 given mass-to-change (m/z) value is measured at one time) can be partially overcome by separating the signal output after the preamplifier into different channels, each having its own mixing frequency and its own low-pass filter.[14] Since the required bandwidth is only around 1 kHz, it is easy to multiplex the channels for the analog/digital A/D converter. The different mixing frequencies must be phase locked to the oscillator that generates the exciting sweep in order for the signal averaging to function properly.

Another method uses the fact that the width of a multiplet at 1 m/z value is often less than 200 Hz. In this case, the heterodyne oscillator simultaneously generates several mixing frequencies that are fed to the same mixer. They are chosen so that the separation between the multiplets is more than 200 Hz. To avoid aliasing, the total bandwidth should not be larger than the frequency difference between the two highest masses measured. By using several channels and a tailored excitation, the detection capacity can be enlarged at will.

The heterodyning method has the advantage of requiring only a very low bandwidth depending on the multiplet measured. Separating multiplets for a 4.7-T magnet and $^{13}C/CH$, D/H_2, and $^{13}CH_5/CD_3$ is less than 100 Hz for m/z values above 60, 30, and 20, respectively. The increase in the signal-to-noise ratio facilitates the high dynamic range often needed in this work. This advantage is somewhat reduced if several multiplets are simultaneously detected; but by using digital filters having steep slopes and a sufficiently high attenuation, bandwidths at any intermediate center frequency can be reduced at will at the expense of some additional computer handling.

The choice of the filter is dictated by the allowable bandpass ripple, the steepness of the slope, and the minimum attenuation desired. Since usually only the noise is to be reduced, the latter value is not very critical. The phases of the mixing frequencies must be carefully locked to the excitation sweep if signal averaging is used.

The fact that the FTICR instrument is completely computer-controlled facilitates such arrangements; however, the existing softwares rarely encourage such experiments! It is often easier to route the signal out of the main computer and to use an external PC with a suitable data acquisition and treatment program (e.g., Matlab$^{®}$, Lab View$^{®}$, etc.). By using two detection channels in parallel, it was possible to detect 24-m/z values between m/z = 41 and 102 simultaneously with a bandwidth of 8 kHz/channel and a very large dynamic range.[15] The resolution was 10^5 at m/z = 102, sufficient to resolve the H_2/D doublet easily at this mass. By using a LeCroy 9112$^{®}$ programmable generator for creating the two mixing frequencies, transients could be averaged directly. However, no observable decrease in the signal-to-noise ratio was observed when the magnitude Fourier transforms were averaged, but there was a slight decrease in the speed of acquisition results. In this case, no special triggering of the signal generator was necessary.

In FTICR, the time between the ionization and the excitation pulses can be increased from milliseconds to several seconds. During this interval, additional pulses can be used to obtain additional information. This scheme adds additional dimensions to the experimental and permits collisions either with neutrals for the study of ion/molecule reactions and collision-included dissociations or with photons for photofragmentation experiments or both. Due to the finite time needed for trapping ions in the middle of the cell, it is not easy to use this time range for quantitative measurements, since several parameters may have an influence.

The present-day ICR cell exhibits very nonlinear electric fields. The orbital radius has been determined by Grosshans and Marshall.[16] Beu and Laude proposed open-cell geometries that allow better pumping without deteriorating too much the cell-geometry.[17] The field in the center of the cell is the most linear, and coupling between the trapping and the cyclo-

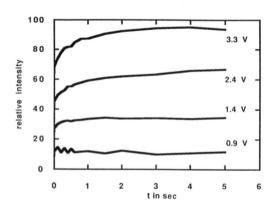

Figure 2. The intensity of the m/z 40 signal of argon as a function of the delay between the end of the ionization pulse and the beginning of the excitation pulse of the detection cycle. The trapping voltage is the parameter. Instrument: Spectrospin CMS 47, 4.7 T kryomagnet; cylindrical cell, 6-cm diameter, 6 cm long.[19]

tron movement is minimized in this region.[18] The influence of the trapping voltage is demonstrated in Figure 2.[19] The instrument is a 4.7 T Spectrospin® instrument with a cylindrical cell with a 6-cm diameter. Argon was chosen to avoid complications due to secondary reactions. The relative intensity of the signal at $m/z = 40$ is plotted as a function of the delay between the end of the ionization pulse and the beginning of the detection cycle. At low-trapping potentials, the signal is independent of time, but its intensity is low. If the depth of the trapping well is increased, the absolute intensity of the signal augments considerably, but it takes about 1 second to obtain stable conditions. Ions formed near the trapping plates are now concentrated in the middle of the cell, a location of the highest sensitivity, instead of being lost as is the case with low-trapping potentials. The influence of the gas pressure is shown in Figure 3.[19] The trapping voltage is held constant (2.4 V), but the emission current is adjusted to obtain approximately equal signal intensities for a delay time of 2 msec, at higher pressures, nearly 2 sec are needed to obtain stable conditions. This result might be considered astonishing, since we would assume that collisions improve the concentration process. We think that even at this trapping voltage, there is some ion loss and the signal intensity at 2 sec is roughly proportional to the concentration of the neutral argon. By normalizing our measurement at a delay time of 2 msec, we had a large space charge effect at higher pressures and a corresponding initial loss. Without going into details, this measurements may serve as a caveat that it is not easy to obtain meaningful time-dependent measurements in FTICR. However, if we use the time between ionization and detection as an additional dimension, this is a problem whose importance is sometimes underestimated. Whenever multiple pulsing and thus some additional reaction time is needed, cell conditions, the number of ions, and pressure conditions have to be carefully adjusted if reproducible and meaningful results are to be obtained.

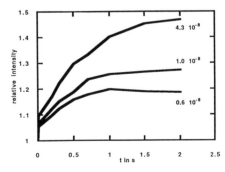

Figure 3. Same experiment as in Figure 2 but with the argon pressure as parameter. Trapping voltage 2.4 V.[19]

2. EXCITATION

The FTICR is an emission spectroscopy, and ions must be excited before the emission so that their rotational movement within the ion source can be detected. Rotational excitation pulses can be used for several other purposes, and they form the basis for multidimensional experiments. The linear response theory of ion excitation has been treated by Guan.[20] Ions can be selectively ejected by increasing their rotational energy to a limit where they are discharged on the excitation and detection plates or in the case of high masses, on trapping plates because of field inhomogeneity.[21-23] Ions can also be only slightly excited to study, for example, collision-induced decomposition (CAD). Phase-dependent excitation and desexcitation form the basis of two-dimensional experiments, but this phase dependence for multiple pulsing can lead to systematically incorrect results if the phase relationship between two consecutive pulses is not destroyed by either multiple collision or the introduction of a random delay together with signal averaging. Quadrupolar excitation[24] can be used to refocus ions.[25]

Excitation-detection is often assumed to be a linear process, however, this is not true because of the nonlinear field distribution within the ICR cell, which has been calculated in detail by Grosshans et al.[26, 27] and Hearn et al.,[28] who also reviews earlier work. The form of the characteristic detection line depends on many parameters, such as cell geometry, the number of ions, the trapping potential, characteristics of the excitation pulse, to name just a few.[29] An elongated cell[30, 31] and field corrections[32] have been proposed to improve the situation. In the classical cell, trapping plates form a short circuit for excitation, resulting in a highly nonlinear field distribution that leads to z-ejection[30, 33, 34] and a combined ICR and trapping frequencies.[35]

With a symmetric trapping potential, the resonance frequency $f_{ICR} + 2 \cdot f_{trap}$ is the most important additional frequency, but with increasing asymmetry of the trapping potential (corresponding to an oscillator in an asymmetric potential) $f_{ICR} + f_{trap.}$ gains in importance. The influence of the trapping field can be diminished by using "guarding grids" at ground potential near the trapping plates, which concentrate the trapping action at the short volume between grids and plates;[36, 37] however, nonlinearity of the excitation field still persists. An important proposal for linearization of the excitation field within the ion source is the "infinity cell" of Caravatti and Allemann,[38] who divided trapping plates into several segments at the same trapping (DC) potential. Due to their shape, these trapping plates simulate the distribution of the (AC) excidation field in the center of the cell and thus possess electrical properties of an infinitely long cell in the

z-direction of the magnetic field. The intensities of harmonic ICR frequencies can be used to quantify the improvement.[39-41] Although there is a definite improvement, the problem of nonlinear trapping field still persists.

Another disadvantage of the present-day spectrometers is the rectangular form of the pulse in time. The resultant frequency distribution corresponds to a $\sin x/x$ distribution with important wiggles far from the center frequencies.[42] This is particularly true for ejection pulses where large amplitudes up to several hundreds volts are used. Even a fraction of a percent can excite a neighboring ion to an energy that falsifies results of ion/molecule reactions by collisionally induced fragmentations whose origin are not easily detected. Singular frequencies can be conveniently apodized by a suitable filter function on a programmable function generator;[43] a modular system has been described by Beu and Laude.[44]

Because of the finite duration of the frequency sweep[15] and the necessity of having phase coherence among ion packets of the same m/z-ratio,[45, 46] it is not easy to produce a mass-independent excitation. Introduction of the SWIFT technique by Marshall *et al.*, where the desired signal is synthesized in the frequency domain and subsequently Fourier transformed into the time domain, is a definite improvement.[47-51] The limited amplitude resolution of the digital-to-analog (D/A) converter requires an optimum choice of phase distribution among the swept frequencies.[52, 53] Even then, experience shows that an amplitude resolution of 8 bits for the D/A converter is not sufficient; 12 bits seems to be a minimum. The limited number of points in the time domain results in the usual limitations of the discrete Fourier transform, such as a limited mass resolution and problems when a complex ejection program is performed. A potential disadvantage is that in addition the complex inverse FT is used, or in this case, a complex function has to be transformed. Standard computer programs often do not allow complex transformations of a large number of points. Although the software for ICR instruments can transform large files with high accuracy, it generally does not allow the transformation of complex numbers. Another possibility is using a direct current pulse of such a short duration that the necessary frequency range is well within the first lobe of a transformed delta function.[54-56]

For frequency chirps, the problem is more complicated. Since existing softwares are not very flexible, programmable generators are a practical way of improving the performance of experiments, but a few points have to be considered. A minimum number of points has to be used per period of the highest frequency. The Nyquist criterion (at least two sample points per period) is not very useful because the synthesizer also forms the frequency that corresponds to aliasing upward around the Nyquist frequency. Careful filtering is necessary, last not least, because of the glitching effects.

In general, between five and eight points per period are indicated, thus reducing the maximum length of a waveform for a given memory capacity and allowing adaequate filtering. Since the waveform is usually periodic, this does not necessarily present an obstacle, but the presence of several frequencies within the waveform introduces the additional condition that at the end of every block, every frequency must terminate with the correct phase. A waveform generator that uses direct digital synthesis for generating a periodic waveform or a variable clock frequency allows the generation of any frequency within its range and the stability of the time base; programming a complex mixture of frequencies is not however easy.

Figure 4 demonstrates the linear relation that can be obtained between the excitation and detection signals for a given mass. This is particularly true with an infinity cell (see Reference 38). Besides a small offset (which could be corrected for), a linear relation is observed up to a rather well-defined ejection amplitude. Because of the nonlinearity at large excitation amplitudes, the ejection region is rather extended even for an infinity cell. Large amplitudes have to be used for complete ejection, which can influence neighboring peaks; however, such linear behavior cannot be easily obtained for a wider range of frequencies, i.e., broadband detection.

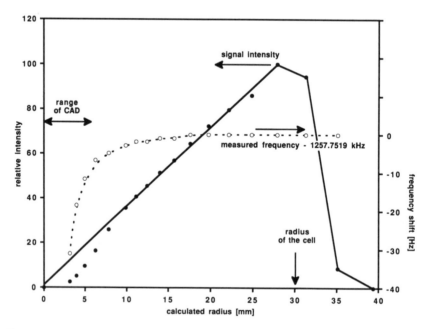

Figure 4. Dependence of the signal intensity and resonance frequency of the ion on the excitation amplitude. Data are given for an infinity cell.[38] Values are in arbitrary units.[19]

Because of nonlinear field distributions, it is not astonishing that the linear range decreases with increasing length of the transient, but the situation is by no means clear. As long as we concentrate on a given amplitude of the excitation and length of the resulting transient, good results are obtained; however, if we try to use the whole range of amplitudes, as is necessary for more complex experiments, better ICR cells must be constructed.

Figure 4 shows an effect that can present an additional difficulty in high-resolution multidimensional spectroscopy: When ions are only slightly excited, as is the case in CAD, the frequency is shifted to lower values. At the same time, relative intensity deviates from a linear relation, because the excited ion is still within the cloud formed by the other ions that shield it electrically (decreasing the excitation field) and magnetically (shifting the ICR frequency).[19] This effect depends on the number of ions; Marshall *et al.* have proposed a method to determine this number.[57]

3. COLLISIONAL ACTIVATION

Ions of a given m/z ratio can be isolated and subsequently rotationally excited by a pulse of small amplitude, corresponding to a radius of a few millimeters; thus, the excited ions are still within or near the cloud of the initially formed ions (see Figure 4).[58] At a higher pressure, the ions collide with a neutral gas and fragment, thus producing a CAD spectrum;[56–61] an example is shown in Figure 5, where the molecular ion of methyl-

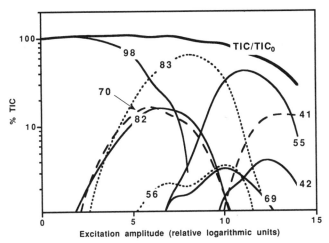

Figure 5. The CAD of m/z = 98 of methylcyclohexane pulsed to a pressure of 2.10^{-8} Torr during ionization; N_2 to $6 \cdot 10^{-7}$ Torr for 1 sec for collision[62] as a function of the excitation amplitude.

cyclohexane is isolated and collides with a N_2 pulse at a pressure of $6 \cdot 10^{-7}$ Torr for 1 sec. The amplitude of excitation is given on the abscissa in a logarithmic scale. The ions $m/z = 83$, 82, 70, and 69 are primary collision fragments. A fragmentation onset can be determined that allows estimation of the energy of the fragmentation process. With increasing excitation amplitude, additional secondary fragments are formed, but the total ion current (TIC) decreases because ions become randomly distributed throughout the ion source and coherent detection excitation is no longer possible even with a linear excitation curve of the type shown in Figure 4. In a CAD experiment, ions collide with a greater energy than necessary to overcome the trapping barrier, and they can be axially ejected along the magnetic field lines. The CAD can easily be performed with high-resolution detection when pressure of the collision gas is pulsed. By using a programmable signal generator, high resolution for the excitation is also feasible.[62] CAD is a case when the two-dimensional detection techniques are a great advantage. Figure 6 shows an example of a CAD experiment involving two peaks of the mass 102 that are selectively excited. In this case, a rectangular pulse is used; the resulting separation is better than 40,000. This is possible in a CAD experiment, since given that the excitation amplitude is small, the other peak of the doublet can be placed

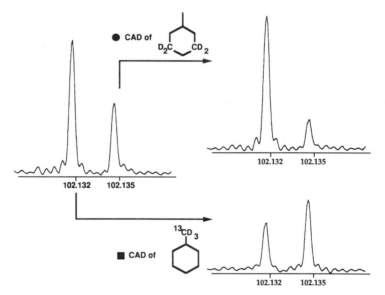

Figure 6. The CAD spectra for two methylcyclohexanes labeled in 3,3,5,5-D4 and 13CD3.[62] This example shows that in the less-demanding CAD experiment, a high resolution can be obtained with rectangular-shaped excitation pulses.

in a minimum of the $\sin x/x$-function, and a small residual excitation on this peak is not critical, contrary to a selective ejection.[62]

A particularly interesting extension of CAD has been proposed by Gauthier *et al.*:[63] An ion is excited a few hundred Hz beside its resonance frequency for about 1 sec [the authors call this a sustained off-resonance irradiation (SORI)]. In this case, the ion is excited and deexcited with a frequency corresponding to the difference between the resonance and excitation frequencies. In the normal CAD case, the excitation pulse is short, but the ion rotates at a given radius until it collides with the neutral gas. In the SORI case, the ion rotates only during a short period ($\sim$msec) on an enlarged radius before it is deexcited by its phase shift with the sustained excitation frequency unless a fragmenting collision takes place. Since this process is repeated for seconds, many "tickling" trials can be conducted. With a sufficiently small excitation amplitude, the lowest energy fragmentation can be sorted out with a high yield and its onset determined with high accuracy; high-mass resolution can be obtained in this way.[64] A somewhat related technique demonstrated by Boering *et al.* involves repetively exciting and deexciting a selected ion by a 180° phase jump (similar to the notch ejection technique) several times during a few hundred milliseconds.[65]

4. EJECTION

The time between ionization and detection can be used to eject one or several masses; unfortunately, FTICR is too slow a method to allow ejection of ions before total fragmentation occurs. (This would be a convenient way of studying very slow metastable fragmentation.) However, no measurable metastable fragmentations have been detected after a few milliseconds. Ejection is primarily used to isolate ions of a given m/z value to study their reaction. The procedure is basically simple: The mass range below the m/z value to be retained is swept with an radio-frequency field covering the required frequency range and an amplitude large enough to excite ions sufficiently to touch the electrodes. A second analog sweep covers the mass range above the chosen value. Sweep time is short compared to the reaction time. Due to the $\sin x/x$ behavior of a rectangular sweep,[42] wiggles are produced outside the swept frequency range in addition to the combination frequencies previously cited that may excite other ions. It is not easy to detect this excitation, and additional fragmentation by, e.g., CAD may be overlooked. This excitation can be diminished or even avoided by lowering the sweep speed (often to an unpractical value).

To avoid this problem, a rather complicated pulsing scheme may be necessary. The same is true if very high-mass resolution for ejection is needed or if several different ions must be isolated at the same time, as is the case for experiments using the Hadamard transformation. In such a situation, the SWIFT method presents a definite advantage in spite of its basic limitations; nevertheless, it also suffers from the wiggles produced by the rectangular window when a complex ejection scheme is needed. The situation can be improved by apodization,[66–69] again coupled with a subsequent loss in mass resolution for ejection.

For simple experiments, we found the notch ejection to be especially useful.[67, 68, 70] In this method, the phase of the ejection pulse jumps by 180° at the frequency of the ion that is to be retained. A rather high ejection resolution (>3000) can be obtained with a simple arrangement.[71] This experiment is particularly efficient in a dual cell.[72] For very high resolution, an external waveform generator that allows the construction of shaped waveforms is particularly useful, and by using multipole apodization, wiggles can be considerably reduced. An additional advantage is that these pulses are wider on the top; thus, the exact frequency is less critical, and pulse length can be shorter. In a doublet, pulse length can be chosen so that the frequency of the ion that should not be excited is placed at the minimum between two wiggles; here again, a multipole apodization (such as a Parzen filter, for example) has the advantage of much wider "holes" for a give attenuation.

It is not easy to detect the rotational excitation of the isolated ion(s), and its effect is often overlooked. On collision with a neutral molecule, rotation excitation produces CADs that superimpose on the ion molecule reactions under study. In our experience, a good way of detecting CAD is to admit either a short pressure pulse of a rare gas (or to add it permanently to the substance under study), then to wait for 1 or 2 seconds before detection. Even at a pressure of 10^{-8} Torr, a sufficient number of collisions take place to produce CAD and thus make the presence of rotational excitation observable. It is our opinion that many ion/molecule reactions are falsified by this excitation, which could possibly explain some of the strange results observed.

5. PULSING THE PRESSURE

The study of ion/molecule reaction involves a widespread application of ICR because of its ability to isolate a specific ion species and study its time behavior. In such a case, it is necessary to increase the pressure; if however high-mass resolution is needed, the number of collisions during the detection cycle must be smaller than a few collisions per second, corresponding to pressures below 10^{-8} Torr. Although it is possible to use observation times of several tens of seconds, this procedure is very slow and there is an increased risk of unwanted additional reactions in the background of the mass spectrometer. The exclusion of background reactions is increased with pulsing.[73] The background is water and hydrogen in the limiting case. Since ion/molecule reactions are most often studied as a function of time as a second dimension, the experiments tend to be very long, even with a very low background pressure.

Pulsing the pressure may be a convenient way of shortening the time needed for one scan. During a short interval, usually a few milliseconds, a reagent gas is admitted. Two possibilities exist: either a very short but well-defined pressure pulse in the μsec time range is formed by means of a fast valve,[74] or a pulse of a rectangular form of several hundreds of milliseconds or seconds length at a lower pressure is created. The pressure increases during a fraction of a second into the 10^{-7}–10^{-4} Torr range and decreases exponentially to the initial pressure. The time constant of this decrease depends on the construction of the instruments and the available pumping speed at the ion source. With the open construction of instrument using a kryomagnet, in this time constant is usually below 1 sec, allowing a reduction in the time needed for a scan to values between 2–10 sec. An additional advantage is the possibility of pulsing different substances during one experimental scan.

This pressure pulsing, although widespread, is a rather crude technique that does not easily permit the study of time dependence in a reaction as an additional dimension, because the admitted reagent gas must be stored at a high pressure to avoid depletion during signal averaging. Orifices of the available pulsed valves are a fraction of a millimeter, corresponding to "apparent" pumping speeds around 0.01 and 0.1 liter/sec. The pumping speed around the ion source in a modern instrument amounts in the best case to a few hundred liter/sec. In short, the ratio between the inlet pressure of the reagent gas and the pressure in the ion source is often around 10^{10}! This means that the pressure increases during a very short time to a very high value and then decreases exponentially to the starting pressure in a very long transient; thus, no quantitative measurements are possible. To obtain an approximately rectangular pressure pulse, the inlet pressure to the pulsed valve must be reduced to a small fraction of a Torr. This is not too difficult to achieve in itself (see Figure 7), but at least three additional difficulties arise: (1) The available pulsed valves are not very gas tight and a small leakage from atmospheric pressure is observed. To avoid this problem, the pressure drop across the closed valve must be reduced,

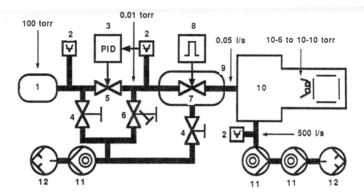

Figure 7. Scheme of the vacuum circuit with a stabilized pulsed valve. (1) Gas to be pulsed; (2) Penning valves, magnetically screened; (3) regulation circuit for the vacuum in front of the pulsed valve; (4) on-off valves; (5) regulating valve for the intermediate pressure; (6) small-leak valve, (7) pulsed valve; (8) timing circuit for the pulsed valve; (9) pumped chamber surrounding the pulsed valve; (10) mass spectrometer; (11) turbomolecular pumps or kryopumps; (12) rotary pumps.

and it is often worthwhile to evacuate the outside of the pulsed valve. (2) There is the possibility that the reagent gas may condense in the capillary present in the needle valve, leading to erratic pressure jumps; this requires heating the pulsed valve. (3) Since the quantity of reagent gas admitted may be of the order of 1–100 µMol/pulse, repetitive reproducible pulsing demands a rather large reservoir volume. It is easier to store the gas to be pulsed at a higher pressure and to use a pressure-regulating

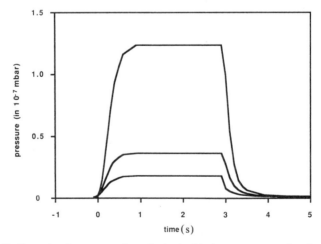

Figure 8. Example of pressure pulses obtained with the arrangement described in text.

device to produce an intermediate pressure. If quantitative results as a function of time are needed, the volume of the intermediate pressure vessel must be sufficiently large, so that during pulsing, the intermediate pressure does not decrease noticeably, since in most cases, the intermediate pressure-regulating valve (a membrane valve) has a time constant of seconds. An additional needle valve produces a small leakage to the vacuum and facilitates pressure regulation. Figure 8 shows an example of the pulse form obtained from such an arrangement.

Although pressure decreases according to a short time constant after closing the valve and permits a measurement even at high-resolution conditions a few hundred milliseconds after closing the valve, the residual pressure has the tendency to build slowly when repeated experiments are performed, as is the case of signal averaging; therefore it may be necessary to wait a few seconds before beginning a new scan.

6. PHOTOFRAGMENTATION

The time between ionization (or ion ejection) and detection can be conveniently used for photofragmentation.[75] For high-intensity pulsed lasers, the possibilities are the same as for a sector, a time-of-flight, or a quadrupole mass spectrometer. If a low-power light source is used, irradiation time is a suitable additional dimension. With light sources that work in the visible and ultraviolet region, the energy of 1 absorped photon is usually sufficient to cause fragmentation. The "rate-constant" obtained from dependence of the fragmentation as a function of length of irradiation corresponds to the absorption coefficient of the ion for fragmentation and can be used to characterize the ion. With a laser that emits photons in the infrared region, this is no longer true, since the energy of 1 photon is no longer enough to induce fragmentation. This is particularly true if infrared lasers in the 10-μm (CO_2 laser) or 5-μm range (CO laser) are used. Since for infrared emission, the lifetime of vibrationally excited ions is most often in the time range of FTICR experimentation, an equilibrium is established between the photons absorption, the distribution of additional internal energy, and photon emission.

The amount of photofragmentation can be plotted as a function of time and often yields as a straight line when $\ln(1-f)$ (f is the fraction of unfragmentated ions) is plotted against irradiation time t. The model assumes a sequential absorption of single photons, and it is based on the following assumptions: (1) The photon is absorbed on resonance between two low lying levels; (2) there is competition between absorption and stimulated emission; and (3) there is a coupling between vibration modes within the molecule (i.e., an excited level can be depopulated by intramolecular vibrational energy transfer and return to its initial state). This is demonstrated in Figure 9, where the decomposition of 5-methyl-2-hexene is shown as a function of laser irradiation time after isolating the molecular ion $m/z = 98$; the only product is the loss of neutral ethylene.[76]

Two parameters are obtained from of the plot of $\ln(1-f)$ against time: the dead time τ and the slope k. The latter is proportional to the number of photons used to reach the limit of fragmentation and the absorption coefficient of the ion. Since the infrared radiation life time is usually not known, this proportionality cannot be used directly to obtain the minimum number of photons necessary for fragmentation. The rate constant k depends on the intensity of the laser pulse:

$$k = \sigma \cdot \Phi^n$$

where σ equals the cross section for photofragmentation, Φ the flux of photons, and n the number of photons absorbed before the absorbed energy is equilibrated within the ion. Very often, a linear dependence is observed with the fluence of the radiation, as demonstrated in Figure 10 for the same system. The constant k is of the order of $0.1-10/\text{sec}$; thus fragmentation is very slow indeed: Fragmentation(s) crossing the lowest energy barrier of the ion with a lifetime of a few milliseconds is observed.

This example has been chosen to remind the reader that even a seemingly simple elimination reaction can be rather complex, since ethylene is expelled from different positions; however, the reaction scheme is much simpler in FTICR than in the ion source of a magnetic instrument or in the metastable range, since it is more probable that in the millisecond range, the ion has reached its stable structure(s), whatever it (these) may be.[77] This possibility corresponds to an ideal metastable decay, and results

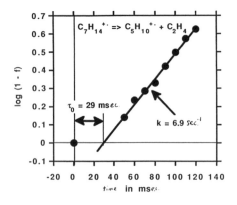

Figure 9. Photofragmentation of 2-dimethyl-2-hexene-3,4^{13}C$_2$. The abscissa represents irradiation time, the ordinate the logarithmic fraction of the decomposed molecular ion $m/z = 100$.[76]

are often similar to metastable fragmentations in a magnetic instrument.[6] The SORI-CAD experiment[63] may yield similar results, but there is not yet enough experimental evidence available. When different isomers cross the same activation barrier, they may have different dead times, although they have similar rate constants, as has been demonstrated for different isomeric butyl ions.[78] The sequential use of two CO_2 lasers has been recommended, with only the first, weaker laser requiring tuning.[79]

7. MULTIPLE DIMENSIONS

Pulsing the pressure or photons is performed to induce a reaction. Ions represented by X produce the ions Y:

$$Y = C(t) \cdot X$$

X and Y can be considered two product vectors, and $C(t)$ is a matrix containing the rate constants and concentrations of neutral products. The concentration list of the reaction ions, X, can be considered as a first dimension; it is obtained by measuring the mass spectrum immediately after ionization. The concentration list of the product ions, Y, corresponds to the mass spectrum obtained after the reaction (or part of it) has taken place. The matrix C introduces the reaction time and the concentration of the neutrals as an additional dimension. Any of the product ions Y_i may have several precursor ions X_i, and it is generally not possible to study the influence of different precursors separately. There are several solutions to solve this problem—either by continuous or discontinuous measurements.

One method consists in measuring Y discontinuously, i.e., to observe

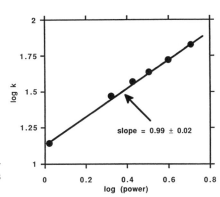

Figure 10. Rate constant of the photofragmentation of 2-dimethyl-2-hexene-3,4$^{13}C_2$ as a function of the power of the IR-laser.[76]

one selected Y_i ion by sweeping X. This method was the standard method with a classical ICR spectrometer, but it is of no practical use with multiplexing advantage of broadband detection in FT instruments. However, if high resolution is needed, this procedure is still valid in the sense that only one product multiplet of one (or few) m/z number(s) is measured at one time. The X also has to be measured discontinuously, which is a time–consuming procedure, since either the pressure must be pulsed or long reaction times at low pressures must be used to obtain the long transients required for high resolution.

The standard procedure is to select a single ion X_i of the X-matrix by ejecting all other ions. The possibility of exciting this isolated ion in the ejection process must be considered. This method is also cumbersome, since one ion after the other must be measured. High-resolution measurements are especially time consuming because of the low pressures needed for long transients.

8. HADAMARD TRANSFORM

This point-by-point measurement technique can be considerably shortened by using the Hadamard technique in mass spectrometry, as proposed by McLafferty *et al.*[80] This statistical practice[81–83] consists in constructing the so-called orthogonal **H**-matrix of experiments uniquely composed of $+1$ and -1. This quadratic matrix has the advantage of being very efficient, but it exists only in the dimensions $N=2$ or $N=0$ (mod 4). The matrix lines correspond to different scans and the columns to different reacting ions. Recall that the quadratic matrix can be used only for a linear model of the first degree; thus, consecutive reactions cannot be detected directly. A $+1$ means that in an experiment, results due to the reaction of the corresponding X_i should be added; a -1 means that these results should be subtracted. The **H**-matrix-problem is easily applied to a double-beam optical spectrometer, where two complementary mechanical masks can be used.[82, 84]

Although it is not *a priori* impossible to use the **H**-matrix in FTICR, it is conceptually more complicated, since there is no direct equivalent of a mechanical mask. To conserve the noise advantage of the Hadamard transform, noise should be detector limited and the additive and the subtractive signals should arrive at the same preamplifier. The easiest way of performing such an experiment is to use a double cell with two identical half-cells, with half of the selected ions excited in one cell and the complementary half in the other cell. Both cells are excited identically for detection, but the signal of one cell is connected with a phase shift of 180° to the detector with respect to the other cell. A similar arrangement has been shown to work by Williams *et al.*[85] This arrangement requires a complicated ejection and excitation scheme, and in reality, the noise advantage cannot be retained. The two half-cells occupy the same place in the homogeneous part of the magnetic field as a larger single cell. However, in a larger cell, a correspondingly larger number of ions can be trapped before a space charge effect begins to degrade the signal. This is a different situation compared to an optical spectrometer. The use of the **H**-matrix with a double cell can be an advantage. In this case, the Hadamard transform may be of a real advantage.

It is conceptually easier to use the so-called **S**-matrix that contains

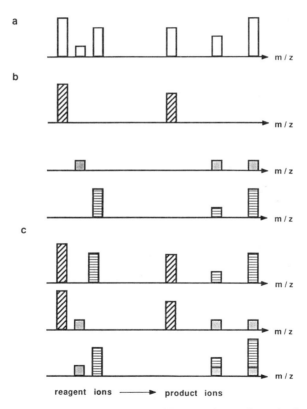

Figure 11. Different ways of measuring the reaction scheme of complex ion molecule reactions, CAD, photofragmentation, and so forth. The three ions at the left are assumed to be the reacting ions; the three on the right, the products. (a) All ions measured simultaneously; (b) single ion measurement; (c) Hadamard.

only +1 and 0; **S** is always one dimension smaller than the corresponding matrix **H**. The simplest one is

$$S = \begin{vmatrix} 1 & 0 & 1 \\ 1 & 1 & 0 \\ 0 & 1 & 1 \end{vmatrix}$$

where a "0" corresponds to ejecting the ion and a "1" to retaining the ion at that place in the matrix. It means that in a first scan (=first line) the masses X_1 and X_3 must be retained (and X_2 ejected); in a second scan, X_3 is ejected and in a third scan, X_1. Thus, in three scans, the influence of every m/z value is measured twice instead of only once, as would be the case in a one-step experiment. This is demonstrated in Figure 11, where we measure the ion/molecule reactions of several ions. Figure 11a corresponds to the case where all X_i are measured simultaneously and all Y_i are observed in the same scan: it gives an overview of all the ions formed. In Figure 11b, the standard procedure, every possible precursor X_i is allowed

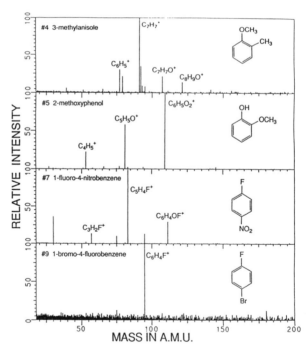

Figure 12. Typical Hadamard-transformed CAD spectra; precursors are the molecular ions shown. Dissociation efficiencies are 80%, 48%, 59%, and 29%, respectively. Spectra obtained by MacLafferty *et al.* (Reprinted with permission of the American Chemical Society.)

to react separately. Figure 11c corresponds to the simplest case of a Hadamard transform: Three of different scans are needed, but every reagent ion is measured twice, thus increasing the signal-to-noise ratio by about $\sqrt{2}$, since the two ion intensities add linearly and the incoherent noise is proportional only to the square root of the number of measurements. In addition, a system of three equations with three unknowns must be solved. The gain in signal-to-noise increases proportional to half of the square root of the dimension of **S**. Since its dimensions are smaller by one less than those of **H**, the next possible set contains of seven reagent ions.

The Hadamard method represents a definitive advantage as long as the reacting ions are known. It presents the following disadvantages given the available software: (1) multiple ejection is not always easy; (2) high-resolution work in the range of reacting ions is difficult; (3) the presence of consecutive reactions is not directly revealed; (4) the gain in signal-to-noise (SNR) is only appreciable with a large number of primary ions requiring a complicated ejection scheme, and (5) this ejection procedure increases the danger of exciting the retained ions rotationally. However, with appropriate software (e.g., SWIFT technique), the Hadamard **S**-matrix is a very promising technique indeed, as demonstrated by Williams *et al.*[85] in a CAD experiment.

Eleven molecular ions were chosen in a mixture of organic compounds in order to test the SNR improvement when an ejection according to an

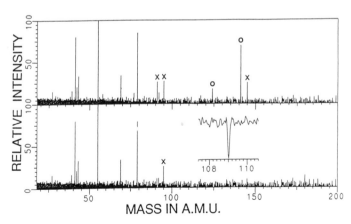

Figure 13. Hadamard-transformed CAD spectra of cyclohexanone (63% dissociation efficiency) using the S-matrix (top) and modified equation coefficients (bottom; inset is negative peak, abundance 38% of that of the base peak). (0:) Other precursor ions; (X:) fragment peaks from other precursors.[84] (Reprinted with permission of the American Chemical Society.)

S-matrix was chosen instead of measuring each molecular ion separately. These molecular ion underwent collision-induced dissociation in N_2 when excited to 25-eV rotational energy. The resulting CAD spectra showed expected improvement of a factor of 1.8 in the SNR (see Figure 12). In a more complex arrangement, an MS/MS/MS experiment was chosen where three daughter ions from three different parents were selected in a double-Hadamard experiment. These experiments clearly showed the versatility of the Hadamard transform.

The ejection program is unfortunately still far from perfect, since an appreciable number of incorrect peaks, partially precursors, partially fragments, showed up in the spectrum (see Figure 13). They could not be completely eliminated even by using a measuring scheme that resembles an H-matrix, probably because of imperfect ejection. A possible improvement is demonstrated in Figure 14. The left side corresponds to the case used by Williams's group: All ions except the selected ones are ejected. The remaining ions are rotationally excited for CAD fragmentation. This experiment suffers from the disadvantage that the ejection pulse may excite the remaining ions and two pulses are needed: the first one to eliminate the ions that are not measured and a second one for the CAD excitation.

Another way of conducting the experiment, which is called the "difference method," is shown on the right of Figure 14: Only ions corresponding to a 0 in the Hadamard experiment are ejected (ions 1, 2, and 6 in A); the other ions ion (ions 3, 4, 5, 7, and all other ions) are allowed to react. This simplifies the ejection problem and very high resolution can be achieved with suitably shaped ejection pulses, since this pulse contains only a few individual frequencies. The disadvantage is that an additional spec-

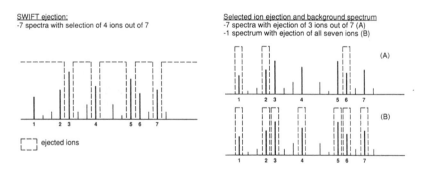

Figure 14. Different ways of performing a Hadamard experiment. *Left*: Ejection of all ions except those marked with a 1 in the S-matrix and excitation of the remaining ions for CAD. *Right*: (A) ejection (for ion/molecule reactions) of the 1 ions; (B) ejection of all Hadamard ions; or (A) excitation (for CAD) of the 1 ions; (B) background spectrum without any excitation.

trum has to be measured (B in Figure 14), where either no ions at all (the "blank" spectrum) or all Hadamard ions are ejected (or excited), corresponding to a small decrease in the Hadamard advantage. In the case of CAD, no ejection, only excitation is used, (1 corresponding to a CAD excitation and 0 to nonexcitation). Since the CAD excitation is of much smaller amplitude than an ejection pulse, the danger of exciting the wrong ion is much smaller and the experiment is less critical. The calculated spectra contain a negative peak (the excited ion) and a series of positive peaks (the product ions).

An experiment where excitation energy is added as an additional variable is easily programmed. It is now possible to use a programmable signal generator to obtain a well-apodized ejection pulse allowing very high resolution of ejection without exciting a neighboring peak. Figure 6 shows that high resolution can be obtained even with rectangular pulses for CAD experiments if correctly applied. The reproducibility of present-day mass spectrometers is sufficiently high as long as pressure pulse and excitation conditions are carefully selected to be linear (see Section 2). The infinity cell[38] represents a definite improvement. In Figures 15–17, the possibility of high resolution and good reproducibility with the method is demonstrated for a few ions in a study of the ion/molecule reaction in the system *n*-hexane-pentanone-3.[19,62] Figure 15 gives the time sequence of the experiment; because ions have to be ejected, shaped pulses are needed for high front-end resolution. Figure 16 demonstrates how the product for three ions is obtained with four spectra; each substance is measured twice. In Figure 17, good reproducibility is shown for a series of experimental points. It can be seen for example that reactions of the $C_4H_9^+$ (B) ion are clearly separated from those of the isobaric $C_3H_5O^+$ (C) that is nearly nonreactive in this mixture. The reaction product $C_5H_{11}O^+$ with ^{13}C in its natural concentration is also seen.

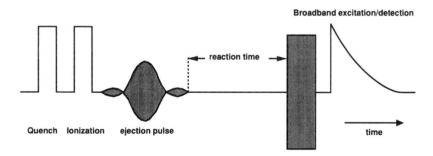

Figure 15. Pulse sequence to perform a high-resolution ion/molecule reaction study shown in Figure 17 with the Hadamard technique.[19]

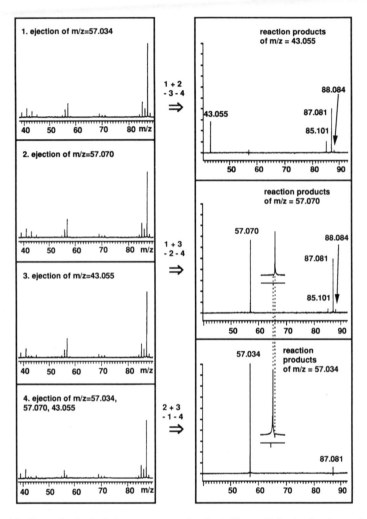

Figure 16. Example of calculating one measured point in Figure 17 (ion/molecule reaction).[19]

Residual excitation that results in either unwanted reactions and fragmentations or an increased rotation radius with a corresponding decreased
detection sensitivity (see Section 2) is easily checked in Hadamard transforms by the appearance of negative peaks in the resulting spectra when
the second method is used. If Hadamard ions are much less reactive than
other ions in the spectrum, the problem of the small difference between
two large numbers limits the reproducibility of all Hadamard experiments.
    In Figure 18, the pulse sequence is shown for a CAD Hadamard

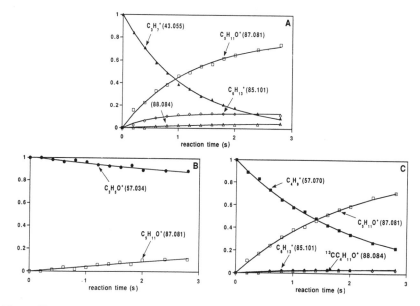

Figure 17. Ion/molecule reactions for three selected ions in the system *n*-hexane-pentanone-3 demonstrating the high resolution and reproducibility possible in Hadamard experiments. Reaction time in this experiment is used as a third dimension.[19, 62]

$$\textbf{A} \quad C_3H_7^+ + C_5H_{10}O \rightarrow C_5H_{11}O^+ + C_3H_6$$
$$C_3H_7^+ + {}^{13}CC_4H_{10}O \rightarrow {}^{13}CC_4H_{11}O^+ + C_3H_6$$
$$C_3H_7^+ + C_6H_{14} \rightarrow C_6H_{13}^+ + C_3H_8$$
$$\textbf{B} \quad C_3H_5O^+ + C_5H_{10}O \rightarrow C_5H_{11}I^+ + C_3H_4O$$
$$\textbf{C} \quad C_4H_9^+ + C_5H_{10}O \rightarrow C_5H_{11}O^+ + C_4H_8$$
$$C_4H_9^+ + {}^{13}CC_4H_{10}O \rightarrow {}^{13}CC_4H_{11}O^+ + C_4H_8$$
$$C_4H_9^+ + C_6H_{14} \rightarrow C_6H_{13}^+ + C_4H_{10}$$

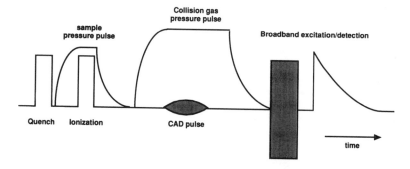

Figure 18. Pulse sequence for performing a CAD experiment with a Hadamard transform.

experiment on three molecular ions of three labeled methylcyclohexanes.[19] Experimental conditions are chosen so that a high back-end resolution can be obtained with the same set up. For a high front-end resolution, the CAD pulse of 20–200 msec length has to precede the collision gas pressure pulse to have a minimum number of collisions during this time period. The spectra in Figure 19 demonstrate the quality of the difference method: Even small differences (little product is formed) give reproducible results, and no false peaks are present. The experimental sequence is rather simple and can be routinely performed.

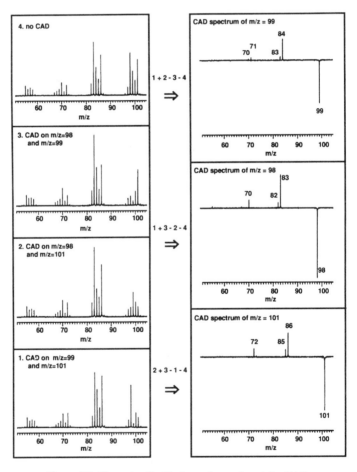

Figure 19. Example of a Hadamard experiment for CAD.

9. TWO-DIMENSIONAL SPECTROSCOPY

It is also possible to use a continuous X-matrix and a continuous Y-matrix. This technique is well known in NMR by the name of two-dimensional spectroscopy.[85] This method has been used in FTICR as well,[87-92] where the name is misleading, since various two- and multi-dimensional arrangements with very different aims exist in mass spectrometry. With N_1 data points in the first dimension and N_2 in the second dimension, the total number of points to be stored in computer memory is $N_1 \cdot N_2$. If high resolution is needed in the X or Y matrix, present-day computer capabilities require the use of heterodyning in one dimension.

The two-dimensional method uses the phase sensitivity of the FTICR technique, where every reacting ion is concentration-modulated by a frequency that is either its own cyclotron frequency or another chosen frequency. The reacting ion transfers this additional frequency information to the daughter ion or even to the granddaughter ion in the case of consecutive reactions. This implies that the lifetime of an activated intermediate complex is short compared to the inverse of the modulation frequency in order to maintain the phase information. This additional information shows up in the second dimension of the product spectra. The advantage of having two continuous dimensions is offset by the fact that all imperfections of the present-day ICR cell, such as nonlinearities (harmonics) and combination frequencies, also appear. The time needed for a two-dimensional experiment is longer and the load on the computer memory is much heavier.

The pulsing sequence for a two-dimensional experiment is shown in Figure 20. After the quench and the ionization pulses, two additional pulses, a and b, are introduced. Pulses a and b are identical for the swept frequency conditions, but not necessarily for the starting phase. The two pulses are separated by the time t_1, which varies during an experiment and defines the so-called first dimension. The time needed for the double pulse is generally a few milliseconds or less. The time between the double pulse and the excitation pulse c is used to perform the processes to be studied (ion/molecule reactions, photofragmentations, collisional-induced dissociations, etc.). The initial ion is transformed, but the frequency and phase information that have been imposed on this ion are transferred to the daughter ions.

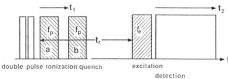

Figure 20. Pulse sequence for a two-dimensional mass spectrum. The double pulse is the excitation for the first dimension (not to scale in *t*).

To understand the effect of the double pulse, let us assume the following experiment: Only one sort of ions is chosen, e.g. argon, and the resulting signal is detected directly after the double pulse without the third excitation pulse f_e. Let us also assume a linear electric field distribution within the cell. After the first pulse a, all ions are excited coherently to the same radius. The effect of the second pulse b depends on its phase dependance with a; when they are in phase, an additional excitation will result (see point B in Figure 21). If they are in antiphase, the $Ar^{+\cdot}$ will return to the central axis of the cell (point A). The phase between the two pulses changes with the delay time between the two pulses which is proportional to t_1, and an amplitude modulation is observed that depends on t_1 [92]

$$\% \text{ TIC} \propto [1 + \cos(2\pi f_1 \cdot t_1)]^{1/2} \tag{1}$$

where f_1 is the cyclotron frequency f_{Ar} of Ar if the two pulses are identical.[89] If the phases of the two pulses are calculated in reference to a frequency f_0, then f_1 corresponds to the difference $f_1 = |f_0 - f_{Ar}|$.[87, 88] When we use t_1 as a second dimension, the radius distribution and thus the signal intensity vary periodically. This period depends on m/z; i.e., it is different and characteristic for ions of different mass. Since the modulation takes place before excitation, we call this dimension the first and the detected mass range the second dimension.

This fact can be used for a series of experiments. Looking at Figure 4, we realize that when we use an excitation amplitude for the two pulses so

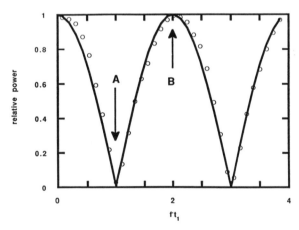

Figure 21. Comparison of the measured (open circles) and calculated (line) signal amplitude for small amplitudes of excitation after the two pulses a and b.[19] Two-dimensional spectroscopy: experimental points and calculated dependence.

that the point B lies near the maximum of the signal intensity, the third pulse, the detection excitation, ejects all ions whose radii exceed a limiting value; thus, we achieve a concentration modulation within the cell. Because of the nonlinearity of potentials within the ICR cell, the detection sensitivity decreases even before reaching this point, and amplitude conditions are less stringent. Ions either fragment or react when colliding with a suitable partner. The product ions have the same periodic concentration dependence and thus their precursor can be seen. When choosing the excitation amplitude according to data in Figure 5, we can optimize our response to a particular primary or secondary CAD fragmentation. When using a laser pulse coaxial with the electron beam, the photofragmentation of all ions excited in the first dimension can be studied simultaneously. Since the ion cloud has practically the same diameter as the electron beam producing it (the additional increase due to the magnetron motion is very small), a well-focused laser beam crossing the cell transversially can be shifted slightly outside the ion cloud, and a selective photoionization experiment can be performed.

After a Fourier transform in two dimensions, a primary ion lies on a diagonal line determined by the frequency f_1 in the first dimension and its cyclotron frequency in the second dimension; the position of a secondary ion is given by the frequency f_1 of its precursor ion in the first dimension and its own cyclotron frequency f_2 in the second dimension. An example for several primary ions is presented in Figure 22 with the ion/molecule reaction of CH_4^+ with methane, where CH_5^+ is formed. The two primary ions CH_3^+ and $CH_4^{+\cdot}$ lie on the diagonal line, and the secondary ion CH_5^+ is on a horizontal line through the position of $CH_4^{+\cdot}$. Note that a small quantity of CH_3^+ is formed by $CH_4^{+\cdot}$.

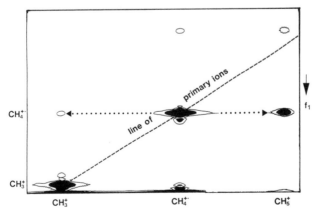

Figure 22. The two-dimensional spectrum of the reaction $CH_4^{+\circ} + CH_4 \rightarrow CH_5^+ + CH_3^{+\cdot}$.[98]

As we have just seen, the electric field distribution within the ICR cell is far from ideal. This influences not only the ease with which ions can be trapped or selectively ejected, but it also complicates the modulation scheme for a two-dimensional experiment. The success of the two-dimensional experiment depends strongly on this characteristic signal-versus-excitation line and the excitation power of the double pulse. If the latter is relatively small, the linear behavior observed is shown in Figure 4. The disadvantage is a small modulation amplitude in f_1 and a large signal without this modulation; the advantages are signal amplitudes proportional to the concentration of the ions, and only small contributions from signals at harmonic frequencies are present. When large amplitudes for double pulses are used, sensitivity and selectivity increase considerably, as does the harmonic content of the signal. It is not impossible that some cross modulation also sets in (as is the case for the complicated ejection program in the Hadamard experiment). Such complications have been demonstrated, e.g., by van der Hart and van de Gucht.[93] There are two possible improvements in this method: One consists in constructing an improved configuration of the cell plates, much progress could be made in this field (see Section 2 and References 94–96). The other improvement depends on the fact that in signal averaging, only constant-phase signals are added linearly. Since only ions at point A (see Figure 21) have lost all phase information about the double pulse, scrambling the time t_r (see Figure 20) within at least a time interval of the period of the lowest cyclotron frequency observed reduce considerably the undesired contribution by ions formed around point B. Note that the phase dependence of the ions at point B can also be used to an advantage: These ions fragment by collision-induced dissociation. When the phase of t_r is correctly chosen, these CAD-induced fragments can be selectively detected. This would be an application of the phase dependence of the two-dimensions experiment.

In the general case, the double pulse is a chirp and a range of frequencies (masses) are covered. Figure 23 illustrates this for methane (also see Figure 21). In this example all three pulses a, b, and c where chosen to be identical, and they covered the mass range from $m/z = 15$–17; usually, the pulses a and b excite the primary whose reactions are to be studied, whereas the pulse c embraces the mass range of the expected product ions. Either one of the pulses can have a narrow frequency range, thus allowing high-resolution work in the range of reacting and/or product ions. A series of transients is recorded with increasing values in t_1. Three of these transients are shown on the left in Figure 23. If they are transformed in the second dimension t_2 and plotted as a cascade of spectra as a function of t_1, the diagram in the middle of Figure 23 results. It is clear that the amplitudes of the three ions change periodically with t_1. The amplitude of

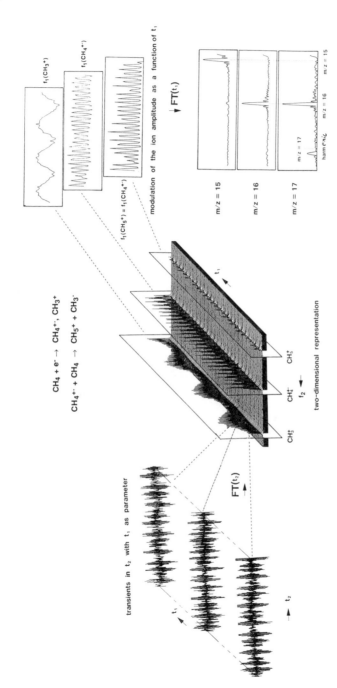

Figure 23. The ion/molecule reaction of CH_4^+ demonstrating various steps in calculating a two-dimensional experiment.[98]

three slices at the three masses of interest in $f_2 = [1/(m/z)_2]$ are depicted at the top right in Figure 23. It is evident that the amplitude of the ion CH_3^+ oscillates at a lower frequency than $CH_4^{+\cdot}$, whereas CH_5^+ has the same frequency as $CH_4^{+\cdot}$. A second transform along t_1 gives the spectra at the bottom right in Figure 23. The CH_3^+ slice presents a major peak at $m/z = 15$ and a very small contribution at $m/z = 16$: CH_3^+ is mainly a primary ion, but part of it is formed in an ion/molecule reaction of $CH_4^{+\cdot}$. The slice $CH_4^{+\cdot}$ yields only one peak at $m/z = 16$. The slice CH_5^+ shows no peak at the position $m/z = 17$ (indicated in Figure 23; the mass scale is linear) but a large contribution at $m/z = 16$ and a smaller peak at twice the distance from the origin to the right—a harmonic peak of $m/z = 16$. Thus, CH_5^+ is not a primary ion; it is formed exclusively from $CH_4^{+\cdot}$. With a Fourier transform program in two dimensions, all these steps are completed in one operation and a presentation such as Figure 24 results.

An important point in a two-dimensional experiments are the smallest step ∂t_1 for a change in t_1 and the number N_1 of discrete values for t_1: The smallest step ∂t_1 should be less than half the period of the highest observed

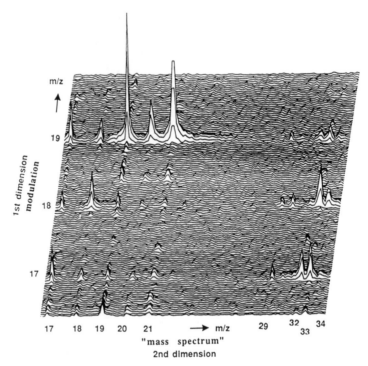

Figure 24. Ion/molecule reactions in CHD_3 in a two-dimensional representation.[98]

cyclotron frequency according to the Nyquist theorem; some oversampling is even indicated. Since no analog filtering is possible in this dimension, aliasing is possible. The minimum number of points should cover at least one period of the lowest frequency to be modulated. The frequency range is determined by the phasing between pulses a and b. If both chirps always start at the same phase, e.g., $0°$, the modulation frequencies in t_1 are equal to the cyclotron frequencies of the ions.[89] To a first approximation, this seems to be the simplest solution, but note that at the present state of the art, there is no analogue filter for f_1 and aliasing can take plase (this can be used as an advantage, see, e.g., Reference 87). Since the resolution can be made quite high, the instruments inherent frequency accuracy can be used to detect aliasing. In a two-dimensional spectrum with a labeled toluene, a charge transfer from water from the background was detected as a minor reaction after 21-fold overs.[97]

If the mass range to be covered in the first dimension is not large, it may be advantageous to use a phasing scheme that resembles heterodyne detection. If the frequency chirp sweeps downward from an upper frequency f_u and the phase of the second chirp is referred to this starting frequency f_u, the modulation frequency of the ions corresponds to the difference between f_u and the cyclotron frequency f_i of this ion.[88] This can be seen in Figure 22, where frequency f_u corresponds to m/z ~ 14.8 m.u. This kind of phasing is standard on a Spectrospin™ CMS 47 instrument, but it can be easily programmed on any instrument that allows the phase of the second pulse to be programmed by varying simultaneously the time t_1 and the starting phase $\Phi_2°$ of the second pulse.[89] By setting the starting phase $\Phi_2°$ of the second pulse equal to $2\pi f_u t_1$, f_u can be considered a virtual frequency that generates the phase continuity during time t_1. When varying t_1, the ion is again modulated by a frequency $(f_u - f_i)$, where f_i is the cyclotron frequency of the ion under consideration. Considerable economy in time and computer memory can be achieved with such a modulation, because the total memory capacity needed for a two-dimensional experiment is proportional to $N_1 \cdot N_2$, where N_2 is the number of data points in the second dimension t_2.

In principal, two dimensional experiments can be conducted at a very high resolution. Since contrary to the Hadamard excitation, the exact frequency of the ions must not be known in a two-dimensional experiment, a considerable advantage is gained, e.g., in a CAD experiment with its low excitation and its corresponding unknown frequency shift (see Figure 4). Such experiments are rather ideal for selective excitation of ions within a mass multiplet. In the case of a CAD excitation with its low excitation amplitudes, however, the ions suffer not only a frequency shift, but also a loss of coherence when the number of ions is too large. This means that the length of the two pulses used for the ion excitation in the first dimension must be as short as possible, which causes some loss in mass resolution. A decrease in the number of ions diminishes this effect, coupled with a decrease in sensitivity. Somehow a compromise has to be found.

An additional problem for all multidimensional spectroscopies is representing the results in a two-dimensional world. The most appropriate way is certainly the presentation by slices shown in the right corner of Figure 23. Such a presentation is possible in mass spectrometry, but has the disadvantage of requiring a lot of space for a complicated case. In addition, a general overlook of the results is not straightforward. Figure 24 demonstrates another possibility for the system CHD_3. Optically, this kind of presentation is probably the easiest to read: Primary ions all lie on the black diagonal line, i.e., the molecular ion $CHD_3^{+\cdot}$ and the two methyl ions CD_3^+ and CHD_2^+. Its daughter ions are located on a horizontal line through a given primary ions. The $CHD_3^{+\cdot}$ forms mainly $CH_2D_3^+$ and CHD_4^+ but also some secondary methyl ions, which subsequently yield

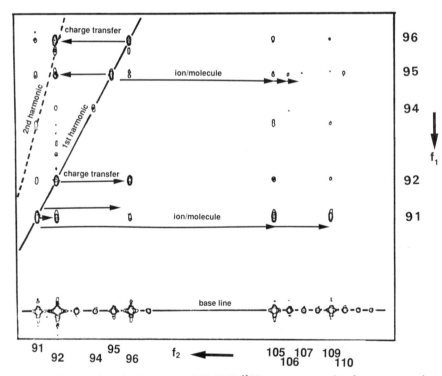

Figure 25. Reactions in the system $C_7H_8/C_6H_5{}^{13}CD_3$ as an example of a more complex reaction system.[98]

ethyl ions in a tertiary reaction. The two primary methyl ions form primarily ethyl ions of different isotopic composition, and excited methyl ions also yield ethenyl ions. The peak height of the ions gives some estimate of their concentration.

When the system is more complex, the preceding presentation is not practical, so it is best to use a mapping system, such as presented in Figure 25 for the system $C_7H_8/C_6H_5.^{13}CD_3$. The estimation of the amplitude of a peak is not so evident, but that is primarily a function of the graphic quality of the computer software available. In Figure 25, the primary ions again lie on a diagonal line (first harmonics). The broken line indicates the position of the second harmonics of the primary ions. Their contribution depends on the amplitude of the two exciting pulses, i.e., the working point on the characteristic line as given in Figure 4. (A compromise must be found between an optimum signal strength and a minimum harmonic content.) Figure 25 also demonstrates the advantage of two-dimensional spectroscopy. In addition to the charge transfer and ion/molecule reactions we would expect, there are a few reactions that are difficult to explain, e.g., the formation of m/z = 92 ($C_7H_8^{+\cdot}$) from $C_7H_7^+$, which is endothermic by at least 1.6 eV. However, these reactions are not two-dimensional artifacts, since they could subsequently be confirmed by classical kinetic and high-resolution FTICR. The results will be published elsewhere.

10. CONCLUSIONS

The matrix C can be considered the transfer function of the system. Several methods are feasible for performing measurements while keeping the multiplex advantage. The Hadamard tranformation has the advantage of a simple experimental design that can be programmed with any instrument. The results, especially for the difference method, are astonishingly well reproducible and make it a very appealing method. An additional advantage is the automatic elimination of background signals. This in itself may be a practical application in high-precision work even when advantages of the Hadamard technique are not needed. Medium mass resolution is easily obtained with the Hadamard transformations; the disadvantage is that the number and the m/z values of the ions being considered at must be known in advance, and if a very high front-end resolution is going to be performed, the small value of total ion current needed in this case means a low sensitivity.

The Fourier transform in two dimensions yields results in a quasi-continuous from. The m/z values of the reacting ion needs not to be known, thus the method is ideal either as a survey with limited mass resolution or for high-resolution work. Contrary to the Hadamard technique, it is relatively critical to find the optimum conditions. The capacity of computer memory and time considerations for the length of the experiment however limit the number of points that can be measured in the first dimension. In such a case, a parametric method may yield a better signal-to-noise ratio.[42] The general theories used in signal identification could be advantageously used,[99] but only very little experience has been gained up to present in this promising field for FTICR.[100,101]

Chapter 4 demonstrated that FTICR is capable of more complex experiments than are usually considered. Although an exact parallelism with NMR is not indicated, (mainly because of the very different bandwidth requirements), quite a number of new experiments can be envisaged. The field is still in its infancy; improvements in cell construction (more linear fields) and the accssibility of computer software are needed, but rapid improvement in this interesting domain is not only necessary, it is expected.

ACKNOWLEDGMENTS

It is a pleasure to thank my present and past collaborators in this field: M. Bensimon, P. Caravatti, F. Clairet, Sophie Haebel, J. Rapin, J. Riveros, F. Verdun, M.-E. Walser, and G. Zhao, for their assistance and many discussions and helpful comments. Financial support from the Swiss National Science Foundation, the Federal Commission for the Encouragement of Research, and Spectrospin/Fällanden is gratefully acknowledged.

REFERENCES

1. Hanson, C. D.; Kerley, E. Q. L.; and Russell, D. H., "Recent Developments in Experimental Fourier Transform Ion Cyclotron Resonance," 2 in *Treatise on Analytical Chemistry* (Wiley: New York, 1989).
2. Rempel, D. L.; Huang, S. K.; and Gross, M. L., *Int. J. Mass Spectrom. Ion Processes* 1986, **70**, 163–84.
3. Huang, S. K.; Rempel, D. L.; and Gross, M. L., *Int. J. Mass Spectrom. Ion Processes* 1986, **72**, 15–31.
4. Aviyente, V.; Shaked, M.; Feinmesser, A.; Gefen, S.; and Lifshitz, C., *Int. J. Mass Spectrom. Ion Processes*, 1986, **70**, 67–77.
5. Lifshitz, C.; Gefen, S.; and Arakawa, R., *J. Phys. Chem.* 1984, **88**, 4242–46.
6. Bensimon, M.; Rapin, J.; and Gäumann, T., *Int. J. Mass Spectrom. Ion Processes*, 1986, **72**, 125–35.

7. Heninger, M.; Fenistein, S.; Durup-Ferguson, M.; Ferguson, E. E.; Marx, R.; and Mauclaire, G., *Chem. Phys. Lett.* 1986, **131**, 439.

8. Mauclaire, G.; Heninger, M.; Fenistein, S.; Wronka, J.; and Marx, R., *Int. J. Mass Spectrom. Ion Processes* 1987, **80**, 99–113.

9. Mauclaire, G., Marx, R.; Heninger, M.; and Fenistein, S., *Adv. Mass Spectrom.* 1989, **11**, 526–27.

10. Heninger, M.; Feninstein, S.; Mauclaire, G.; Marx, R.; and Yang, Y. M., *J. Chem. Soc. Faraday Trans. II* 1989, **85**, 1705–12.

11. Faulk, J. D.; and Dunbar, R. C., *J. Phys. Chem.* 1989, **93**, 7785–89.

12. Huang, F.-S.; and Dunbar, R. C., *Int. J. Mass Spectrom. Ion Processes* 1988, **82**, 17–31.

13. Kuo, C.-H.; Wyttenbach, T.; Beggs, C. G.; Kemper, P. R.; and Bowers, M. T., *J. Chem. Phys.* 1990, **92**, 4849.

14. Rempel, D. L.; Ledford, Jr.; J. B.; Sack, T. M.; and Gross, M. L., *Anal. Chem.* 1989, **61**, 749–54.

15. Walser, M.-E., unpublished measurements.

16. Grosshans, P. B.; and Marshall, A. G., *Int. J. Mass Spectrom. Ion Processes* 1992, **115**, 1–19.

17. Beu, S. C.; and Laude, D. A., Jr., *Int. J. Mass Spectrom. Ion Processes* 1992, **112**, 215–230.

18. Delong, S. E.; Mitchell, D. W.; Cherniak, D. J.; and Harriso, T. M., *Int. J. Mass Spectrom. Ion Processes* 1989, **91**, 273–82.

19. Haebel, S., "Front-End Resolution and Hadamard Transform in Fourier Transform Ion Cyclotron Resonance." *Thesis EPFL No 1132*, 1993, Lausanne. Haebel, S.; and Gäumann, T., *Int. J. Mass Spectrom. Ion Processes*, to be published.

20. Guan, S., *J. Am. Soc. Mass Spectrom.* 1991, **2**, 483–86.

21. Grosshans, P. B.; and Marshall, A. G., *Int. J. Mass Spectrom Ion Processes* 1990, **100**, 347–79.

22. Grosshans, P. B.; Limbach, P. A.; and Marshall, A. G., *Proc. Thirty-Ninth ASMS Conf. Mass Spectrom.* Am. Soc. Mass Spectrom., 1991, 1505–06.

23. Beu, S. C.; and Laude, D. A., Jr., *Int. J. Mass Spectrom. Ion Processes* 1991, **108**, 255–68.

24. Savard, G.; Becker, St.; Bollen, G.; Klage, H.-J.; Moore, R. B.; Otto, Th.; Schweikhard, J.; Stolzenberg, H.; and Wiess, U., *Phys. Lett.* 1991, **A158**, 247–52.

25. Speir, J. P.; Gorman, G. S.; Pitzenburger, C.; Turner, C. A.; Wayne, P. P.; and Amster, A. J., *Anal. Chem.* 1993, **65**, 1746–52.

26. Grosshans, P. B.; and Marshall, A. G., *Anal. Chem.* 1991, **63**, 2057–61.

27. Grosshans, P. B.; Shields, P. J.; and Marshall, A. G., *J. Chem. Phys.* 1991, **94**, 5341–52.

28. Hearn, B. A.; Watson, C. H.; Baykut, G.; and Eyler, J. R., *Int. J. Mass Spectrom. Ion Processes* 1990, **95**, 299–316.

29. Ledford, E. B.; Rempel, D. L.; and Gross, M. L., *Int. J. Mass Spectrom. Ion Processes* 1983/84, **55**, 143–54.

30. van de Guchte, W. J.; and van der Hart, W. J., *Int. J. Mass Spectrom. Ion Processes* 1990, **95**, 317–26.

31. Hofstadler, S. A.; and Laude, D. A., Jr., *J. Am. Soc. Mass Spectrom.* 1990, **1**, 351–60.

32. Hansen, C. D.; Castro, M. E.; Kerley, E. L.; and Russell, D. H., *Anal. Chem.* 1990, 62, 520–26.

33. Wang, M.; and Marshall, A. G., *Int. J. Mass Spectrom. Ion Processes* 1990, **100**, 323–46.

34. Riegner, D. E.; Hofstadler, S. A.; and Laude, D. A., Jr., *Anal. Chem.* 1991, **63**, 261–68.

35. van der Hart, W. J.; and van de Guchte, W. J., *Int. J. Mass Spectrom. Ion Processes* 1988, **82**, 17–31.

36. Wang, M.; and Marshall, A. G., *Anal. Chem.* 1990, **62**, 515–20.
37. Grosshans, P. B.; and Marshall, A. G., *Proc. Thirty-Ninth ASMS Conf. Mass Spectrom.*, Am. Soc. Mass Spectrom., 1991, 40–50.
38. Caravatti, P.; and Allemann, M., *Org. Mass Spectrom.* 1991, **26**, 514–18.
39. Nikolaev, E. N.; Gorshkov, M. V.; Mordehai, A. V.; and Tal'rose, V. L., *Rapid Comm. Mass Spectrom.* 1990, **4**, 144–46.
40. Grosshans, P. B.; Shields, P.; and Marshall, A. G., *J. Am. Chem. Soc.* 1990, **112**, 1275–77.
41. Hofstadler, S. A.; and Laude, D. A., Jr., *Int. J. Mass Spectrom. Ion Processes*, 1990, **101**, 65–78.
42. Marshall, A. G.; and Verdun, F. R. *Fourier Transforms in NMR, Optical, and Mass Spectrometry* (Elsevier: Amsterdam, 1990).
43. Hanson, C. D.; Castro, M. E.; Kerley, E. L.; and Russell, D. D., *Anal. Chem.* 1990, **62**, 1352–55.
44. Beu, S. C.; and Laude, Jr., D. A., *Anal. Chem.* 1991, **63**, 2200–03.
45. Hanson, C. D.; Kerley, E. L.; Castro, M. E.; and Russell, D. H., *Anal. Chem.* 1989, **61**, 2040–46.
46. Hanson, C. D.; Castro, M. E.; and Russell, D. H., *Anal. Chem.* 1989, **61**, 92130–36.
47. Marshall, A. G.; Wang, T.-C. L.; and Ricca, T. L., *J. Am. Chem. Soc.* 1985, **107**, 7893.
48. Lin Wang, T.-C.; Ricca, T. L.; and Marshall, A. G., *Anal. Chem.* 1986, **58**, 2935–38.
49. Marshall, A. G.; Lin Wang, T.-C.; Chen, L.; and Ricca, T. L. "New Excitation and Detection Techniques in Fourier Transform Ion Cyclotron Resonance Mass Spectrometry," in ACS Symp. Rev., Bucharan, M. V., ed., 1987, **359**, 21.
50. Cody, R. B.; Goodman, S. D.; and Campana, J. E., *Proc. Fortieth Pittsburgh Conf. Anal. Chem. Appl. Spectrosc.*, Am. Chem. Soc., (Atlanta, Ga, 1989) abstract 719.
51. Guan, S., *J. Chem. Phys.* 1989, **91**, 775–77.
52. Chen, L.; Lin Wang, T.-C.; Ricca, T. L.; and Marshall, A. G., *Anal. Chem.* 1987, **59**, 449–54.
53. Guan, S.; and McIver, R. T., Jr., *J. Chem. Phys.* 1990, **92**, 5841–46.
54. McIver, Jr., R. T.; Hunter, R. L.; and Baykut, G., *Rev. Sci. Instr.* 1989, **60**, 400–5.
55. McIver, Jr., R. T.; Baykut, G.; and Hunter, R. L., *Int. J. Mass Spectrom. Ion Processes* 1989, **89**, 343–58.
56. McIver, Jr., R. T.; Hunter, R. L.; and Baykutt, G., *Anal. Chem.* 1989, **61**, 489–91.
57. Limbach, P. A.; Grosshans, P. B.; and Marshall, A. G., *Anal. Chem.* 1993, **65**, 135–40.
58. Feng, W.; and Gäumann, T., unpublished results.
59. Cody, R. B.; and Freiser, B. S., *Int. J. Mass Spectrom. Ion Phys.* 1982, **41**, 199–204.
60. Cody, R. B.; Burnier, R. C.; and Freiser, B. S., *Anal. Chem.* 1982, **54**, 96–101.
61. McIver, R. T., Jr.; and Bowers, W. D., in *Tandem Mass Spectrometry*, McLafferty, ed. (Wiley: New York, 1983), p 187.
62. Haebel, S.; Walser, M. E.; Clairet, F.; Feng, W.; and Gäumann, T., to be published.
63. Gauthier, J. W.; Trautman, T. R.; and Jacobson, D. B., *Anal. Chim. Acta* 1991, **246**, 211–25.
64. Heck, A. J. R.; deKoning, L. J.; Pinkse, F. A.; and Nibbering, N. M. M., *Rapid Comm. Mass Spectrom.* 1991, **5**, 406–14.
65. Boering, K. A.; Rolfe, J.; and Brauman, J. I., *Rapid Comm. Mass Spectrom.* 1992, **6**, 303–5.
66. Aarstol, M.; and Comisarow, M. B., *Int. J. Mass Spectrom. Ion Processes* 1987, **76**, 287–97.
67. Noest, A. J., "ICR Studies of Some Anionic Gas Phase Reactions and FTICR Software Design," PLD Thesis, University of Amsterdam, Amsterdam, 1982.

68. Noest, A. J.; and Kort, C. W. F., *Comp. Chem.* 1982, **6**, 111–15.
69. Brenna, J. T.; and Creasy, W. R., *J. Chem. Phys.* 1989, **91**, 775–77.
70. de Koning, L. J.; Fokkens, R. H.; Pinske, F. A.; and Nibbering, N. H. M., *Int. J. Mass Spectrom. Ion Processes* 1987, **77**, 95–105.
71. Zhao, G. H., "Photofragmentation of $C_7H_8^{+\cdot}$ Ions, Phd Thesis EPFL No. 1130, Lausanne, 1993.
72. Farrell, J. T., Jr.; Peiping, Lin; and Kenttämaa, H. I., *Anal. Chim. Acta* 1991, **246**, 227–32.
73. Gardner, J. M.; Lester, M. I.; and Kimock, F. M., *Int. J. Mass Spectrom Ion Processes* 1989, **91**, 199–207.
74. Corat, E. J.; and Trava-Airoldi, V. J., *Rev. Sci. Instr.* 1990, **61**, 1968–71.
75. van der Hart, W. J., *Mass Spectrom. Rev.* 1989, **8**, 237–68.
76. Bensimon, M., "Photodissociation d'ions par laser infrarouge dans un spectromètre de masse à transformation de Fourier," Phd Thesis, EPFL No. 694, Lausanne, 1987. Bensimon, M.; Gäumann, T.; Rapin, J.; Stahl, D.; and Waghmare, S., *Adv. Mass Spectrom.* 1989, **11** 170–71.
77. Gäumann, T., "The mass spectra of alkanes," in *"The Chemistry of Alkanes and Cycloalkanes,"* Patai, S.; Rappoport, Z., eds. (Wiley: Chichester, UK, 1992), pp. 395–454.
78. Bensimon, M.; Gäumann, T.; Guenat, C.; and Stahl, D., *Adv. Mass Spectrom* 1985, **Part B**, 807–8.
79. Watson, C. H.; Zimmermann, J. A.; Bruse, J. E.; and Eyler, J. R., *J. Phys. Chem.* 1991, **95**, 6091–86.
80. Mc.Lafferty, F. W.; Tauffer, D. B.; Loh, S. Y.; and Williams, E. R., *Anal. Chem.* 1987, **59**, 2213–16.
81. Plackett, R. L.; and Burman, J. P., *Biometrika* 1946, **33**, 305–25.
82. Harwit, M., "Hadamard Transform in Analytical Systems," in *Transform Techniques in Chemistry*, Griffiths, P. R., ed. (Plenum: New York, 1978.)
83. Sloane, N. J. A., "Hadamard and Other Discrete Transforms in Spectroscopy," in "Fourier, Hadamard, and Hilbert Transforms in Chemistry," Marshall, A. G., ed. (Plenum: New York, 1982).
84. Treado, P. J.; and Morris, M. D., *Anal. Chem.* 1989, **61**, 723A–734A.
85. Williams, E. R.; Loh, S. Y.; McLafferty, F. W.; and Cody, R. B., *Adv. Mass Spectrom.* 1989, **11**, 1878–79. *Anal. Chem.* 1990, **62**, 698–703.
86. Ernst, R. R.; Bodenhausen, G.; and Wokaun, A., *Principles of Nuclear Magnetic Resonance in One and Two Dimensions* (Clarendon Press: Oxford, U.K., 1987).
87. Pfändler, P.; Bodenhausen, G.; Rapin, J.; Houriet, R.; and Gäumann, T. J., *Chem. Phys. Lett.* 1987, **138**, 195–200.
88. Pfändler, P.; Bodenhausen, G.; Rapin, J.; Walser, M. E.; and Gäumann, T., *J. Am. Chem. Soc.* 1988, **110**, 5625–28.
89. Bensimon, M.; Zhao, G.; and Gäumann, T., *Chem. Phys. Lett.* 1989, **157**, 97–100.
90. Guan, S.; and Jones, P.-R., *Rev. Sci. Instr.* 1988, **59**, 2573–76.
91. Guan, S., *J. Chem. Phys.* 1989, **91**, 775–77.
92. Guan, S.; and Jones, P. R., *J. Chem. Phys.* 1989, **91**, 5291–95.
93. van der Hart, W. J.; and van de Guchte, W. J., *Int. J. Mass Spectrom. Ion Processes* 1988, **82**, 17–31.
94. Nikolaev, E. N.; and Gorshkov, M. V., *Int. J. Mass Spectrom. Ion Processes* 1985, **64**, 115–25.
95. Wang, M.; and Marshall, A. G., *Anal. Chem.* 1989, **61**, 1288–93.
96. Pan, Y.; Ridge, D. P.; and Rockwood, A. L., *Int. J. Mass Spectrom. Ion Processes* 1988, **84**, 293–304.

97. Bensimon, M.; and Gäumann, T., unpublished results.
98. Bensimon, M.; Gäumann, T.; Rapin, J.; Verdun, F.; Walser, M.-E.; and Zhao, G. H., *Proc. Thirty-Seventh ASMS Conf. Mass Spectrom.*, Miami Beach, 1989, 65–67.
99. Ljung, L. *System Identification: Theory for the User* (Prentice Hall: Englewood Cliffs, 1987) and literature cited therein.
100. Meier, J. E.; and Marshall, A. G., *Anal. Chem.* 1991, **63**, 551–60.
101. Williams, C. P.; and Marshall, A. G., *Proc. Thirty-Seventh ASMS Conf. Mass Spectrom*, Miami Beach, 1991, 1503–04.

5

Experimental Fourier Transform Ion Cyclotron Resonance Mass Spectrometry

David A. Laude, Jr. and Steven C. Beu

1. INTRODUCTION

The field of ion cyclotron resonance (ICR) mass spectrometry was established in 1949 with the invention of the omegatron mass spectrometer by Hipple et al.[1, 2] This new type of mass spectrometer employed a combination of static electric and magnetic fields to trap ions that were mass selected by applying radio frequency (rf) excitation at their cyclotron frequency and then detected on striking a collector electrode. Unfortunately, the many intriguing features of this spectrometer were countered by its poor mass resolution and mass range, and it was to find limited analytical utility. In the 1960s, ICR was investigated as a detector for studying gasphase reactive and nonreactive collisions. Wobschall constructed an ICR spectrometer that differed from the omegatron in that ionization and detection occured in separate regions of the vacuum chamber and an rf bridge circuit was used to measure ICR power absorption.[3] This instrument matured into the multisection drift ICR spectrometer[4] that combined double resonance techniques[5] with marginal oscillator detection.[6] Drift cell ICR found great utility in ion/molecule studies in the laboratories of Baldeschweiler et al.,[7] Beauchamp,[8] and Bowers et al.,[9] among

David A. Laude, Jr. and Steven C. Beu • Departments of Chemistry and Biochemistry, The University of Texas at Austin, Austin, Texas 78712.

Experimental Mass Spectrometry, edited by David H. Russell. Plenum Press, New York, 1994.

others, although the technique continued to exhibit the unattractive low-performance features of its ICR predecessors.

It was not until the early 1970s that two advances were made in ICR instrument design that provided the foundation for the modern high-performance ICR experiment discussed in Chapter 5. In 1970, McIver reported on a trapped ion cell designed to constrain ion motion in the applied magnetic and electric fields of an ICR spectrometer.[10] The trapped ion cell was distinguished from the drift cell by a one hundredfold increase in ion storage time and by a pulsed rather than continuous mode of operation. These two features were exploited by Comisarow and Marshall, who in 1974 described a nondestructive multiplex detection scheme for ICR that dramatically improved performance.[11] In this pioneering work, a time domain ICR signal was acquired by monitoring the ion image charge developed on electrodes as ions orbited within the cell. This signal was decoded with the Fourier transform to yield a frequency domain representation of a conventional mass spectrum. The technique is today known synonymously as Fourier transform ion cyclotron resonance (FTICR) or Fourier transform mass spectrometry (FTMS). Modern FTICR continues the rich history of ICR as a detector for gas-phase ion/molecule reaction studies,[12] but in addition, it is acknowledged as a high-performance spectrometer capable of unrivaled mass resolution and mass precision.[13–15]

Chapter 5 begins with a description of the FTICR experiment as conceived by Comisarow and Marshall,[11, 16, 17] then presents more recent innovations in FTICR instrument design and experimental capabilities, providing the reader with a general overview of the present status of experimental FTICR and its prospects for future development.

2. SINGLE-SECTION CELL FTICR EXPERIMENT

In Section 2, we review the basic FTICR experiment in which ions are formed by electron ionization in a single-section cubic-trapped ion cell, driven to coherent, large amplitude orbits by resonant rf excitation, and then detected. Elementary equations describing ion physics in the cubic cell accompany this presentation of the experiment. A brief theoretical development of the FTICR line shape, and some aspects of data acquisition and processing are also discussed.

2.1. Simple Ion Motion in the Trapped Ion Cell

Figure 1 shows a cubic FTICR-trapped ion cell of length c. The cell is oriented in a strong homogeneous magnetic field along Cartesian coor-

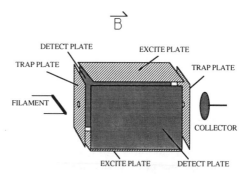

Figure 1. Single-section cubic FTICR-trapped ion cell as configured for electron ionization.

dinates with origin in the center of the cell and the *z*-coordinate aligned with the magnetic field *B*. The three orthogonal sets of parallel plates include the receiver and transmitter plates, which are positioned collinear with the magnetic field, and the trap plates, which are positioned perpendicular to the magnetic field. An electron filament assembly and collector are positioned at opposite ends of the trap plates along the *z*-axis centerline of the cell ($x=y=0$). To initiate the FTICR experiment, the filament assembly is gated on for a few milliseconds to pass accelerated electrons through the cell and form ions by electron ionization (EI). As will be described, these ions assume stable trajectories within the trapped ion cell, and they are subsequently mass analyzed.

Within the confines of a strong homogeneous magnetic field, ion motion in the *xy*-plane assumes circular orbits with cyclotron frequency f_c. This frequency is related to *B* (T) and to the mass *m* (kg) and charge *q* (C), of the ion by Equation 1, which describes the fundamental relationship in ICR mass analysis.

$$f_c = \frac{qB}{2\pi m} \qquad (1)$$

Typical values of f_c range from a few kHz to several MHz, and for example, for $B=3.0$ T, an ion of 100 Da per elementary charge (Da/z) has a frequency of 461 kHz. The radius of an ion's cyclotron orbit *r* depends on the radial kinetic energy of the ion E_k (J), as shown in Equation 2.

$$E_k = \frac{q^2 B^2 r^2}{2m} \qquad (2)$$

Ions in thermal equilibrium at room temperature assume an average kinetic energy of $3kT/2$, where k is Boltzmann constant and T is the absolute temperature. Two-thirds of this energy will be in the *xy*-plane, and for example, at 300 K and 3.0 T, a singly charged ion of 100 Da establishes a radius of about 77 µm, which is small with respect to cell dimensions.

Ion motion in the z-direction along magnetic field lines is restricted by applying a continuous potential V_t to trapping electrodes positioned at right angles to the magnetic field. Ions oscillate in the potential well established between the electrodes as long as the maximum z-axis kinetic energy E_o achieved at the bottom of the well is less then $V_t - V_o$, where V_o is the potential in the center of the well. For example, in the cubic cell with trapping potential $V_t = 1.0$ V, V_o is 0.33 V and E_o must be less than 0.67 eV to retain the ion. In the cubic cell, the trapping electric field is approximately quadrupolar, and the z-axis, or trapping motion, is effectively harmonic. Equation 3 shows the solution of the equation of motion for trapping frequency f_t.

$$f_t = \frac{1}{2\pi} \left(\frac{4q\alpha V_t}{mc^2} \right)^{1/2} \tag{3}$$

Here, α is a cell geometry factor equal to 1.3869 for the cubic cell[18], and c is the cell length. Typical values of f_t are tens of kHz, and for example, the 100 Da/z ion in a 2.5-cm cubic cell with $V_t = 1.0$ V has a trapping frequency of 14.7 kHz. From Equations 1 and 3, the combined motion of a single ion along the center line in an ICR-trapped ion cell assumes a trajectory that can be described as a helical motion of changing pitch.

One important consequence of the applied trapping electric fields is the generation of a second form of radial motion in the cell, termed magnetron motion.[20] Magnetron motion arises from the interaction of a charged particle with crossed electric and magnetic fields. Although the trapping electric field is applied parallel to the magnetic field, the finite electrode dimensions and adjacent grounded plates result in significant field curvature. This curvature introduces a radial electric field component of the trapping field that is orthogonal to the magnetic field. The resulting $E \times B$ motion of the ion may be described as a slow precession of the cyclotron orbit center along isopotential contours around the center of the radial electric field. The frequency of this precession f_m can be approximated in a cubic cell by Equation 4.

$$f_m = \frac{V_t \alpha}{\pi c^2 B} \tag{4}$$

For example, with $B = 3.0$ T and $V_t = 1.0$ V, a value of 235 Hz is obtained for the magnetron motion. Note that unlike f_c, f_m is very small and does not vary with mass, so it is of little analytical use. The motion is important, however, in that it can modulate the observed amplitude of cyclotron motion. Poor quantitative behavior results.[21] The forces responsible for the magnetron motion also induce a negative shift in the observed cyclotron frequency.[20] Additional complications resulting from the radial electric field include reduced ion storage time, decreased mass resolution and mass accuracy, and restricted mass range.[22-25]

In the FTICR experiment, radial electric fields arise from both ion space charge and nonideal trapping fields. Thus, to minimize the effects of such fields, the FTICR experiment should be conducted with a small ion population formed along the center line of the cell where the radial electric field is minimized and with as low a trapping potential as possible. The example just given of ions formed by electron ionization approximates such an experiment. However, in many applications, careful control of ion population and spatial distribution in the cell is not possible, and the adverse effects of radial electric fields must be considered. Much of the research in FTICR instrument development has been directed toward an improved understanding of these effects and toward the development of cells in which such effects are minimized to improve analytical performance. A brief discussion of these efforts is presented in Section 5.

2.2. Ion Excitation and Detection

The cyclotron motion of the initially acquired ion ensemble is of low amplitude and random phase and therefore does not induce a detectable ICR signal on the receive plates. To create a condition conducive to detection, an alternating electric field can be applied to the transmitter electrodes that are collinear with the magnetic field. Ions with a cyclotron frequency equal to the frequency of the applied field will fall into resonance and absorb energy from the field. The resonant energy absorption results in phase coherent ion cyclotron motion of increasing amplitude. The ions spiral outward from their initial position until the applied field is terminated or until the ions strike the cell electrodes and are lost. Equation 5 relates the increased radial kinetic energy of the ions to excitation voltage E_{rf} and event time t.

$$E_{tr} = \frac{q^2 E_{rf}^2 t^2}{8mc^2} \qquad \text{(5)}$$

Combining Equation 5 with Equation 2 yields the final radius of the ion cloud, shown in Equation 6.

$$r = \frac{E_{rf} t}{2Bc} \qquad \text{(6)}$$

For example, applying a 15-V_{pp} rf pulse for a 100-μsec interval with $B = 3.0$ T and $c = 2.5$ cm produces a cyclotron orbital radius of a 1.0 cm.

Once an isomass ion packet assumes its higher amplitude spatially coherent orbit, detection can be accomplished in several ways. In the omegatron experiment, the ions gained sufficient amplitude at resonance to strike a collector plate, thus generating an ion current.[1] In the drift ICR experiment, power absorption by the ion cloud during excitation was detected with a marginal oscillator.[6] In contrast, in the FTICR experiment, an alternating image current is induced in parallel receiver plates as the ions orbit the cell at their cyclotron frequency.[11] This detected current is amplified, digitized at a rate that satisfies the Nyquist criterion (which requires a sinusoidal signal to be sampled at least twice per period if the true frequency is to be represented by the digitized data), and stored for data processing. Fourier transformation of the acquired signal yields a frequency spectrum in which the frequency of a peak is indicative of ion mass, and the area of the peak is proportional to the ion population. One important feature of Equation 5 is that the orbital radius achieved by the coherent ion bundle is independent of mass and thus is the same for all ions. Because signal intensity varies inversely with the orbital radius, this feature simplifies any quantitative comparison of relative ion populations. The FTICR derives its primary advantages over previous ICR experiments because of this nondestructive detection procedure.

2.3. FTICR Line Shape Signal Processing

In the absence of collisions or other damping mechanisms, the time domain signal S for a single isomass ion packet is given by Equation 7.

$$S(t) = A \sin(2\pi f_c t + \phi) \tag{7}$$

Here, A is a constant proportional to the ion abundance, and ϕ is the phase angle. With Fourier transformation, Equation 7 yields the frequency domain sinc function shown in Equation 8.

$$S(f) = \frac{A \sin(2\pi f_c)}{2\pi f_c} = A \sin(2\pi f_c) \tag{8}$$

Equation 8 describes the zero pressure (i.e., collisionless) frequency domain FTICR mass spectral line shape. In practice, this absorption mode line shape is approximated if pressures below 10^{-8} Torr are maintained.

The effects of pressure on the FTICR experiment are of far greater consequence than in other types of mass spectrometry, because during the detection event, reactive and nonreactive collisions between neutrals and ions reduce the spatial coherence of a detected ion packet.[26] If collision processes are first order, then a decaying exponential adequately defines the damping of the signal and a high-pressure line shape of the form shown in Equation 9 replaces Equation 7 in defining the time domain signal.

$$S(t) = A \sin(2\pi f_c t + \phi) \exp(-t/\tau) \qquad (9)$$

Here τ is an overall relaxation time analogous to the spin-spin relaxation time, T_2, found in Fourier transform nuclear magnetic resonance (FTNMR). Fourier transformation of Equation 9 now yields the Lorentzian line shape, shown in Equation 10, that approximates FTICR peak shapes at pressures above 10^{-7} Torr.

$$S(f) = \frac{A\tau^2}{1 + (2\pi f)^2 \tau^2} \qquad (10)$$

One of the fundamental advantages of FTICR is its ability to detect ions over a broad mass range simultaneously. This follows from the fact

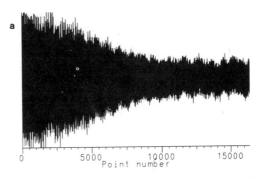

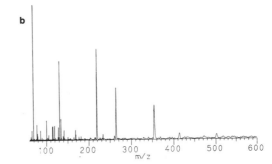

Figure 2. (a) The FTICR time domain signal and (b) FTICR mass spectrum obtained following electron ionization of perfluorotributylamine.

that all ions within a given mass range may be excited prior to detection. The resulting multiple isomass ion bundles that orbit within the cell induce a net image current that is the superposition of image currents of different amplitude, frequency, and phase, each defined by Equation 6, 8, or some combination of the two. This time domain response can be uniquely decoded by Fourier transformation to yield multiple frequency peaks, each corresponding to one of the isomass bundles. An example is found in Figure 2, where the FTICR time and frequency domain signals generated for the electron ionization mass spectrum of perfluorotributylamine are presented.

One complication in FTICR signal processing is that the phase relationship between individual oscillators being detected is a complex function of excitation conditions.[27-28] An inability to account for these phase differences in the transformed absorption mode mass spectrum easily results in distorted peak shapes. As a simple solution to the problem, a magnitude Fourier transform is usually generated as an alternative to the absorption mode transform because it yields phase-independent nonnegative frequency domain line shapes.[29] However, disadvantages of magnitude line shapes include a reduction in resolution at the half-height, accompanied by significant broadening at the baseline, which can interfere with detecting nearby low-intensity peaks. Nevertheless, magnitude spectra remain the standard mode for presentation, and for this reason, some form of apodization is usually applied to improve FTICR spectral appearance.[30]

3. THE MODERN FTICR EXPERIMENT

3.1. Magnet and Vacuum Chamber Considerations

The FTICR experiment differs from conventional mass spectrometers in that ion formation, manipulation, and detection can be accomplished in the same region of the instrument. Consequently, the FTICR instrument can be quite different in appearance from other mass analyzers. Essential components include a high field magnet, which provides a homogeneous field over a large volume, an ultrahigh vacuum (UHV) chamber, a trapped ion cell, ionization source, and appropriate controlling electronics. The two basic instrument configurations in use are dictated by the type of magnet used and specifically by design constraints imposed by the orientation of the magnetic field with respect to the magnet. For electromagnet or permanent magnet-based instruments, the z-axis assigned in Figure 1 to the ICR trapped ion cell aligns with the magnetic field in the narrow region between the magnet pole faces. Two important limitations of this design are a maximum magnetic field strength of about 2 T, which limits mass resolution, mass accuracy, and mass range and difficult external ion injection due to a restricted z-dimension length that accomodates only the cell. One advantage of electromagnets is that access from the sides of the cell provides more convenient interfacing of gas chromatography inlets, solids probes, pulsed valves, and laser optics; thus for applications requiring these peripheral devices but not ultrahigh performance, electromagnet-based systems are arguably superior. For example, the preponderance of Freiser's early studies of gas-phase metal ion chemistry were performed with a small electromagnet.[12]

Superconducting solenoid magnets are increasingly the magnets of choice for analytical FTICR because of the markedly improved performance achieved at higher fields.[31] Another advantage, convenient ion trajectory access to the cell along the z-axis, has prompted development of new spectrometer configurations and z-axis ion manipulation pulse sequences not possible with electromagnets. Examples of these include the external injection source[32–35] and tandem mass analyzers,[36, 37] multiple[38] and elongated trapped ion cells,[19] and suspended trapping pulse sequences.[39, 40] The primary disadvantages of superconducting magnets are initial capital, subsequent operating expense, and restricted mechanical access to the cell because of its position in the magnet solenoid. More elaborate and costly interfaces for chromatography, desorption, and laser experiments are also required because of the restricted access. Nevertheless, because high-field magnets offer superior performance and increased flexibility for ion transport, they are a standard component of modern FTICR spectrometers. Subsequent descriptions in Chapter 5 assume their use, although in principle, electromagnet systems may be employed for those experiments that do not require z-axis translation of ions.

In selecting a magnet, field strength is a primary consideration because figures of merit for mass resolution,[41] mass accuracy,[42] and upper mass limit[43] can be shown both theoretically and experimentally to improve with increasing field strength. Unfortunately, the cost of magnets increases exponentially with field strength, and presently, magnetic fields strength above about 7.0 T are prohibitively expensive.

A second parameter to consider in selecting the magnet is field homogeneity. Several theoretical discussions of the effects of field homogeneity on line shape are presented in the literature,[44, 45] with the most comprehensive by Laukien,[44] suggesting that radial magnetic gradients ultimately limit mass resolution to about 10^8 independent of pressure. Any additional increase in resolution would require shimming of the field. The z-axis gradients in the magnetic field were also found to be less important than radial field gradients. In practice, the magnetic field homogeneity should be as uniform as possible, and commercial cryoshimmed super-conducting magnets generally exhibit field variation of less than 1 part in 10^6 within the volume occupied by the trapped ion cell. However, exceptional performance can be achieved in much poorer fields; for example, a resolution of 77,000 at 5900 Da/z was demonstrated on a 7.0-T magnet, even though homogeneity was only 1 part in 10^4 over the cell volume.[46] We have obtained a mass resolution in excess of 10^6 at mass 77 on a 2.0-T magnet, even though homogeneity over the trapped ion cell volume was only 1 part in 10^3. Explanations for such good performance in relatively inhomogeneous fields include the likelihood that the z-amplitude achieved by ions in these experiments was a small fraction of the cell length and motional averaging of radial inhomogeneities occurs when detecting the cyclotron frequency.[44] Whatever the explanation, for all but the highest mass resolution requirements, magnetic field homogeneity is not the primary factor limiting mass resolution.

Another factor of importance in designing the FTICR spectrometer is the vacuum system. FTICR instruments differ from other mass spectrometers in that UHV conditions are essential to optimum performance. Low pressures are desirable in FTICR because the collision of excited spatially coherent ions with neutrals during data acquisition is primarily responsible for the continuous decay in signal intensity. Vacuum system components capable of low 10^{-9} Torr pressures are common in most analytical spectrometers. These pressures are most often achieved with diffusion or turbomolecular pumps, but vacuum chambers have been designed with more elaborate pumping schemes when pressures below 10^{-9} Torr are required for ultrahigh resolution[41] or surface chemistry experiments.[47] As a rule of thumb, the ion cloud exhibits detectable spatial coherence for about one second after excitation at 10^{-8} Torr, and if a corresponding signal acquisition time is employed, a mass resolution exceeding 10^5 for ions below 1000 Da can be achieved. For most analytical application, this resolution is more than adequate, and pressures in the low 10^{-8} Torr range are an accepted standard for high-performance measurements. Of course, with more extreme conditions, the transient signal can be observed for a much longer period, for example, Wanczek demonstrated a mass resolution in excess of 10^8 for H_2O^+ at 4.7 T by using a titanium sublimation pump in a liquid nitrogen trap to obtain a pressure of 8×10^{-11} Torr and measuring a 51 sec transient.[41]

3.2. The Trapped Ion Cell

The heart of the FTICR experiment is the trapped ion cell, which consist of three orthogonal sets of parallel electrodes. These electrodes may be arranged in a variety of geometries, including variations of the hyperbolic Penning trap,[48] and mechanically simpler geometries, including cubic,[49] orthohombic,[19] and cylindrical cells.[50, 51] Unfortunately, no ideal trap configuration is possible due to mutually incompatible objectives for the shape of trapping and excitation electric fields. It is desirable for trapping fields to exhibit quadrupolar geometry, since this yields cyclotron frequency shifts independent of ion position within the cell. However, excitation fields should be spatially homogeneous to yield ion excitation independent of ion position; no current electrode geometry achieves these objectives simultaneously. In practice, cubic and cylindrical cells are most commonly used because they are easy to construct, and performance is still exceptional. That high-resolution detection can be achieved in such poorly shaped traps might seem surprising given the great lengths to which physicists will go to create perfect hyperbolic electrodes for Penning traps.[52] However, even with simple cubic cell geometries, experimental conditions can be created so that the ions experience only the approximately quadrupolar trapping fields near the center of the cell.

Materials for cell construction should be selected with considerations given to minimal magnetic susceptibility, ease for machining, and chemical inertness. Stainless steel is the most popular choice despite unfavorable paramagnetic properties, and in fact, it is the material used in all cells with which performance standards have been achieved (see Section 4). However, in those experiments, ions were formed in the cell; in external injection experiments, where ions must traverse a trap plate orifice, magnetic susceptibility in the plate must be minimized to avoid significant perturbation of ion trajectories.[53] Other desirable aspects of cell design require the surface area to be minimized to achieve good conductance in the cell and to reduce charging that occurs on unclean trap plates; for these reasons, highly transmissive mesh or perforated plate is often used. However, for many typical applications, the properties of the FTICR cell components are often not critical and homebuilt cells are a cost-effective alternative to expensive commercial cells.

In selecting the size of the trapped ion cell, dimensions of the vacuum chamber and the spatial homogeneity of the magnetic field are limiting factors. Experimental factors to consider are mass resolution, which is optimized in smaller cells (due to greater magnetic field homogeneity over the small cell volume); dynamic range, which increases in elongated cells (due to reduced space charge effects at lower ion density); and maximum excitation energy, which increases with radial dimension. For example, the original 2.5-cm cubic cell marketed by Nicolet Analytical Instruments afforded excellent mass resolution but had a limited dynamic range. The move to longer cells, including elongated rectangular and cylindrical cells,

significantly increased cell capacity and also minimized the proportion of
the cell over which a significant radial component of the trapping electric
field was experienced. For access to higher ion excitation energies, the
radial dimension of the cell should also be increased, because as seen from
Equation 4, the maximum translational kinetic energy of an ion increases
with the square of the cyclotron orbital radius. Collision-activated ion dissociation measurements may then be performed by FTICR in both low-
and high-energy collision regimes. For example, Russell and Bricker used
a larger diameter cell to investigate the keV collisional processes for the
dissociation of benzene molecular ion.[54] In practice, cells with unit aspect
ratios and dimensions of about 5 cms are a practical compromise with
respect to dynamic range, mass resolution, and range of collision energies.
For external injection experiments, however, the use of longer cells
increases trapping efficiencies.[35]

3.3. Differential-Pumping Configurations

Much effort has been expended to develop FTICR instrumentation in
which the ionization region is separated from the analyzer-trapped ion cell.
Although this detracts from the simplicity of the Comisarow and Marshall
experiment, optimum FTICR performance is obtained at low pressures
that are incompatible with most ion sources and inlet system. To date, two
approaches have been taken to inject ions from a high-pressure region into
the FTICR analyzer cell. They are distinguished by proximity of the source
of the analyzer cell. If ions are formed near the analyzer cell in the
magnetic field, they can be constrained radially and easily injected along
field lines. One such design, shown in Figure 3, is the differentially pumped

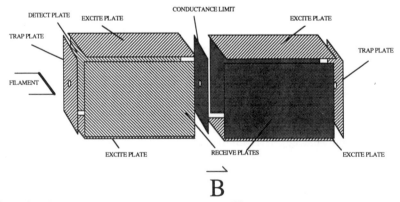

Figure 3. The FTICR dual-section trapped ion cell[55] used for ion transfer from a high-
pressure source cell to a low-pressure analyzer cell along the magnetic field axis, $\vec{B}$.

dual-trapped ion cell.[55, 56] In this example, a common trap plate shared by source and analyzer cells also serves as a conductance limit. Ions formed in the source trap are transported to the analyzer region by grounding the common trap plate and establishing an electric field to direct ions of the appropriate charge. In this way, an ion population proportional to the source region neutral population is selectively transferred to the analyzer region, yielding sensitivity improvements exceeding two orders of magnitude. This type of interface is particularly effective for direct introduction of effluent from a gas chromatograph (GC). Other tandem cell arrangements include multielectrode assemblies that permit ions to be transferred to any desired region of the vacuum chamber within the strong magnetic field,[38] and Wanczek's dual-cell assembly in which the source cell is positioned in the fringing field to take advantage of TOF mass selection capabilities.[57]

Multiple-trap assemblies, such as the dual cell, offer advantages of simplicity and cost effectiveness because they exploit the existing strong magnetic field as a means of confining and focusing source-generated ions. In addition, they permit a versatile assortment of ion injection and ion manipulation pulse sequences to be implemented. For several types of relatively low-gas-load experiments, including chemical ionization and GC, they are the method of choice for solving the source/analyzer pressure conflicts in FTICR; However, they are not a general solution to the source/analyzer pressure conflict. Given the physical constraints of superconducting magnet systems, it is not practical to implement more than two pumping stages with sufficient space to accommodate a fully functional trapped ion cell in each stage. Thus, source cell pressures only 10^2–10^3 times higher than the analyzer cell pressure can be achieved, and the installation of such high-pressure sources as electrospray or atmospheric pressure ionization is not possible.

It has been demonstrated in our laboratory that an alternative to the tandem trap configuration can allow additional stages of differential pumping to be employed between the source and analyzer region while preserving the ion-focusing benefits of ionization within the strong magnetic field.[58] In this design, a multiple concentric tube vacuum chamber is employed to provide pumping to each of several adjacent conductance limits located near the source cell of a dual-trapped ion cell. No trapped ion cells are employed within these additional stages of differential pumping, and thus, they can be staggered within a relatively small distance along the vacuum chamber. Electrospray ionization (ESI) was achieved at atmospheric pressure, well within the magnet bore while maintaining analyzer cell presures in the 10^{-9} Torr range. Low-resolution spectra of bovine albumin dimer at 132,532 Da were acquired, which is the largest molecule observed to date by FTICR. Such a pumping arrangement can be extended to any high-pressure ionization process that benefits from the radial-focusing properties of the strong magnetic field.

Although the benefit of strong ion focusing may be derived from source placement within the strong magnetic field, not all sources are compatible with such an arrangement. For example, sources employing charged particle beams must direct the beam along the z-axis because transverse orientations generate poorly controlled particle trajectories in the magnet. This can place severe constraints on the geometry of ion extraction and complicate the process of generating a spatially well-defined ion population. Additionally, high-voltage sources, such as Cs^+ secondary ion mass spectrometry (SIMS), are not easily implemented in magnetic fields.[59] Thus, to accomodate the full complement of available mass spectrometry sources and inlets better, several types of external source instruments have been constructed. The common aspect of all such designs is that ions are formed well away from the strong magnetic field to avoid undesirable magnetic effects. Such configurations have the further advantage that any desired source pressure can be accommodated by employing multiple stages of differential pumping without resorting to complex vacuum chamber designs. The fundamental problem with the external source then becomes the efficient injection of externally generated ions through the fringing magnetic field into the magnet bore, with subsequent retention in the analyzer cell. Fairly elaborate combinations of focusing and velocity adjustment optics are required to overcome the magnetic mirror effect, a physical phenomenon in which charged particles with radial kinetic energy are decelerated as they move through a large magnetic field gradient. Excellent discussions of this effect are provided elsewhere.[32-35] For this discussion, it is sufficient to state that the principal requirement for efficient injection is minimization of the ratio of radial to axial velocity of injected ions. This has been accomplished with both electrostatic and rf quadrupole lenses, both described in the following paragraphs.

The first succesful demonstration of external source FTICR was by McIver et al.[32] Two sets of rf quadrupole lenses were used, with the first set guiding ions from the source to the fringing magnetic field and the second set focusing ions onto field lines with a minimal radial velocity. In this way, the magnetic mirror effect was avoided, and a large fraction of the ions reached the trapped ion cell. This type of interface has been used succesfully with Cs^+ SIMS to introduce ions of large biomolecules and clusters into the cell. Lebrilla et al. have detected a CsI cluster ion with a resolution of 53,000 at 9745 Da/z,[60] while Hunt et al. have detected ions beyond 12,000 Da/z for the protein cytochrome c but with much poorer resolution.[61] Henry et al. have demonstrated external source ESI with rf quadrupole injection and achieved resolution in excess of 60,000 for the +16 charge state of cytochrome c.[62, 63]

A second external source configuration uses electrostatic lenses to inject ions with high efficiency. In work by both Alford *et al.*[35], Smalley,[64] and Kofel *et al.*[34, 36] two sets of simple electrostatic lenses are used to inject ions into the magnetic field. Again, the approach is to focus ions onto field lines with a minimal radial component of motion. A distinct advantage of the electrostatic focusing system is improved control over ion kinetic energies. This allows more elaborate types of mass selection to be performed and simplifies the tailoring of ion kinetic energy to match the trapping potential.[37]

A necessary element of all experiments in which ions are introduced to the analyzer from outside the cell is a technique for trapping the injected ions. However, this aspect of the experiment is more poorly understood than the injection process. One difficulty is a confusion over the actual mechanisms by which ions are trapped, and at least two broad categories

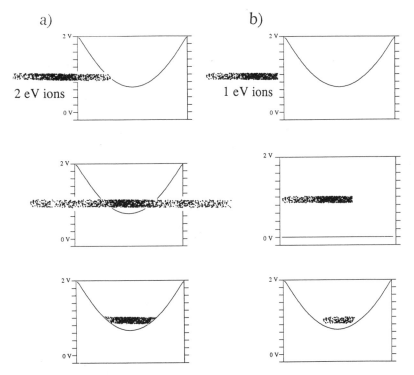

Figure 4. Depiction of (a) accumulated trapping and (b) gated trapping of externally generated ions arriving at the FTICR-trapped ion cell.

of trapping techniques, termed accumulated trapped and gated trapping, can be delineated. Figure 4 distinguishes the two approaches. In accumulated trapping, ions are injected through the cell with axial kinetic energies sufficient to penetrate the trapping potential. Despite the apparent incompatibility of kinetic energies and trapping potentials, ions are found to accumulate in the trapped ion cell, and FTICR spectra are readily generated. To trap high-energy ions, a substantial energy loss or redirection must occur during the brief time ions are in the cell. Suggested mechanisms include ion-neutral and ion-ion collisional stabilization,[32] as well as redistribution of axial-to-radial kinetic energy by electric and magnetic field inhomogeneities[57] associated with cell electrodes. Alternatively, detected ions can arise from charge exchange or other ion/molecule reactions in the cell[65] or by electrostatically shielding low-energy ions in a dense ion population.[66] Whatever the mechanism, accumulated trapping is often a simple and effective technique for acquiring spectra of externally generated ions. For example, it is the routine method by which Lebrilla *et al.*[60] and Hunt *et al.*[61] trap ions in their quadrupole systems, and Brucker Spectrospin employs a patented version of accumulated trapping with their commercial external source FTICR interface.[67] As discussed in Section 3.5.3., it is also one mechanism by which laser-desorbed ions are trapped. However, inefficient and nonselective trapping are negative features of accumulated trapping that are not likely to be improved until mechanisms for accumulation are better understood.

The alternative to accumulated trapping is gated trapping, a potentially more efficient and better characterized approach. Here, trap potentials are maintained below the axial kinetic energy of a desired ion until it enters the cell. At this time, the trapping potential is increased to trap the ion. Gated trapping can be applied to either a continuous ion beam, in which case an equilibrium distribution of the ion population is acquired, or a pulsed ion beam,[36, 37] in which case time-of-flight dispersion permits mass-selected ions to be captured. The usefulness of this latter experiment may be extended to allow arbitrary mass range selection using dynamic trapping techniques.[37]

3.4. Sample Introduction

Because the FTICR experiment is performed at pressures that are typically a factor of 10^3 lower than those for conventional mass analyzers, sample introduction methods also differ from those used in conventional mass spectrometers. In general, every effort is made in the FTICR experiment to minimize the sample or matrix pressure, especially in single-section cell systems. This effort is necessary not only because of low-pressure requirements for high-resolution detection, but also to promote cleanliness of the cell and vacuum chambers. Contaminants on cell trap plates may result in charging, which reduces the efficiency of ion transfers and quenches and increases undesirable space charge effects. Chemical noise is also a problem when rapid sample turnaround is required.

3.4.1. Volatile Liquids and Gases

Volatile species are introduced through either pulsed valves[68-70] or high-precision leak valves. Typical analyte pressures in the mid to low 10^{-8} Torr range are adequate for generating high signal-to-noise EI spectra. What is not generally recognized is that even compounds thought to exhibit thermal instability or insufficient vapor pressures to be introduced through a volatile inlet or on a heated probe with conventional mass spectrometers can generate sufficient room temperature vapor pressure to be sampled by FTICR.

3.4.2. Sampling Solids

The practice of coupling a solids inlet probe with thermal desorption and a subsequent ionization event is not in wide use with FTICR for several reasons. Cells are inconveniently positioned in the magnet bore, thus requiring large and cumbersome solids probes. Care must be exercised to avoid substantial pump-down times to eliminate water and other background species introduced with the probe. Also, because the neutral population varies during desorption, it is often difficult to select ionization parameters that avoid space charge effects. Poor quality spectra result. Using a differentially pumped dual cell alleviates some of the difficulty because direct contamination of the analyzer cell by the probe is avoided and greater control over ion transfer to the analyzer cell is afforded.

The most common technique for detecting nonvolatile solids (often used as a replacement for thermal desorption even when unnecessary) is laser desorption/ionization (LDI). The LDI/FTICR is one of the better techniques for generating ions from otherwise intractable samples. Because LDI is executed with a pulsed laser, a gas plume similar to that generated by a pulsed valve is created and quickly pumped away prior to detection. Thus LDI is rare among the many new ionization and sample introduction methods developed for mass spectrometry in that it is fully compatible with the single-section trapped ion cell.

3.4.3. Chromatographic Sources

As might be expected, high-pressure chromatographic sources are incompatible with single-cell FTICR, and for example, early GC interfaces featuring direct splits of the effluent exhibited reduced sensitivity.[71, 72] Fortunately, the development of pulsed valves and the differentially pumped dual cell now make direct introduction of capillary GC effluent routine.[68, 73-76] With reports of low-picogram-detection limits for broadband EI spectra,[73, 74] FTICR is competitive with other mass analyzers for GC/MS; however, as the associated gas load increases, for example, with supercritical fluid chromatography (SFC) and high performance liquid chromatography (HPLC), the two stages of pumping possible with the dual-cell interface quickly become inadequate. Examples of dual-cell SFC/FTICR have been demonstrated,[77, 78] but an additional pumping stage is still required to achieve optimum performance: Two pumping stages are entirely inadequate for liquid introduction, and for example, in the only successfully demonstration of HPLC, thermospray ionization was interfaced through four pumping stages to an external source FTICR.[79] Spectrospin developed a continuous-flow FAB interface for their external source FTICR, and this should improve the likelihood for routine HPLC/FTICR.[80]

3.5. FTICR Ionization

As previously discussed, the low-pressure and strong magnetic field conditions under which sample introduction and ionization occur in FTICR are compatible with few sources without modifying the original single-section cell design. Nevertheless, it is essential for FTICR to accomodate the many new ionization techniques that have so enhanced the capabilities of mass spectrometry in detecting large polymers and biomolecules. Considerable effort has been made to create effective interfaces for these sources, specifically with the differentially pumped external source instruments, and now, most techniques have at least been demonstrated. Three of the more common, EI, chemical ionization (CI), and LDI, are discussed here; Others, including Cs^+ (SIMS),[33] fast atom bombardment (FAB),[80] plasma desorption (PD),[81] field desorption (FD),[82] electrospray ionization (ESI),[58, 62, 63] multiphoton ionization (MPI),[73, 83] and thermospray[79] have been demonstrated but are not yet in widespread use. Of these, Cs^+ deserves special mention as the source for high-mass analysis, which is common to external source quadrupole FTICR spectrometers constructed by Hunt, Lebrilla, and McIver. Furthermore, ESI is gaining favor as a means for analyzing very high-mass ions with FTICR.

3.5.1. Electron Ionization

The EI is compatible with the low-pressure requirements of a single-section trapped ion cell with a single stage of pumping, and it is therefore readily implemented in FTICR. Electron gun assemblies, which consist of the filament and appropriate gating and accelerating grids, are positioned to emit electrons along magnetic field lines that converge at the center line of the cell. The filament assembly can be mounted on the cell, but if mounted in the fringing field, the beam will be focused in proportion to the ratio of magnetic fields it experiences. In either case, electrons traverse centered holes or grid openings in the trap plates during the beam event and impinge on a positively biased collector on the opposite side of the cell. The collector is employed to prevent low-energy scattered or secondary electrons from reentering the cell. Filament materials are typically thin rhenium or tungsten ribbon, although Russell has suggested the use of a lanthanum hexaboride crystal to generate a better defined source of electrons.[84]

In the FTICR EI experiment, a narrow collimated beam of ions is formed within the flight path of the electrons. These ions are constrained within a volume similar to that of the electron beam by the magnetic field and the trapping potentials. The number of ions produced depends on the electron current, sample pressure, and beam event duration as in conventional mass spectrometers, but experimental requirements are quite different. Because it is necessary to avoid space charge, the total ion population is usually kept between 10^4 and 10^6. Thus nanoampere to low microampere electron currents are appropriate for beam times on the order of a few milliseconds duration if there are neutral analyte pressures in the mid 10^{-8} Torr range.

The EI should be an essential part of all FTICR instruments regardless of desired application, because the ionization technique is easily operated at very low pressures and quickly and inexpensively implemented. The electron beam is also an invaluable diagnostic tool for evaluating the condition of a cell with respect to alignment, cleanliness, and electrical connections. The disadvantages of EI include poor z-axis spatial resolution and mechanical interference with other sources, probes, and ion beams that must be introduced along the z-axis center line of the cell.

3.5.2. Chemical Ionization

In contrast with electron ionization, conventional CI is not directly compatible with FTICR because of the high-reagent-gas-load requirements; CI is performed in conventional sources at a pressure of 1 Torr, eight orders of magnitude above acceptable pressures for FTICR detection. Fortunately, by taking advantage of increased reagent ion storage time in the cell, it is possible to compensate for reduced reagent ion concentration by increasing experiment time to perform the desired CI experiment. This concept was first implemented by Ghaderi et al.[85] who used reagent gas pressures of 10^{-6} Torr, analyte pressures of 10^{-7} Torr, and reaction times of several hundred milliseconds to generate sufficient CI product ions for FTICR detection. Methane, isobutane, and ammonia reagent gases were demonstrated to produce spectra similar to those obtained by conventional methods. Also demonstrated was a procedure by which chemical ionization, specifically protonation of the analyte, was accomplished by reaction of EI fragment ions with analyte neutrals during a reaction delay. This process is termed self-CI.

Several problems arise from maintaining a constant CI reagent gas pressure in the analyzer cell; for example, the experiments just described at 10^{-6} Torr are several orders of magnitude above optimum analyzer cell pressures. Other problems include contamination from EI fragments that may be acquired along with CI data, as, for example, during alternate EI/CI GC/MS detection. Another problem is that during lengthy reaction delays, the neutral background reagent gas may undergo reaction with product ions and contaminate or quench the CI mass spectrum. To overcome these problems, either pulsed introduction of CI reagent gas or differential pumping is required. The former has been demonstrated for both positive methane CI and negative ammonia CI of GC effluent with FTICR detection,[86, 87] and Cody has demonstrated the advantage of using the differentially pumped dual cell to perform ammonia CI while avoiding contamination from unwanted ion/molecules reactions.[88]

It should be mentioned that substituting increased experiment time for reagent gas pressure is not always satisfactory for two reasons: first, the rapid-scanning capabilities of FTICR are sacrificed; second, any FTICR experiment that includes a lengthy delay between ion formation and event detection, and this necessarily includes any form of pulsed gas introduction, increases the likelihood of competing ion/molecule reactions. To avoid this problem, the vacuum system where the analyzer cell resides must be kept scrupuloulsy clean. The alternative is to interface a conventional high-pressure CI source to the FTICR so that lengthy CI reaction times are unnecessary.

3.5.3. Laser Desorption/Ionization

By a wide margin, the most prevalent ionization technique applied to FTICR of non-volatile materials is LDI. Its success is due to both its near universal applicability and its compatibility with the pulsed and low-pressure aspects of FTICR detection. LDI/FTICR was first demonstrated for organic molecules by McCreary *et al.*[89] and successfully extended to organic molecules in the 1000–5000 Da range as described by Wilkins *et al.*[90] and Nuwaysir.[91] Compounds analyzed by LDI/FTICR include the low-molecular-weight biomolecules peptides,[92] carbohydrates,[93, 94] and oligonucleosides,[95] and also porphyrins,[96] polymers,[97–99] polymer additives,[100, 101] carbon clusters,[35, 102] and clinical drugs.[103] In addition to routine analytical evaluation of these materials, LDI also provides a source of metal ions for ion/molecule reactions,[12] and is used to interrogate surfaces as part of a laser microprobe apparatus.[104, 105]

The most common procedure for LDI/FTICR of organic and biological molecules borrows from LDI/TOF work as described by Van Peyl *et al.*[106] Spectra are dominated by a cation-attached species representative of the molecular ion. The observed cation often originates from adventitious sodium or potassium salts, but more commonly, a salt such as KBr is mixed directly with the sample and deposited on a stainless steel probe tip.

Pulsed CO_2 radiation at 10.6 μm or Nd:YAG radiation at 532 nm or 1064 nm are used in most LDI/FTICR experiments. In a thermal rather than resonance absorption mechanism for ionization, the primary laser requirement is application of mJ-level energies to small (less than 1-mm^2) spot sizes within a period of tens of nanoseconds. These conditions provide the laser power densities of 1×10^7 to 1×10^9 W/cm^2 necessary for ionization. The traditional interface for LDI uses conventional optics, including windows, mirrors, and lenses of appropriate material. Laser light is introduced through a UHV window with a long focal length lens to a second lens in the vacuum chamber that focuses light to a small-sized spot at the sample probe. Precise positioning of the probe with respect to the focused light is essential to obtain necessary power densities; the probe is usually positioned along the center line within a few centimeters of the cell to maximize signal intensity. Because it is possible to maintain trapping potentials throughout the desorption event yet still observe large FTICR signals, some mechanism analogous to accumulated trapping must be occurring to retain ions. Van Peyl *et al.*[106] have shown that gas-phase cation attachment to intact desorbed neutrals is the likely mechanism. This is consistent with the idea of cation attachment in the cell between energetic alkalai ions that penetrate the trapping fields and the desorbed neutrals of the sample compound. Moreover it has been demonstrated in our laboratory that a laser-generated cation neutral salt adduct may be responsible for transferring the cation to the desorbed neutral analyte molecule.[107]

As an alternative to conventional optical assemblies for LDI/FTICR, we now employ a probe-mounted fiber optic interface to a Nd:YAG laser to generate spectra.[108] Light is focused into the fiber but exits as unfocused light within 1 mm of the sample surface. Spot sizes of less than 1 mm^2 are still obtained, and power densities in excess of 2×10^8 W/cm^2 are achieved. Advantages of the fiber optic assembly are optical design simplicity and ease of alignment. In addition, because the assembly is probe mounted, it is possible to translate the LDI interface throughout the vacuum chamber to evaluate ionization and trapping mechanisms in LDI/FTICR.

3.6. Ion Excitation

Excitation of coherent ion motion is the most important aspect of the FTICR experiment, since it is necessary not only for ion detection, but also for mass selection, ion ejection, and collisional activation. Generating selective, broadband, high-power excitation waveforms is a subject of great interest in FTICR. Desired properties in the excitation pulse are simultaneous, rapid excitation of desired ions with sufficient power to achieve desired kinetic energies, homogeneous excitation over bandwidths of several megahertz, and tailored excitation to achieve high-resolution mass discrimination. Many of these features are mutually incompatible, and thus no ideal excitation process exists. The oldest forms, pulsed and swept radiofrequency excitation[109] suffer from many limitations. Two novel approaches, impulse excitation[110] and stored waveform inverse Fourier transform excitation (SWIFT),[111] have been implemented; the latter approach promises significant improvements over older techniques.

3.6.1. Pulsed Radiofrequency Excitation

The simplest excitation method employs a fixed rf sinusoidal electric field applied with the transmitter plates at the cyclotron frequency of interest. The excitation bandwidth varies inversely with pulse duration and is a few hundred hertz for a typical duration of a few milliseconds. The technique is the optimum method for selective irradiation of a single ion or selective narrow bandwidth; however, broadband excitation is not practical. For example, Marshall *et al.* have calculated that a 30 nsec, 7.5-kV rf pulse would be necessary to encompass a 2-MHz bandwidth.[112]

3.6.2. Frequency Sweep

For the past decade, swept, or chirp, radiofrequency excitation has been used to generate broadband FTICR spectra. With this technique, the frequency of the applied field is varied linearly during the application interval. The advantage of swept excitation is that the electric fields have a much lower amplitude than required for wide band pulsed excitation. Typical fields of less than 100 V_{pp} are applied at rates of 10^2–10^3 Hz/μsec over megahertz bandwidths; thus, swept excitation is relatively fast, with broadband excitation accomplished in a few milliseconds. Unfortunately, several undesirable features of swept excitation reduce FTICR spectral performance. One problem is that ions are not simultaneously excited, which complicates frequency domain phasing. In addition, the use of slow ejection sweeps during double-resonance experiments allows chemical noise in the form of ion molecule reactions to complicate the spectrum. A second limitation of swept excitation is its nonuniform power spectrum, which, especially at the low- and high-frequency limits of the sweep, creates nonquantitative relative peak areas in the spectrum. The most significant limitation, however, is poor mass selectivity. When employed in double-resonance experiments for ejecting ions within a selected mass range, nominal mass separation is possible below 100 Da but deteriorates at high mass due to compression of the frequency scale. Moreover, ions of interest near termination points of the frequency sweep undergo substantial off-resonance excitation. Complications include unwanted kinetic energy effects in ion chemistry studies and nonquantitative signal intensities in the subsequent detection process. The poor front-end mass resolution of swept excitation conflicts with high-resolution FTICR detection and consequently reduces the overall analytical performance of FTICR for double-resonance experiments. Several modifications of the simple linear sweep have been examined in an effort to improve mass selectivity; these include notch ejection[113] and the combination of hard and soft sweeps,[114] which can be effective but are time consuming to implement.

3.6.3. Impulse Excitation

McIver *et al.* have developed impulse excitation as an alternative to linear rf sweeps for broadband, non selective excitation.[110] Impulse excitation is conceptually similar to pulsed broadband rf excitation in that very short pulses are employed to achieve a large excitation bandwidth. However, instead of an rf pulse, a high-voltage rectangular DC pulse is applied to transmitter electrodes, effectively overwhelming the magnetic field and accelerating all ions toward excitation plates of opposite polarity. Equation 11 approximates the final radius of gyration for impulse excitation.[115]

$$r = \frac{2E_{dc}t}{Bc} \tag{11}$$

For example, typical values are $E_{dc} = 120$ V, $t = 50$ nsec, $B = 1$ T, and $c = 0.03$ m, leading to an orbit of 0.4 cm. Importantly, this equation indicates that all ions are accelerated to similar orbits, independent of mass and charge. An essential requirement of the technique is that impulse duration be very short compared to the time of a cyclotron orbit. This places restrictions on quantitative detection of low-mass ions. Advantages of impulse excitation include simplicity in theory and execution, simultaneous and essentially instantaneous excitation of all ions, which simplifies phase correction, and homogeneous excitation for higher mass ions. However, there are two potentially serious limitations of the technique: (1) Because excitation is nonselective, it cannot be used for double-resonance ion excitation or ejection experiments. However, should it prove superior for quantitative high-mass detection, it can be used in conjunction with other more selective excitation techniques. (2) The high-amplitude nature of the excitation results in large displacements of the center of cyclotron gyration.[104, 106] In other words, excited ions exhibit large magnetron radii, which limits the maximum excitation radius and incurs additional problems, as discussed in Section 2.1.

3.6.4. SWIFT

None of the approaches to ion excitation presented so far adequately address the need for high-resolution mass selection in double-resonance experiments. Given the ultrahigh mass resolution capabilities of FTICR for detection, it would be ideal for achieving corresponding levels of resolution in all aspects of ion manipulation. In 1985, Chen and Marshall first reported on an ion manipulation technique with just such promise,[111] the SWIFT technique,[117] which employs waveforms constructed by the experimenter to facilitate selective quantitative excitation of any ion. Highly selective excitation is determined by the duration of the excitation event.

Figure 5 presents various stages of the SWIFT excitation process used for selective broadband excitation of perfluorotributylamine. Initially, an idealized frequency domain representation of the desired excitation waveform is constructed from which a time domain excitation waveform is obtained, as in Figure 5a. The frequency domain representation of this waveform, shown in Figure 5b, facilitates SWIFT excitation over a mass range of 50–600 Da, with two isomass suppression windows. At the time of the experiment, the time domain waveform is converted into an analog signal and applied to the transmitter plates during the excitation event; the resulting electron ionization FTICR spectrum is shown in Figure 5c. Much higher mass resolution than this can be achieved; for example, Marshall *et*

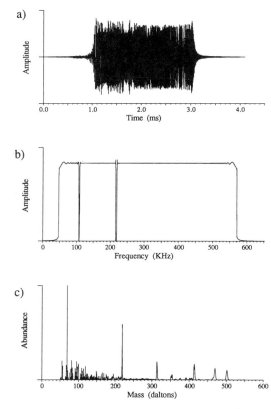

Figure 5. (a) Time domain excitation waveform for selective SWIFT excitation in the mass range from 50–600 Da with suppression of ions at 131 Da and 264 Da; (b) frequency domain representation of the time domain waveform; (c) electron ionization FTICR spectrum of perfluorotributylamine generated by application of the time domain waveform to the trapped ion cell excite plates.

al. have performed a SWIFT heterodyne-mode experiment that illustrates the exceptional mass selection capabilities of SWIFT. A mass selectivity of 0.01 Da is achieved to isolate the N_2^+ from neighboring CO^+ and $C_2H_4^+$ by applying a tailored 4-msec duration waveform.[117]

The principle advantages of SWIFT are derived from its ability to generate complex broadband excitation waveforms with essentially flat power spectra. In principle, a series of consecutive hard- and soft-mass selective frequency sweeps could accomplish the same task with similar mass selection capability. In practice, however, precise control of frequency, amplitude, and phase in this collection of sweeps is not possible, and resulting spectra exhibit the poor performance features of swept excitation. In contrast, SWIFT permits precision control of frequency, amplitude, and phase selection. Real performance benefits include improved quantitation of peak areas, increased dynamic range in complex mixture analysis, reduced event time compared to the equivalent frequency sweeps, compression of FTICR event sequence by combining excitation events, and improved mass selection for double-resonance experiments. Disadvantages include significant increases in cost and complexity, but the primary disadvantage is the trade-off between mass selection and event time. For example, to achieve a resolution approaching 1 Hz, it is necessary to apply the waveform for nearly 1 sec. In practice, resolution cannot increase indefinitely; it is instead bounded by the same constraints of pressure on ion coherence that limit mass resolution during the detection event. Of course, routine SWIFT experiments can be performed with much shorter excitation event times if mass selection requirements are not extreme.

3.7. Ion Dissociation

Ion dissociation techniques for modern analytical mass spectrometry are well established, and as a result, considerable effort has been made to achieve similar success with FTICR. As with other tandem mass spectrometers, FTICR ion dissociation must include precursor ion selection, a dissociation process, and product ion detection as general features of the experiment. However, FTICR differs fundamentally from other tandem instruments in that these events all occur in a single analyzer-trapped ion cell rather than in spatially distinct mass analyzers and dissociation cells. This temporal rather than spatial separation of FTICR ion dissociation events permits multiple stages of ion dissociation to be performed without increasing instrument complexity or cost. For reasons to be discussed, quantitative, high-yield ion dissociation is not routine in the FTICR-trapped ion cell. To date, the four dissociation mechanisms that have been demonstrated for FTICR are collision-activated dissociation (CAD) (also referred to as collision-induced-dissociation or CID),[118] surface-induced dissociation (SID),[119] electron-induced dissociations (EID) (also referred to as electron impact excitation of ions from organics, or EIEIO),[120] and photodissociation.[121]

3.7.1. Collision-Activated Dissociation

The CAD is the oldest and most common form of ion dissociation applied to FTICR. In the trapped ion cell, a sequence of events selects a precursor ion of interest, kinetically excites this ion to undergo dissociative collisions with neutrals, then detects the product ions produced. Figure 6 shows a CAD example for the dissociation of the phenylether molecular ion. Electron ionization yields the spectrum in Figure 6a from which the molecular ion at Da/z 170 is selected by applying ejection sweeps to remove unwanted low-mass ions from the cell. The isolated ion, shown in Figure 6b, is then kinetically excited by a resonance rf pulse at its cyclotron frequency to permit dissociative collisions with neutrals. As calculated from Equation 2, typical values for these laboratory collision energies are in the range of 10–200 eV, with conversion efficiencies of 5–10%. To increase the likelihood of collisions, an inert target gas, such as argon, is present in the cell at pressures of 10^{-6}–10^{-5} Torr at the time of excitation. The dissociation products detected by standard FTICR excitation and detection events are shown in Figure 6c.

Target gas pressures necessary to increase collision frequency are incompatible with high-performance detection. To counter this problem in a single-section cell, pulsed introduction of the target gas is followed by a several hundred millisecond delay for base pressures to return prior to detection.[122] This was how data in Figure 6 were collected. Alternatively, the differentially pumped dual cell permits high-pressure source cell dis-

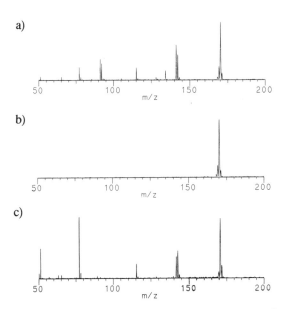

Figure 6. The FTICR spectra acquired at various stages in a CAD experiment. (a) Broadband electron ionization spectrum of phenyl ether; (b) selective ejection of ions to isolate the precursor ion at 170 Da; (c) product ion spectrum acquired following resonance excitation of the precursor ion in the presence of 5×10^{-7} Torr Ar.

sociations in the presence of steady-state argon pressures, with subsequent transfer of product ions to the analyzer-trapped ion cell for detection.[123]

Several aspects of the CAD experiment are of note. First, because the entire experiment takes place in the trapped ion cell, it is possible to perform consecutive dissociation experiments.[122] Also unique to FTICR is the ability to accelerate the precursor ion to collision energies ranging from low eV energies typical of quadrupole-based CAD work to KeV processes observed in sector spectrometers. This energy range is possible because as shown in Equation 2, increases in magnetic field and cell radius extend the range of ion energies that can be achieved within the confines of the trapped ion cell. For example, Russell and Bricker[54] and Bricker et al.[124] have investigated KeV dissociation of benzene but also discussed limitations that arise because of target gas ionization.[125]

Despite the advantages just described, CAD is executed in the FTICR-trapped ion cell with neither the generality nor the efficiency of CAD experiments performed on tandem sector or quadrupole instruments. The necessity of overcoming the high-collision gas pressure is one reason, since optimization of experiment conditions for either the pulsed valve or dual-cell transfer experiment is not trivial. Another possible reason is that the FTICR detection scheme is inherently ill-suited to monitoring collisional dissociation products when the cyclotron radius increases for both collisional excitation and detection purposes. This is because the radial excitation of the precursor ion generates product ions in regions of the cell that are radially displaced from the z-axis. Subsequent detection of these off-axis product ions, especially for higher energy dissociations, become increasingly inefficient and nonquantitative.

3.7.2. Electron Dissociation

Cody and Freiser have demonstrated that collisions between electrons and ions generate dissociation products in the FTICR-trapped ion cell.[120, 126] Low-energy electrons, typically a few eV below the ionization potential of neutral species in the trapped ion cell, are directed through the cell along the center line. Product ions from previously trapped and mass-selected precursor ions are then detected. The EID differs from the CAD experiment in that electrons rather than ions provide the kinetic energy in the collision complex. This is an advantage for FTICR, since electrons are accelerated along the z-axis in the center of the cell, preventing departure of product ions from the center line. A disadvantage of EID is the competing interference arising from electron ionization of analyte and background neutrals. It is probably for this reason that the technique has not found widespread application.

3.7.3. Surface-Induced Dissociation

Mabud *et al.* have demonstrated that ion collision with surfaces generate dissociation products in high yields. This SID process was demonstrated with a variety of tandem spectrometers including BEQ[127] and TOF/TOF,[128] among others. The many compelling reasons for implementing SID with FTICR include unit collision efficiency, retaining product ions closer to the *z*-axis, low-pressure requirements, and rapid analysis time, all of which recommend SID over CAD.

The first demonstration of SID in an FTICR dual cell was by Ijames and Wilkins;[119] in that experiment, a precursor ion was first selected from the source cell and transferred to the analyzer cell. The precursor ion was then provided with sufficient collision energy by exciting both axial and radial motion. An excitation pulse at the resonance frequency of the ion of interest was applied, and then using the conductance limit trap plate as the collision surface, a 20-V potential was applied to the exterior analyzer trap plate with the conductance limit was set to ground potential. Given this combination of electric fields, ions struck the plate in grazing collisions and were dissociated. Although for a perfectly smooth surface only the ion energy normal to the collision surface should be deposited in such collisions, the actual surface is apparently rough enough on the molecular scale to provide collision sites that allow deposition of the radial energy component. Thus, collision energies to several hundred electron volts could be deposited in this fashion. Estimates of internal energy deposition were as high as 8 eV for a total ion kinetic energy of 140 eV. Dissociation spectra were generated for a variety of precursor ions formed by electron ionization or laser desorption. Product spectra were similar to those obtained by low-energy dissociation on tandem spectrometers constructed for SID. Apparent dissociation efficiencies ranged from 1–30%, which are comparable to those achieved by Mabud *et al.*[127]

Williams *et al.* recently demonstrated SID with Hunt's external source quadrupole FTICR.[129] In this work, mass-selected peptide ions struck the screened front-trap plate of the trapped ion cell and dissociated. Product spectra for peptide up to Da/z 3055 resembled 193-nm photodissociation spectra, which suggests an internal energy deposition of about 6 eV.

3.7.4. Photodissociation

Ultimately, the best approach for ion dissociation in the FTICR-trapped ion cell may be photodissociation. In addition to the advantages enjoyed by all mass analyzers, including high-dissociation efficiency and quantum-state selection, photodissociation in the FTICR experiment

(1) permits product ions to be formed along the center line of the cell, (2) low pressures are maintained, (3) experiment times can be rapid, and (4) multiple dissociation events are possible. Another often cited justification for the development of photodissociation techniques is that CAD is ineffective for large ions because of decreasing internal energy transfer resulting from collisions with target gases.

Dunbar pioneered the application of photodissociation in ICR-trapped ion cells[130] and extended the technique to FTICR.[131] Since then, photodissociation in the FTICR-trapped ion cell has been successfully implemented with a variety of lasers and optical arrangements.[132–137] Examples include 193-nm[133] and 308-nm[137] light from excimer lasers, two-photon visible light,[134] and infrared radiation from pulsed and continuous wave CO_2 lasers at 10.6 μm.[135, 136] More elaborate experiments include consecutive photodissociation experiments[133] and two laser experiments in which photodissociation is performed on laser desorbed ions.[136, 137]

The benefits of photodissociation are unfortunately often outweighed by the added expense of the laser and associated optics, so the method has not enjoyed the widespread use that might be expected. Although high-power pulsed lasers are a standard feature with many FTICR spectrometers, their primary use has been in LDI. Adding a second laser can be prohibitively expensive and difficult to justify if non-selective dissociation is required and can be accomplished at no additional cost by CAD, EID, or SID.

3.8. FTICR Pulse Sequences

The FTICR experiment is governed by a series of discrete events, executed under computer control, that permit any desired mass spectrometry experiment to be conducted. A collection of events is termed an experimental pulse sequence and is analogous to the FTNMR pulse sequence. Although pulse sequences may be quite elaborate, see, for example, the two-dimensional FTICR pulse sequences in Chapter 4, individual events can fall into three classes. The first class apply DC or rf excitation to increase the amplitude of the cyclotron motion. Depending on cell pressure and translational energy, these are used to facilitate collisional activation, establish coherent high-amplitude ion packets for detection, or eject ions from the cell. The second class of pulses alter trap plate potentials to inject or eject ions from the cell along magnetic field lines and to manipulate the z-amplitude of the ion, for example to accomplish SID[119] or ion cloud compression.[138] A final category of pulses are delay events during which electric fields are not changed. These passive events allow time for non-reactive homogeneous[139] and inhomogeneous relaxation, reactive and dissociative collisions, and image current detection.

The simplest pulse sequence consist of four events common to every experiment. Executed first in every pulse sequence is a quench period during which both positively and negatively charged particles are removed from the trapped ion cell, typically by establishing a large electric field gradient between trap plates. The quench event is followed by ionization or ion introduction event during which ions are either formed in the cell or injected from an external source. Examples of these two ion formation modes are given in Sections 3.3 and 3.5. The final common elements of the FTICR sequence are the excitation event, described in Section 3.6, and the image current detection event.

The pulse sequence may be expanded beyond the basic sequence to perform more complex experiments. For example, Figure 7 depicts the standard pulse sequence used for CAD discussed in Section 3.7. Here, the basic pulse sequence includes three additional resonance excitation events and two additional delay events. The first two resonance events are ejection sweeps used to remove all ions above and below the precursor ion of interest from the cell. This is done by ensuring that energy delivered by the ejection pulse is sufficiently large to induce orbital radii greater than the

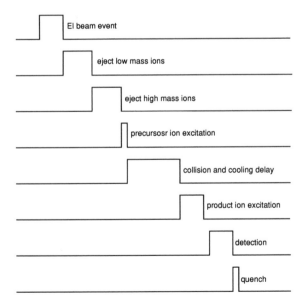

Figure 7. Depiction of FTICR pulse sequence events associated with the CAD experiment performed in Figure 6.

cell dimensions. The third resonance pulse is used to provide a specific amount of kinetic energy to the precursor ion. In the presence of high-pressure collision gas introduced during a delay, ions undergo collisions and dissociate. The final delay before excitation allows time for system base pressures to be reestablished and for fragment ions to stabilize prior to detection. If it were also desirable to dissociate a selected product ion prior to detection, the appropriate resonance excitation and delay events could be inserted.

4. FTICR PREFORMANCE

This section reviews the current status of FTICR performance, paying particular attention to aspects of FTICR performance that set it apart from conventional spectrometers. Although the ultimate performance standards as of this writing are documented, a discussion of the routine capabilities of the spectrometer is also included.

4.1. Ion Trapping

Once formed and trapped in the FTICR cell, ions may be retained for extended periods of time. For example, Alleman *et al.* demonstrated that over a 13-hour period, 30% of the initial ion population remained stored in a 4.7-T magnet at 1×10^{-8} Torr.[41] In practice, there are many ways in which an ion can be lost from the cell, and certain precautions should be taken when forming and trapping ions to avoid these effects. These precautions include operating at a low pressure, forming ions along the center line of the cell where the radial electric field is smallest, creating a sufficiently small ion population so that ion ion interactions are negligible, and minimizing the axial energy of the ions. Among the loss mechanisms that become important if any of these conditions are not satisfied are reactive collisions or collisional diffusion,[140] ion evaporation,[138] and resistive-wall destabilization of magnetron motion.[141] If these effects are minimized, then more subtle effects, such as magnetic field gradients, unbalanced electrostatic forces (such as those due to contact potentials), and gravity can induce radial drifts that ultimately reduce ion storage time.[142]

Extended ion trapping is exploited in FTICR experiments that require ultrahigh resolution measurements or when gas-phase ion chemistry is studied. The popularity of FTICR for the study of ion/molecule chemistry arises because the ultralow pressure conditions under which reactions occur provide a unique opportunity for examining intrinsic properties of reaction dynamics and energetics in the absence of solvation effects. The extended trapping times are required in such low-pressure experiments because reactive encounters are infrequent and long reaction times may be required to accumulate detectable quantities of product ions. Temporal profiles of reaction intermediates and products can be generated by varying the reaction delay prior to detection. Freiser,[12] in particular, has exploited the use of FTICR to examine reactions of metal ion species with neutral organic molecules, although dozens of laboratories now routinely describe novel ion chemistry performed in a low-pressure-trapped ion cell and detected by FTICR.

4.2. Mass Resolution

The FTICR is perhaps best identified as an ultrahigh resolution spectrometer capable of mass resolution for low-mass ions that can be orders of magnitude higher than that achieved with any other type of mass analyzer. Resolution is defined for FTICR by differentiating Equation 1 to obtain with rearrangement,

$$R = \frac{m}{dm} = \frac{-qB}{2\pi m \, df_c} \tag{12}$$

and substituting into Equation 1,

$$R = \frac{m}{dm} = -\frac{f_c}{df_c} \tag{13}$$

Resolution is observed to vary proportionally with magnetic field, and it is a primary justification for developing the 6-T and 7-T superconducting magnet systems used in many laboratories. Equation 13 also indicates the inverse relationship between resolution and mass. Thus, while Alleman *et al.*[41] and more recently Laukien[143] have obtained a resolution in excess of 10^8 for Da/z 18 of H_2O, equation 13 suggests upper mass resolution values of only 10^6 at Da/z 1000 and 10^5 at Da/z 10,000 are possible. Thus, the impressive resolution of 60,000 at Da/z 5900 for an ion of PEG 6000 obtained by laser desorption/ionization on a 7-T magnet is also indicative of the eventual inferiority of FTICR for resolution of ultrahigh mass ions.[46] Nevertheless, for ions with mass-to-charge ratios below a few thousand Da/z, FTICR is unrivaled for high-resolution mass analysis. This feature provides a strong motivation for developing ESI sources for FTICR because the multiple charging of ions obtained with this source allow high-mass ions to be analyzed within this optimum mass-to-charge range.

We briefly discuss now the experimental procedure by which ultrahigh resolution measurement is made. As stated earlier, the mass resolution achieved depends on the amount of time available to define the frequency. For the 10^8 mass resolution of H_2O just described, a peak width of 0.04 Hz was generated at 4.7 T. Based on the uncertainty principle, it is necessary for the transient be sampled for about 50 sec, and in fact, a 51 sec acquisition time was used. This points to an experimental difficulty in FTICR: If this resolution measurement were to be accomplished with direct detection, then according to sampling theory, data would have to be digitized at 4 MHz for 50 sec, which corresponds to a transient consisting of 2×10^8 data points. This is an impractically large amount of data for storage and processing. One solution to this impediment to high-resolution detection is the "mixer" mode or heterodyne experiment in which the much lower difference frequency between a synthesizer output and the cyclotron frequency is measured. Detection of this lower frequency allows correspondingly smaller digitization rates and thus results in smaller data sets for equivalent observation periods. Although this approach is effective in minimizing the problem of manipulating and storing large numbers of data points, it comes at the expense of reducing the detectable bandwidth.

As just described, mass resolution is often purposely limited by data acquisition parameters selected to obtain broadband spectra. Ultimately, however, mass resolution is limited by duration of the transient signal, which in turn depends on homogeneous and inhomogeneous relaxation effects in the coherent ion cloud.[144] Homogeneous relaxation involves processes that do not result in loss of phase coherence; such processes include radiative energy loss and resistive damping with external circuits but also the special case of collisional damping in which the coherence of the ion packet is not disturbed. This latter case arises when large ions collide with low-mass neutrals, and it has been exploited by Williams et al. to achieve multiple remeasurement of the same ion population.[139] Inhomogeneous relaxation includes processes resulting in loss of phase coherence such as Doppler broadening, and magnetic field inhomogeneities.[128]

4.3. Mass Measurement Accuracy

To a first approximation, the frequency of an ion is related to its mass by Equation 1, which accounts only for the applied magnetic field. In practice, any radial electric or magnetic fields in the cell alters the cyclotron frequency. The most severe contributions are from radial electric fields due to the radial component of the electric-trapping field and the local electric field generated by the space charge of the ion population, as discussed in Section 2. Several research groups have developed calibration equations to relate mass to frequency better by accounting for radial electric field terms. Ledford *et al.* first evaluated the contribution of trapping fields to the radial electric field in the FTICR-trapped ion cell and established a mass-frequency relationship based on this effect.[145] Following from the theory developed by Jeffries *et al.* to describe the evolving ion space charge in the cell,[146] Francl and coworkers developed the first calibration equation to approximate the ion cloud radial electric field contribution.[147] Later, Ledford *et al.* derived an algebraically improved mass frequency relationship from Jeffries's space charge model with the form shown in Equation 14.[148]

$$m = \frac{qB}{2\pi f_c} - \frac{2qGtV_{eff}}{4\pi^2 f_c} = \frac{A}{f_c} + \frac{B}{f_c^2} \tag{14}$$

Here, G_t is a trap geometry factor and V_{eff} describes the effective trapping potential experienced by ions.

To perform mass calibration, a calibrant such as perfluorotributyl-amine (PFTBA) is introduced to the spectrometer and a calibration table is generated from the electron ionization spectrum. If sufficient data points are used to define peaks within the bandwidth,[149] calibration curves with less than 1 Hz of uncertainty can be obtained with calibration functions derived from Equation 14. Broadband mass measurements typically exhibit a few part per milion error if measured at the calibration-trapping potential and for ion populations subject to minimal space charge distortion.[87, 88, 149–152] White *et al.* have shown that given the stability of modern superconducting magnets, by carefully controlling trap potentials, low-ppm error mass measurement accuracy can be achieved with a single FTICR calibration table for more than 1 week.[151] This is a clear advantage over accurate mass measurement performance with high-resolution double-sector mass analyzers. The stability of the FTICR mass calibration has been borne out experimentally with the demonstration of low-ppm errors for both chromatography,[87] laser desorption,[92] and CAD spectra.[88]

The presence of an ion-density-dependent term in the frequency expression becomes an important concern if it is not possible to match calibrant and analyte ion populations. Typically, a lengthy tuning process is required to achieve this goal, and unfortunately, for experiments in which the detected neutral population varies or is poorly controlled, this is not possible. Frequency shifts of hundreds of Hz are observed for this mismatch of ion populations. Efforts have been made to generate calibration expressions that account for such different ion populations,[152] but in an uncontrolled sampling environment, they are of limited utility. As an alternative, suspended-trapping techniques have been employed to obtain a self-regulating ion populations for detection below the cell's space charge limit. Trap plate potentials are altered to facilitate the efflux of excess ions[40] and low-ppm errors from a single calibration table have been obtained for GC/FTICR spectra differing by several orders of magnitude in initial neutral analyte concentration.[74, 75]

4.4. Mass Range

As recently as the early 1980s, one advantage of FTICR that was stressed was a supposedly unlimited mass range; in fact, several theoretical constraints to the FTICR experiment now have been shown to bound the upper mass limit. The first one to be recognized was the critical mass of the ion trap beyond which the outward radial electric field force exceeds the Lorentz force.[153] Another limitation to high-mass detection is recognized from Equation 4, which shows that as mass increases, the cyclotron orbit at thermal energies increase, so that an upper limit based on radial dimensions of the cell can be determined.[153, 154] Of these two limiting factors the radial component of the trapping field imposes the more severe upper mass limit; for example, the upper mass limit due to the radial electric field is about 40,000 Da for a singly charged ion in a 2.5-cm cell at 1.0-V trapping potential in a 3.0-T magnetic field. In contrast, the limit due to the thermal radius alone is greater than 10,000,000 Da for the same cell dimension and magnetic field strength. Experimentally, much lower practical mass limits are incurred, because a much smaller initial radial orbit is necessary if the orbit assumed after excitation is to achieve adequate spatial and phase coherence.

Given the strictures placed on acceptable ion energies, spatial distribution, and the size of the radial electric field for high-mass analysis, it is not surprising that high-mass analysis efforts have fallen short of initial predictions. Summarizing the results for singly charged ions to date, Amster *et al.* used FTICR to detect the 16241 Da CsI cluster ion resulting from bombarding CsI with C_s^+.[155] The Lebrilla *et al.* Cs^+ SIMS external-source quadrupole FTICR recently yielded CsI cluster ions to 31830 Da.[60] Hunt *et al.*, who routinely use an external-source quadrupole FTICR, have observed a spectrum for horse cytochrome c at 12384 Da.[61] The largest ion detected by direct infrared LDI/FTICR is at mass 9,700 for an oligomeric unit of PEG-8000,[59] however, Solouki and Russell[156] and Castoro *et al.*,[157] who observed ions from myoglobin dimer at 34000 Da,[156] have now obtained spectra from matrix-assisted LDI. It should be mentioned that beyond about 2000 Da/z, the mass resolution and sensitivity for organic ions often deteriorate for as yet experimentally undetermined reasons.[158] Thus, as previously mentioned, there is currently much effort directed at developing ESI interfaces for the FTICR, and the best high-mass results to date with ESI/FTICR are for multiply charged ions from bovine albumin dimer at 132,532 Da detected in the mass-to-charge range of about 1800–3000 Da/z.[58]

5. EXPERIMENTAL LIMITATIONS TO THE FTICR MEASUREMENT

Section 4.4 described several of the extraordinary results generated by FTICR. In most cases, these experiments were performed after considerable tuning to create precisely controlled ion populations in optimal applied yields. Unfortunately, attempts to implement common mass spectrometry experiments, such as ion dissociation, chemical ionization, or chromatographic detection, routinely result in inappropriate conditions for such high-performance detection, and spectral performance is disappointing. Problems that arise include pressure broadening and space charge distortion, which have already been discussed. In addition, ion motion in non-ideal applied fields has been shown to result in ion ejection, complications to mass calibration, and limitations on mass range. Thus, much of the research in instrument development has been directed at finding solutions to these problems with the intention of increasing the reliability of FTICR in routine application. In Section 5, we briefly summarize problems with the FTICR experiment and their effect on instrument performance.

5.1. Pressure Broadening

As described in Section 4, mass resolution is inversely related to pressure because collisions between ions and background neutrals during detection result in dephasing and decay of the coherent ion motion, thereby damping the detected signal. Even when ultrahigh resolution is not required, it is always desirable to operate at the lowest pressure possible to maximize sensitivity and minimize chemical interferences. However, many useful mass spectrometry experiments must be performed at much higher pressures than those required for optimum FTICR performance. Without some effort to rectify this pressure conflict, the generality of the FTICR measurement is severely compromised, especially for high-performance analytical applications. Section 3.3 described remedies in the form of pulsed valves and differential pumping that have for the most part provided effective solutions to the pressure problem for more routine analytical measurements.

5.2. Space Charge

A problem endemic to all ion storage devices is a limited dynamic range imposed by a finite ion capacity beyond which ion-ion coulombic repulsion alters the analytical lineshape. As described in Section 4.3, an ion-density-dependent decrease in effective cyclotron frequency arises due to the radial electric field that is created. This shift is observed almost immediately as the ion population in the cell increases before other line shape distortion is observed, and it can be hundreds of Hz in a severe space charge environment. For ion populations about two orders of magnitude above the trapped ion cell detection limits, other effects are observed, including coulombic line broadening[159] and peak shape distortion. The problem is especially severe for closely spaced peaks and often interferes with high-resolution measurements. A variety of effective means to reduce space charge effects are routinely employed, including increasing cell length, using selective ionization techniques (for example, performing electron ionization at an electron energy below the ionization potential of helium in GC experiments), applying selective ejection pulse sequences, physically partitioning ions,[160] and reducing the total ion population with suspended trapping pulse sequences.[40]

5.3. Radial-Trapping Fields

Ideally, the trapping electric field applied at electrodes normal to the applied magnetic field should be parallel everywhere to the magnetic field. In fact, as discussed in Section 2.1, a significant radial electric field is generated due to finite cell dimensions and the presence of the adjacent excite and receive electrodes. There has been an increased awareness of the adverse effects of these fields on the FTICR signal within the last few years. The field has been shown to cause a trapping-potential-dependent shift in effective cyclotron frequency, which is corrected in standard mass calibration equations, and were also found to limit the mass range of the trapped ion cell. Efforts to minimize these effects include increasing the z-axis dimensions of the cell, employing improved cell geometries, adding electric field shims to the cell,[161, 162] and placing grounded screen electrodes just inside the trapping electrodes.[154]

Besides the radial-trapping fields, a second type of nonhomogeneous electric field is generated in the cell when excitation potentials are applied to the excite electrodes. Again, due to the finite dimension of the cell and the presence of the adjacent trapping electrodes, there is significant inhomogeniety in the excitation field. This results in a corresponding reduction in the effective excitation field as ions move away from the center of the cell and approach the trapping electrodes. Thus, ions with sufficient z-axis amplitude to encounter this reduced field experience a reduced average excitation. If the ion cloud is not localized at the center of the cell at the time of excitation, nonquantitative behavior occurs in the detected FTICR signal.[163] In addition to reduced amplitude, the significant field curvature near the trap plates results in an axial component to the excite field that can increase the amplitude of trapping motion. This aggravates the problem of inhomogeneous excitation and can further result in axial ejection.[164] As was the case with the trapping field, various types of electric field shimming have been proposed as a means of reducing these excite field inhomogeneities.[161, 162, 165] However, a mechanically simpler and potentially more effective approach has been recently developed in our laboratory,[166, 167] employs an open trapped ion cell geometry[166] with capacitive coupling of the excitation electrodes to the trapping electrodes[167], thereby eliminating the axial component of the excitation field over the effective volume of the trapping potential well.

6. CONCLUSION

The last decade has seen tremendous growth in the field of ICR mass spectrometry as initial speculation and early disappointments concerning the analytical capabilities of FTICR have in more recent years been replaced by earnest theoretical and experimental efforts to evaluate and develop the technique. The FTICR is today recognized as an indispensable tool for low-pressure gas-phase ion/molecule and ion chemistry studies. As a high-performance analytical mass spectrometer, it exhibits unrivaled mass resolution and accuracy below 2000 Da. The inherent compatibility of FTICR with pulsed desorption/ionization techniques makes it an increasingly attractive choice for materials and surface analysis. In looking to the future, as better ion injection and trapping techniques are developed, FTICR will be increasingly integrated into tandem mass spectrometers when high-performance ion selection and detection are required. The continuing development of ion-source-interfacing techniques for the FTICR will make available a wide variety of ionization techniques and significantly improve the experimental versatility of the technique. Finally, as work continues to understand better and control ion motion in the trapped ion cell, improved trapped ion cells will be constructed. Significantly improved analytical performance, especially at higher mass, will be the ultimate consequence of these efforts.

REFERENCES

1. Hipple, J. A.; Sommer, H.; and Thomas, H. A., *Phys. Rev.* 1949, **76**, 1877–78.
2. Sommer, H.; Thomas, H. A.; and Hipple, J. A., *Phys. Rev.* 1951, **82**, 697–702.
3. Wobschall, D., *Rev. Sci. Instr.* 1965, **36**, 466–75.
4. Syrotron mass spectrometer, Varian Associates, Palo Alto, CA.
5. Beauchamp, J. L.; and Armstrong, J. L., *Rev. Sci, Instr.* 1969, **40**, 123–28.
6. McIver, R. T., Jr., *Rev. Sci. Instr.* 1973, **44**, 1071–74.
7. Baldeschweiler, J. D.; and Woodgate, S. S., *Acc. Chem. Res.* 1971, **4**, 114–20.
8. Beauchamp, J. L., *Annu. Rev. Phys. Chem.* 1971, **22**, 527–61.
9. Bowers, M. T.; and Su, T., *Adv. Electron. Electron Phys.* 1973, **34**, 223–79.
10. McIver, R. T., Jr., *Rev. Sci. Instr.* 1970, **41**, 555–58.
11. Comisarow, M. B.; and Marshall, A. G., *Chem. Phys. Lett.* 1974, **25**, 282–83.
12. Freiser, B. S., *Talanta* 1985, **32**, 697–708.
13. Laude, D. A., Jr.; Johlmann, C. L.; Brown, R. S.; Weil, P. A.; and Wilkins, C. L., *Mass Spectrom. Rev.* 1986, **5**, 107–66.
14. Wanczek, K. P., *Int. J. Mass Spectrom. Ion Processes* 1989, **95**, 1–38.
15. Nibbering, N. M. M., *Adv. Phys. Org. Chem.* 1988, **24**, 1–55.
16. Comisarow, M. B.; and Marshall, A. G., *Chem. Phys. Lett.* 1974, **26**, 489–90.
17. Comisarow, M. B.; and Marshall, A. G., *J. Chem. Phys.* 1975, **62**, 293–95.
18. Sharp, T. E.; and Eyler, J. R., *Int. J. Mass Spectrom. Ion Phys.* 1972, **9**, 421–39.
19. Hunter, R.; Sherman, M.; and McIver, R. T., Jr., *Int. J. Mass Spec. Ion Phys.* 1983, **50**, 259–74.
20. Dunbar, R. C.; Chen, J. H.; and Hays, J. D., *Int. J. Mass Spectrom. Ion Processes* 1984, **57**, 39–56.
21. Honovich, J. P.; and Markey, S., *Int. J. Mass Spectrom. Ion Processes* 1990, **98**, 51–68.
22. White, R. L.; Ledford, E. B., Jr.; Ghaderi, S.; Spencer, R. B.; Kulkarni, P. S.; Wilkins, C. L.; and Gross, M. L., *Anal. Chem.* 1980, **52**, 463–68.
23. Wang, T. L.; and Marshall, A. G., *Int. J. Mass Spectrom. Ion Processes*, 1986, **68**, 287–301.
24. Wang, M.; and Marshall, A. G., *Anal. Chem.* 1989, **61**, 1288–93.
25. Huang, S. K.; Rempel, C. L.; and Gross, M. L., *Int. J. Mass Spectrom. Ion Processes*, 1986, **72**, 15–31.
26. Marshall, A. G., *Anal. Chem.* 1979, **51**, 1710–14.
27. Comisarow, M. B.; and Marshall, A. G., *J. Chem. Phys.* 1976, **64**, 110–19.
28. Ledford, E. B., Jr.; White, R. L.; Ghaderi, S.; Gross, M. L.; and Wilkins, C. L., *Anal. Chem.* 1980, **52**, 1090–94.
29. Comisarow, M. B., *Transform Techniques in Chemistry*, Griffiths, P. R., ed. (Plenum: New York, 1978).
30. Giancaspro, C.; and Comisarow, M. B., *Appl. Spectrosc.* 1983, **37**, 153–66.
31. Alleman, M.; Kellerhals, Hp.; and Wanczek, K., *Chem. Phys. Lett.* 1980, **75**, 328–31.
32. McIver, R. T., Jr.; Hunter, R. L.; and Bowers, W. D., *Int. J. Mass Spectrom. Ion Processes* 1985, **64**, 67–77.

33. Hunt, P. F.; Shabanowitz, J.; McIver, R. T., Jr.; Hunter, R. L.; and Syka, J. E. P., *Anal. Chem.* 1985, **57**, 765–68.
34. Kofel, P.; Alleman, M.; Kellerhals, Hp.; and Wanczek, K. P., *Int. J. Mass Spectrom. Ion Processes* 1985, **65**, 97–103.
35. Alford, J. M.; Williams, P. E.; Trevor, D. J.; and Smalley, R. E., *Int. J. Mass Spectrom. Ion Processes* 1986, **72**, 33–51.
36. Kofel, P.; Alleman, M.; Kellerhals, Hp.; and Wanczek, K. P., *Int. J. Mass Spectrom. Ion Processes* 1986, **72**, 53–61.
37. Beu, S. C.; and Laude, D. A., *Int. J. Mass Spectrom. Ion Processes* 1991, **104**, 109–27.
38. Hofstadler, S. A.; and Laude, D. A., Jr., *Int. J. Mass Spectrom. Ion Processes* 1990, **97**, 151–64.
39. Dunbar, R. C.; and Weddle, G. H., *J. Phys. Chem.* 1988, **92**, 5706–09.
40. Laude, D. A., Jr.; and Beu, S. C., *Anal. Chem.* 1989, **61**, 2422–27.
41. Alleman, M.; Kellerhals, Hp.; and Wanczek, K. P., *Int. J. Mass Spectrom. Ion Processes*, 1983, **46**, 139–42.
42. Ledford, E. B., Jr.; Rempel, D. L.; and Gross, M. L., *Anal. Chem.* 1984, **56**, 2744–48.
43. Gross, M. L.; and Rempel, D. L., *Science* 1984, **226**, 261–68.
44. Laukien, F. H., *Int. J. Mass Spectrom. Ion Processes* 1986, **73**, 81–107.
45. Schuch, D.; Chung, K. M.; and Hartmann, H., *Int. J. Mass Spectrom. Ion Processes* 1984, **56**, 109–21.
46. Ijames, C. F.; and Wilkins, C. L., *J. Am. Chem. Soc.* 1988, **110**, 2687–88.
47. Sherman, M. G.; Kingsley, J. R.; Hemminger, J. C.; and McIver, R. T., Jr., *Anal. Chim. Acta* 1985, **178**, 79–89.
48. Rempel, D. L.; Ledford, E. B., Jr.; Huang, S. K.; and Gross, M. L., *Anal. Chem.* 1987, **59**, 2527–32.
49. Comisarow, M. B., *Int. J. Mass Spectrom. Ion Processes* 1981, **37**, 251–57.
50. Lee, S. H.; Wanczek, K. P.; and Hartmann, H., *Adv. Mass Spectrom.* 1980, **8B**, 1645–49.
51. Kofel, P.; Alleman, M.; Kellerhals, Hp.; and Wanczek, K. P., *Int. J. Mass Spectrom. Ion Processes* 1986, **74**, 1–12.
52. Schnatz, H.; Bollen, G.; Dabkiewicz, P.; Egelhof, P.; Kern, F.; Kalinowsky, H.; Schweikhard, L.; Stolzenberg, H.; and Kluge, H. J., *Nucl. Instrum. Methods Phys. Res.* 1986, **A251**, 17–20.
53. Kerley, E. L.; Hanson, C. D.; Castro, M. E.; and Russell, D. H., *Anal. Chem.* 1989, **61**, 2528–34.
54. Russell, D. H.; and Bricker, D. L., *Anal. Chim. Acta* 1985, **178**, 117–24.
55. FTMS-2000, Nicolet Analytical Instruments, Madison, WI; now marketed by Extrel, Inc.
56. Hanson, C. D.; Kerley, E. L.; and Russell, D. H., *Anal. Chem.* 1989, **61**, 83–85.
57. Kofel, P.; Alleman, M.; Kellerhals, Hp.; and Wanczek, K. P., *Int. J. Mass Spectrom. Ion Processes* 1989, **87**, 237–47.
58. Hofstadler, S. A.; and Laude, D. A., Jr., *Anal. Chem.* 1992, **64**, 572–75.
59. Ijames, C. F.; and Wilkins, C. L., *J. Am. Soc. Mass Spectrom.* 1990, **1**, 208–12.
60. Lebrilla, C. B.; Amster, I. J.; and McIver, R. T., Jr., *Int. J. Mass Spectrom. Ion Processes* 1989, **87**, R7–R13.
61. Hunt, D. F.; Shabanowitz, J.; Yates III, J. R.; Zhu, N. Z.; Russell, D. H.; and Castro, M. E., *Proc. Natl. Acad. Sci. USA* 1987, **84**, 620–23.
62. Henry, K. D.; Williams, E. R.; Wang, B. H.; McLafferty, F. W.; Shabanowitz, J.; and Hunt, D. F., *Proc. Natl. Acad. Sci. USA* 1989, **86**, 9075–78.
63. Henry, K. D., Quinn, J. P.; and McLafferty, F. W., *J. Am. Chem. Soc.* 1991, **113**, 5447–49.

64. Smalley, R. E., *Analytical Instrum.* 1988, **17**, 1–21.
65. Beu, S. C.; and Laude, D. A., Jr., *Int. J. Mass Spectrom. Ion Processes* 1990, **97**, 295–310.
66. Beu, S. C.; and Laude, D. A., Jr., *Int. J. Mass Spectrom. Ion Processes* 1992, **113**, 59–79.
67. United State Patent No. 4,924,089.
68. Carlin, T. J.; and Freiser, B. S., *Anal. Chem.* 1983, **55**, 571–74.
69. Sack, T. M.; and Gross, M. L., *Anal. Chem.* 1983, **55**, 2419–21.
70. Johlman, C. L.; Laude, D. A., Jr.; Brown, R. S.; and Wilkins, C. L., *Anal. Chem.* 1985, **57**, 2726–28.
71. Ledford, E. B., Jr.; White, R. L.; Ghaderi, S.; Wilkins, C. L.; and Gross, M. L., *Anal. Chem.* 1980, **52**, 2450–51.
72. White, R. L.; and Wilkins, C. L., *Anal. Chem.* 1982, **54**, 2443–47.
73. Sack, T. L.; McCreary, D. A.; and Gross, M. L., *Anal. Chem.* 1985, **57**, 1290–95.
74. Hogan, J. D.; and Claude, D. A., Jr., *Anal. Chem.* 1990, **62**, 530–35.
75. Hogan, J. D.; and Laude, D. A., Jr., *J. Am. Soc. Mass Spectrom* 1990, **1**, 431–39.
76. Hogan, J. D.; Hofstadler, S. A.; and Laude, D. A., Jr., "Gas Chromatography/FTMS," in *Analytical Applications of Fourier Transform Ion Cyclotron Resonance Mass Spectrometry*, Asamoto, B., ed. (VCH Publishers: New York, 1991).
77. Lee, E. D., Jr.; Henion, J. D.; Cody, R. B.; and Kinsinger, J. A., *Anal. Chem.* 1987, **59**, 1309–12.
78. Laude, D. A., Jr.; Pentoney, S. L.; Griffiths, P. R.; and Wilkins, C. L., *Anal. Chem.* 1987, **59**, 2283–88.
79. Miller, W. G.; Meek, J. T.; Stockton, G. W.; Thompson, M. L.; and Wayne, R. S., *Proc. Thirthy-Sixth ASMS Conf. Mass Spectrom. and Allied Tropics* (San Francisco, 1988), pp. 1247–48.
80. Kruppa, G. H.; Caravatti, P.; Radloff, C.; Laukien, F.; Watson, C.; and Wronka, J., "Applications of the CMS 47X External Ion Source FTICRMS," in *Analytical Applications of Fourier Transform Ion Cyclotron Resonance Mass Spectrometry*, Asamoto, B., ed. (VCH Publishers: New York, 1991).
81. Williams, E. R.; and McLafferty, F. W., *J. Am. Soc. Mass Spectrom.* 1990, **1**, 427–30.
82. Radloff, Ch.; Grossmann, P.; Caravatti, P.; and Alleman, M., *Proc. Thirty-Eight ASMS Conf. Mass Spectrometry and Allied Tropics* (Tuscon, 1990), pp. 834–35.
83. Williams, E. R.; and McLafferty, F. W., *J. Am. Soc. Mass Spectrom.* 1990, **1**, 361–65.
84. Kerley, E. L.; Hanson, D. C.; and Russell, D. H., *Anal. Chem.* 1990, **62**, 409–11.
85. Ghaderi, S.; Kulkarni, P. S.; Ledford, E. B., Jr.; and Wilkins, C. L., *Anal. Chem.* 1981, **53**, 428–37.
86. Laude, D. A., Jr.; Johlman, C. L.; Brown, R. S.; and Wilkins, C. L., *Fresenius Z. Anal. Chem.* 1986, **324**, 839–45.
87. Johlman, C. L.; Laude, D. A., Jr.; Brown, R. S.; and Wilkins, C. L., *Anal. Chem.* 1985, **57**, 2726–28.
88. Cody, R. B., *Anal. Chem.* 1989, **61**, 2511–15.
89. McCreary, D. A.; Ledford, E. B., Jr.; and Gross, M. L., *Anal. Chem.* 1982, **54**, 1431–35.
90. Wilkins, C. L.; Weil, D. A.; Yang, C. L.; and Ijames, C. F., *Anal. Chem.* 1985, **57**, 520–24.
91. Nuwaysir, L. M.; and Wilkins, C. L., "Lasers Coupled with Fourier Transformation Mass Spectrometry", in *Lasers and Mass Spectrometry*, Lubman, D. M., ed. (University Press: New York, 1989).
92. Wilkins, C. L.; and Yang, C. L. C., *Int. J. Mass Spectrom. Ion Processes* 1986, **72**, 195–208.
93. Coates, M. L.; and Wilkins, C. L., *Biochem. Mass Spectrom.* 1985, **12**, 424–28.

94. Coates, M. L.; and Wilkins, C. L., *Anal. Chem.* 1987, **59**, 197–200.
95. McCreary, D. A.; and Gross, M. L., *Anal. Chim. Acta* 1985, **178**, 91–103.
96. Brown, R. S.; and Wilkins, C. L., *Anal. Chem.* 1986, **58**, 3196–99.
97. Brown, C. E.; Kovacic, P.; Wilke, C. E.; Cody, R. B.; and Kinsinger, J. A., *J. of Polymer Sci. Lett.* 1983, **23**, 456–63.
98. Brown, R. B.; Weil, D. A.; and Wilkins, C. L., *Macromolec.* 1986, **19**, 1255–60.
99. Shomo II, R. E.; Marshall, A. G.; and Lattimer, R. P., *Int. J. Mass Spectrom. Ion Processes* 1986, **72**, 2099–2107.
100. Asamota, B.; Young, J. R.; and Criterin, R. J., *Anal. Chem.* 1990, **62**, 61–70.
101. Johlman, C. L.; Wilkins, C. L.; Hogan, J. D.; Donovan, R. L.; Laude, D. A., Jr.; Youseffi, M. J., *Anal. Chem.* 1990, **62**, 1167–72.
102. So, H. Y.; and Wilkins, C. L., *J. Phys. Chem.* 1989, **93**, 1184–87.
103. Shomo II, R. E.; Marshall, A. G.; and Weisenberger, C. R., *Anal. Chem.* 1985, **57**, 2940–44.
104. Brenna, J. T.; Creasy, W. R.; McBain, W.; and Soria, C., *Rev. Sci. Instr.* 1988, **59**, 873–79.
105. Brenna, J. T., *Microbeam Anal.* 1989, 306–10.
106. Van Peyl, G. J. Q.; Haverkamp, J.; and Kistemaker, P. G., *Int. J. Mass Spectrom. Ion Processes* 1982, **42**, 125–41.
107. Hogan, J. D.; and Laude, D. A., Jr., *Anal. Chem.* 1991, **63**, 2105–09.
108. Hogan, J. D.; Beu, S. C.; Laude, D. A., Jr.; and Majidi, V., *Anal. Chem.* 1991, **63**, 1452–57.
109. Comisarow, M. B.; and Marshall, A. G., *Chem. Phys. Lett.* 1974, **26**, 489–90.
110. McIver, R. T., Jr.; Baykut, G.; and Hunter, R. L., *Int. J. Mass Spectrom. Ion Processes* 1989, **89**, 343–53.
111. Chen, L.; and Marshall, A. G., *Int. J. Mass Spectrom. Ion Processes*, 1987, **79**, 115–25.
112. Marshall, A. G.; and Roe, D. C., *J. Chem. Phys.* 1980, **73**, 1581–90.
113. Noest, A. J.; Kort, C. W. F., *Comput. Chem.* 1983, **7**, 81–86.
114. deKoning, L. J.; Fokkens, R. H.; Pinske, F. A.; and Nibbering, N. M. M., *Int. J. Mass Spectrom. Ion Processes* 1987, **77**, 95–105.
115. McIver, R. T., Jr.; Hunter, R. L.; and Bahkut, G., *Anal. Chem.* 1989, **61**, 489–91.
116. Grosshans, P. B.; and Marshall, A. G., *Int. J. Mass Spectrom. Ion Processes* 1990, **100**, 347–79.
117. Marshall, A. G.; Wang, T.-C.; and Ricca, J. L., *J. Am. Chem. Soc.* 1985, **107**, 7893–97.
118. Cody, R. B.; and Freiser, B. S., *Int. J. Mass Spectrom. Ion Processes* 1982, **41**, 195–204.
119. Ijames, C. F.; and Wilkins, C. L., *Anal. Chem.* 1990, **62**, 1295–99.
120. Cody, R. B.; and Freiser, B. S., *Anal. Chem.* 1979, **51**, 547–51.
121. Cassady, C. J.; and Freiser, B. S., *J. Am. Chem. Soc.* 1984, **106**, 6176–79.
122. Cody, R. B.; Burnier, R. C.; Cassady, C. J.; and Freiser, B. S., *Anal. Chem.* 1982, **54**, 2225–28.
123. Wise, M. B., *Anal. Chem.* 1987, **59**, 2289–93.
124. Bricker, D. L.; Adams, T. A., Jr.; Russell, D. H., *Anal. Chem.* 1983, **55**, 2417–18.
125. Russell, D. H.; and Bricker, D. L., *Anal. Chim. Acta* 1985, **178**, 117–24.
126. Cody, R. B.; and Freiser, B. S., *Anal. Chem.* 1987, **59**, 1056–59.
127. Mabud, M. A.; DeKrey, M. J.; and Cooks, R. G., *Int. J. Mass Spectrom. Ion Processes* 1985, **67**, 285–94.
128. Bier, M. E.; Amy, J. W.; Cooks, R. G.; Syka, J. E. P. ; Ceja, P.; and Stafford, G., *Int. J. Mass Spectrom. Ion Processes* 1987, **77**, 31–47.
129. Williams, E. R.; Henry, J. D.; McLafferty, F. W.; Shabanowitz, J.; and Hunt, D. F., *J. Am. Soc. Mass Spectrom.* 1990, **1**, 413–16.

130. Dunbar, R. C. "Tropics in Ion Photodissociation" in *Ion Cyclotron Resonance Spectrometry II*, Hartmann, H.; and Wanczek, K. P., eds. (Springer Verlag: Berlin, 1982), pp. 1–26.

131. Chen, J. H.; Hays, J.; and Dunbar, R. C., *J. Chem. Phys.* 1984, **88**, 4759–64.

132. Cassady, C. J.; and Freiser, B. S., *J. Am. Chem. Soc.* 1984, **106**, 6176–79.

133. Bowers, W. D.; Delbert, S. S.; and McIver, R. T, Jr., *Anal. Chem.* 1986, **54**, 969–72.

134. Dunbar, R. C.; and Ferrara, J., *J. Chem. Phys.* 1985, **83**, 6229–33.

135. Baykut, G.; Watson, C. H.; Weller, R. R.; and Eyler, J. R., *J. Am. Chem. Soc.* 1985, **107**, 8036–42.

136. Watson, C. H.; Baykut, G.; and Eyler, J. R., *Anal. Chem.* 1987, **59**, 1133–38.

137. Nuwaysir, L. M.; and Wilkins, C. L., *Anal. Chem.* 1989, **61**, 689–94.

138. Gross, M. L.; Huang, S. K.; and Rempel, D. L., *Int. J. Mass Spectrom. Ion Processes* 1986, **70**, 163–84.

139. Williams, E. R.; Henry, K. D.; and McLafferty, F. W., *J. Am. Chem. Soc.* 1990, **112**, 6157–62.

140. Francl, T. J.; Fukuda, E. K.; and McIver, R. T., Jr., *Int. J. Mass Spectrom. Ion Phys.* 1983, **50**, 151–67.

141. Beu, S. C.; and Laude, D. A., Jr., *Int. J. Mass Spectrom. Ion Processes* 1991, **108**, 255–68.

142. Chen F. F., *Introduction to Plasma Physics* (Plenum Press: New York, 1974).

143. Laukien, F. H., *Proc. Thirty-Fifth ASMS Conf.* Spectrometry and Allied Topics (Denver, 1987), pp. 781–82.

144. Comisarow, M. B., "Signals, Noise, Sensitivity, and Resolution in Ion Cyclotron Resonance Spectroscopy", in *Ion Cyclotron Resonance Spectrometry II*, Hartmann, H.; and Wanczek, K. P., eds. (Springer-Verlag: Berlin, 1978), pp. 484–513.

145. Ledford, E. B., Jr.; Ghaderi, S.; White, R. L.; Spencer, R. B.; Kulkavic, P. S.; Wilkins, C. L.; and Gross, M. L., *Anal. Chem.* 1980, **52**, 463–68.

146. Jeffries, J. B.; Barlow, S. E.; and Dunn, G. H., *Int. J. Mass Spectrom. Ion Processes* 1983, **54**, 169–87.

147. Francl, I. J.; Sherman, M. G.; Hunter, R. L.; Locke, M. J.; Bowers, W. D.; and McIver, R. T., Jr., *Int. J. Mass Spectrom. Ion Processes* 1984, **54**, 189–99.

148. Ledford, E. B., Jr.; Rempel, D. L.; and Gross, M. L., *Anal. Chem.* 1984, **56**, 2744–48.

149. Shomo II, R. E.; Marshall, A. G.; and Weisenberger, C. R., *Anal. Chem.* 1985, **57**, 2940–44.

150. Rempel, C. L.; Ledford, E. B., Jr.; Sack, T. M.; and Gross, M. L., *Anal. Chem.* 1989, **61**, 749–54.

151. White, R. L.; Onyuvinka, E. C.; and Wilkins, C. L., *Anal. Chem.* 1983, **55**, 339–43.

152. Francl, I. J.; Sherman, M. G.; Hunter, R. L.; Locke, M. J.; Bowers, W. D.; and McIver, R. T., Jr., *Int. J. Mass Spectrom. Ion Processes* 1984, **54**, 189–99.

153. Grossman, P. B.; Wang, M.; Ricca, T. L.; Ledford, E. B., Jr.; Marshall, A. G., *Proc. Thirty-Sixth ASMS Conf.* Mass Spectrometry and Allied Tropics (San Francisco, 1988), pp. 592–93.

154. Wang, M.; and Marshall, A. G., *Anal. Chem.* 1989, **61**, 1288–93.

155. Amster, I. J.; McLafferty, F. W.; Castro, M. E.; Russell, D. H.; Cody, R. B., Jr.; and Ghaderi, S., *Anal. Chem.* 1986, **58**, 483–85.

156. Solouki, T.; and Russell, D., *Proc. Natl. Acad. Sci. USA* 1992, in press.

157. Castoro, J. A.; Koster, C.; and Wilkins, C., *Rapid Comm. Mass Spectrom.*, in press.

158. Hanson, C. D.; Castro, M. E.; Russel, D. H.; Hunt, D. F.; and Shabanowitz, J., American Chemical Society series, "FTMS of Large (m/z 5000) Biomolecules" 1977, **359**(FTMS), 100–15.

159. Wang, T. L.; and Marshall, A. G., *Int. J. Mass Spectrom. Ion Processes* 1986, **68**, 287–301.
160. Kerley, E.; and Russel, D. H., *Anal. Chem.* 1989, **61**, 53–57.
161. Wang, M.; and Marshall, A. G., *Anal. Chem.* 1990, **62**, 515–20.
162. Hanson, C. D.; Castro, M. E.; Kerley, E. L.; and Russell, D. H., *Anal. Chem.* 1990, **62**, 520–26.
163. Riegner, D.; Hofstadler, S. A.; and Laude, D. A., Jr., *Anal. Chem.* 1991, **63**, 261–68.
164. Huang, S. K., Rempel, D. L.; and Gross, M. L., *Int. J. Mass Spectrom. Ion Processes* 1986, **72**, 15–31.
165. Caravatti, P.; and Allemann, M., *Org. Mass. Spectrom.* 1991, **26**, 514–18.
166. Beu, S. C.; and Laude, D. A., Jr., *Int. J. Mass Spectrom. Ion Processes* 1992, **112**, 215–30.
167. Beu, S. C.; and Laude, D. A., Jr., *Anal. Chem.* 1992, **64**, 177–80.

6

Elucidation of Protein Structure and Processing Using Time-of-Flight Mass Spectrometry

Amina Woods, Rong Wang, Marc Chevrier, Tim Cornish, Cathy Wolkow, and Robert J. Cotter

1. INTRODUCTION

Traditionally, structural analysis by mass spectrometry has been limited to small organic molecules that are volatile and thermally stable, since the initial step involves heating the sample in a vacuum. During the past several years, new ionization methods have been developed that can produce gas-phase molecular and fragment ions from nonvolatile samples; these are presented to the mass spectrometer as liquids (solutions) or solids. As a consequence of these developments, mass spectrometry has come to play an increasing role in the biological sciences.

Fast atom bombardment (FAB), introduced by Barber *et al.*[1] in 1981, has made it possible to analyze routinely intact peptides and small proteins weighing up to 5000 Da. The steady, long-lasting sample ion currents produced by this technique made this an ideal ion source for scanning instruments: double-focusing magnetic sector and quadrupole mass spectrometers. The demand for higher mass ranges stimulated the

Amina Woods, Rong Wang, Marc Chevrier, Tim Cornish, Cathy Wolkow, and Robert J. Cotter • Middle Atlantic Mass Spectrometry Laboratory, Department of Pharmacology and Molecular Sciences, The Johns Hopkins University School of Medicine, Baltimore, Maryland 21205.

Experimental Mass Spectrometry, edited by David H. Russell. Plenum Press, New York, 1994.

subsequent development of high-field-magnetic analyzers and extended mass range quadrupoles, while the need to obtain structural information about peptides (amino acid sequence) and other biopolymers led to the development of tandem mass spectrometers. These include high-performance four-sector instruments and triple quadrupoles in which fragmentation is carried out between the two mass analyzers by collision-induced dissociation (*CID*) with an inert gas. When combined with *array detection*,[2] four-sector mass spectrometers provide a highly sensitive approach to protein sequencing.

In 1976, Macfarlane *et al.*[3] introduced plasma desorption mass spectrometry (PDMS), which is used almost exclusively with time-of-flight (*TOF*) mass analyzers; normally, they have considerably lower mass resolution than either sector or quadrupole instruments. Because ions are extracted and analyzed in a time frame that is often shorter than that required for fragmentation,[4,5] plasma desorption mass spectra generally reveal only molecular ions. However, because TOF mass analyzers do not scan the mass range but record ions of every mass after each ionization event, PDMS can have a significant advantage in sensitivity. In theory, the TOF analyzer has an unlimited mass range. Molecular ions have been recorded by PDMS up to 35 kDa,[6] while a practical limit for routine use appears to be about 15 kDa.

The first laser desorption (LD) mass spectra of relatively large, thermally labile molecules was reported in 1978 by Posthumus *et al.*[7] using CO_2 (10.6-μm) and Nd:YAG (1.06-μm) lasers on a sector instrument. Because the short (generally 20–40 nsec) laser pulses produce ions in a very short time frame, TOF[8, 9] and Fourier transform mass spectrometers (FTMS)[10] provided the most appropriate means for observing ions of every mass from a single laser pulse. In our laboratory (and others), LD has been used for the structural analysis of industrial polymers,[11, 12] oligosaccharides,[13, 14] and glycolipids.[15, 16] Its succesful application to peptide and protein analysis began with the introduction of matrix-assisted laser desorption (MALD) in 1988 by Karas and Hillenkamp.[17] This technique is currently capable of molecular weight measurements for proteins in excess of 100 kDa.[18]

Electrospray ionization (ESI)[19] provides the same high-molecular-weight information as MALD. The predominance of multiply-charged ions places the measured mass/charge ratios within the limits of relatively inexpensive quadrupole and triple quadrupole instruments[20, 21] and *ion traps*,[22] although ESI has been developed for magnetic sector instruments as well.[23] In addition, ESI provides an excellent interface for *on-line* microbore HPLC and capillary zone electrophoresis (CZE) mass spectrometry.

In our laboratory, we have been able to address many of the structural problems related to protein structure (glycosylation and phosphorylation sites, carbohydrate structure and heterogeneity, disulfide bond locations, point mutations, and cleavage sites in protein processing) using plasma desorption (PD) and MALD. Both are TOF techniques from which (generally) only molecular weight information is available, and they are used because of their high sensitivity. These instruments are readily available commercially and/or are easily constructed. When used in combination with appropriate chemical and enzymatic digestion steps, they provide a practical approach to the structural analysis of proteins and an alternative to more expensive tandem instruments that are not likely to be available in protein laboratories.

Chapter 6 describes PD and LD mass spectrometry (MS), and reviews a number of our own recent applications of these two TOF methods to elucidate protein/peptide structure and processing. In so doing, we attempt to illustrate the basic strategies that can be used to determine the location of disulfide bonds, cleavage sites of precursor proteins, and chemical and post translational modifications, using molecular weight measurements alone. We conclude with our current efforts to improve the mass resolution of the TOF mass analyzer and to develop a practical, inexpensive tandem (TOF/TOF) instrument for peptide sequencing.

1.1. Plasma Desorption Mass Spectrometry

In PDMS, ionization is carried out by the highly energetic fission fragments from a sample of $^{252}Cf_{98}$. This radioactive isotope decays with

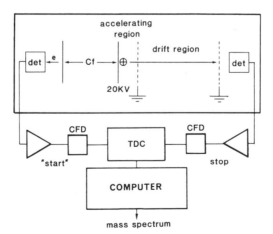

Figure 1. Block diagram of the PDMS.

a half-life of 2.65 years: 97% as α-particles and 3% as two simultaneous, multiply charged fission fragments emitted in opposite directions. A typical decay involves the simultaneous emission of $^{106}Tc_{43}$ and $^{142}Ba_{56}$ with energies of 104 and 79 MeV, respectively.[24] In a PDMS (see Figure 1), one of the fission fragments is recorded by a *start* detector, which consists of a thin nickel foil that converts the energy of the incoming particle to a burst of secondary electrons, a dual-channel plate detector, a preamplifier, and a constant fraction discriminator (CFD) that can distinguish between the pulse height from a 6.1-MeV α-particle and a fission fragment. The start pulse then initiates the timing cycle in a time-to-digital converter (TDC). The other fission fragment penetrates an aluminized mylar foil to which the sample has been applied, releasing from 1–10 sample ions. These ions are accelerated by the electrical potential placed across the sample foil and an extraction grid, traverse a *drift* region, and are then recorded by a *stop* detector. The arrival times are compared with the time of the start pulse, and (to a first approximation) mass depends on the time (t) spent in the drift region (D):

$$\frac{m}{e} = \frac{2Vt^2}{D^2} \tag{1}$$

where V is the accelerating voltage. For a 10-µCi source, from 1200–1800 such events occur each second. Mass spectra can be obtained for many samples by accumulating the ion counts from 1–3 million fission events (10–30 min). For weaker samples or enzymatic digests (where several molecular ion peaks are observed in the mass spectrum), mass analysis may be carried out over several hours, or overnight. The mass spectrum is calibrated from the measured flight times of H^+ and Na^+ appearing in the positive ion spectrum or H^- and CN^- in the negative ion spectrum.

Peptide samples are generally dissolved in glacial acetic acid or aqueous trifluoroacetic acid and electrosprayed[25] onto the aluminized mylar foil. Alai *et al.*[26] noted that an ion signal could be greatly improved if peptides were dissolved in aqueous TFA solutions containing an equimolar mixture of oxidized and reduced glutathione. Thiols have been shown to preserve the folded form of proteins,[27] so it is thought that this decreases the contact area between the peptide and the aluminium surface. A more widely used approach is to first coat the sample foil surface with nitrocellulose,[28] which also reduces the interaction between peptides and protects the sample surface from weak hydrophobic attractions; in addition, this approach enables the researcher to rinse off ionic species (primarily Na^+) with deionized water. These contaminants accompany most biological matrices, the buffers used to solublize and stabilize proteins, and the solutions used to extract and purify peptides; and they can interfere with the quality of the mass spectrum. At the same time, mass spectra of mixtures (for example, unfractionated tryptic digests used for peptide mapping) may reveal only the most hydrophobic peptides.[29, 30] This problem can be alleviated by partial fractionation by reversed-phase HPLC, so that resultant mixtures contain peptides of similar hydrophobicity.

The strength of PDMS lies in its ability to carry out enzymatic reactions directly on the hydrophobic surface used to hold the sample in the ionization source so that the same sample can be mass analyzed before, during, and after digestion. The PDMS has a practical mass range of about 15 kDa. It has been suggested that the effective surface *interaction area* resulting from absorption of an MeV particle is about 20 Å in diameter and this places a limit on the size of peptides that can be desorbed.[31]

A commercial instrument, the BIN-10K™ PDMS developed by Bio-Ion, Nordic (Uppsala, Sweden), is used in the experiments that follow. A more recent version of this instrument, the Bio-Ion 20™, is available from Applied Biosystems Incorporated (Foster City, CA).

1.2. Matrix-Assisted Laser Desorption

In their first report, Karas and Hillenkamp[17] dissolved protein samples in solutions containing nicotinic acid (a strong ultraviolet absorber) and carried out their analyses on a Leybold-Hereaus (Köln, Germany) LAMMA-1000™ reflecting TOF mass spectrometer, using the fourth harmonic (266 nm) from a pulsed (10-nS), Q-switched Nd:YAG laser. Similar results were achieved by Tanaka *et al.*[32] using the 337-nm line from a pulsed (15-nS), 4-mJ nitrogen laser with a coaxial, reflecting TOF mass analyzer and a matrix composed of 300-Å diameter cobalt powder in glycerol. This instrument was later commercialized (LAMS-50K™, Shimadzu Corporation, Kyoto, Japan). Subsequently, Beavis and Chait[33] introduced a 2-meter linear TOF instrument in which the sample probe tip was placed at a voltage of 20 kV, and ions were extracted by a two-stage grid system. Spengler and Cotter[34] used a frequency-quadrupled, Q-switched Nd:YAG laser with a CVC Products (Rochester, NY) model 2000™ TOF mass spectrometer for the desorption of proteins in excess of 100 kDa and oligodeoxyribunucleotides.[35] Salehpour *et al.*[36] used a modified Bio-Ion, Nordic (Uppsala, Sweden) BIN-10K™ plasma desorption mass spectrometer with the 308-nm line from a Xe/HCl excimer laser in combination with a Coumarin 153/307 dye laser system and a second harmonic generation crystal to generate the 266-nm radiation.[37]

Subsequent development has generally focused on the use of longer wavelengths. Beavis and Chait[38] used the third harmonic (355 nm) from a Nd:YAG laser in combination with cinnamic acid derivatives (ferulic, caffeic, or sinapinic acid)[39] as matrices. Nelson *et al.*[40] employed the 581-nm wavelength from an excimer-pumped dye laser for desorption of peptides and nucleic acids from frozen aqueous solutions. Pulsed IR laser radiation at 2.94 µm (from a mechanically Q-switched Erbium:YAG laser)[41] and 10.6 µm (from a TEA-CO$_2$ laser)[42] have also been used succesfully by Overberg *et al.*[41] Commercial instruments using a 5-nsec pulsed N$_2$ laser (337 nm) are available from Vestec Corporation (Houston, TX) and Finnigan (Hemmel-Hempstead, UK). An instrument employing a 1.5-mJ, 600-psec pulsed nitrogen laser, 20-kV ion extraction, and a 1 m flight tube was reported by Chevrier *et al.*[43] and is used in the experiments that follow.

2. INSTRUMENTATION AND STRATEGIES

Mass resolutions for simple, linear TOF mass spectrometers are generally in the range of one part in 300–600. For peptides in the mass

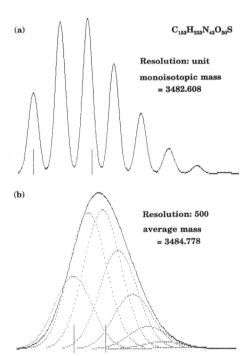

(a)

$C_{153}H_{225}N_{42}O_{50}S$

Resolution: unit

monoisotopic mass
= 3482.608

(b)

Resolution: 500

average mass
= 3484.778

Figure 2. Molecular ion distributions for the peptide glucagon at (a) unit mass resolution and (b) at a resolution of 500, calculated by the EXMASS program described by Yergey.[44] Monoisotopic and average masses are indicated by the vertical centroids.

range of 1,000–4,000 Da, it is not possible to resolve isotopic contributions (D, ^{13}C, ^{15}N, ^{18}O, etc.) to the molecular ion signal. Thus, the isotopically averaged mass (*average mass*), rather than the *monoisotopic mass*, is measured (see Figure 2).[44] For mass spectra in which only molecular ions are observed, this does not present a particular problem. Mass accuracies of 0.1%–0.01% can easily be achieved at relatively low mass resolution when there are no interfering species. For larger peptides and proteins, mass measurement accuracy may be compromised by the inability to distinguish between the isotopic clusters of MH^+ ions, MNa^+ ions, and (in the case of matrix-assisted LD) adducts formed between the sample and matrix. In addition, Beavis and Chait[45] and Pan and Cotter[46] have shown that proteins desorbed by matrix-assisted LD carry velocities comparable to those of the expanding plume of matrix ions in which they are entrained and (as a result of their larger mass) considerable kinetic energy in the direction of their flight. This results in an apparent shift to lower mass, as well as peak broadening and lower mass resolution. However, molecular weight measurements by matrix-assisted LD are generally far more accurate than SDS-PAGE (sodium dodecyl sulfonate polyacrylamide gel electrophoresis) measurements. Molecular weights form an approximate upper bound from which one can account for all of the enzymatic fragments.

For linear TOF mass spectrometers, mass resolution (to a first approximation) is given by:

$$\frac{\Delta m}{n} = \frac{U_o}{eV} \tag{3}$$

where U_o is the initial kinetic energy distribution of desorbed ions.[47] Mass resolution is therefore improved using high-accelerating voltages V. In the two instruments described later, accelerating voltages of 20 kV are used. Including a *reflectron*[48] in a TOF mass spectrometer can improve mass resolution considerably, since the reflectron provides corrections for differences in ion kinetic energy. At the same time, the simple linear arrangement currently provides the best sensitivity for observing molecular ions of high-molecular-weight peptides and proteins.

2.1. The BIO-ION Plasma Desorption Mass Spectrometer

Plasma desorption mass spectrometry was carried out on a Bio-Ion (Uppsala, Sweden) BIN-10K^{TM} TOF mass spectrometer. This instrument is equipped with a 10-μCi ^{252}Cf source, a 15-cm flight tube, a 1-nS time resolution time-to-digital converter, and a PDP 11/73-based data system.

Samples were deposited on nitrocellulose-backed aluminized mylar foils. The nitrocellulose sample foils were prepared by electrospraying a solution (10 µl) containing approximately 2-µg/µl nitrocellulose (Bio-Rad 0.45-µm nitrocellulose membrane) in acetone onto an aluminized mylar foil. Alternatively, nitrocellulose-treated foils could be obtained from Applied Biosystems (ABI). Peptide solutions, ranging in concentration from 100 fmol/µl–1nmol/µl, were prepared by dilution in aqueous 0.1% TFA or in 0.1% TFA in 95% water and 5% acetonitile. Peptides soluble in basic pH were prepared in solutions of 25-mM ammonium bicarbonate (pH 7.5) or 25-mM Tris (pH 8.5). 2-10 µl of sample solution was applied to a nitrocellulose foil mounted on a spin-device (built in-house) and spin dried so that soluble salts migrate with the solvent front to the edge of the sample foil. Excess salts can be further remove by additional washing with 0.1% aqueous TFA and spin drying.

2.2. Matrix-Assisted Laser Desorption Using a 600-psec N_2 Laser

Laser desorption mass spectra were carried out on an instrument designed in-house (see Figure 3).[43] The ion source housing was constructed from a single (7in. × 7in. × 8in.) aluminium block. Electrical feed-through flanges, ion gauge, laser bean window, and final focusing lens, sample probe, turbomolecular pump and 1-m flight tube are all attached to this block using standard ASA flanges with viton O-rings. The ion source

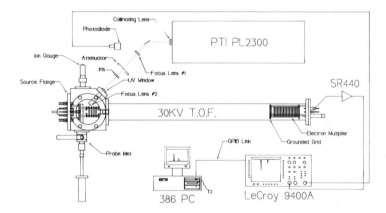

Figure 3. Block diagram of the linear TOF LD instrument used in this study. (Reprinted with permission from Reference 43.)

and ion extraction optics are mounted directly on the electrical feed-through flange. The sample probe is stainless steel with a vespel tip containing five stainless steel disks so that up to five samples can be loaded simultaneously. As the probe is inserted, each of the sample disks in turn makes electrical contact with the 20-kV feed-through flange. Ions are extracted by a grounded grid followed by an Einsel lens, *X*- and *Y*-deflectors and a grounded flight tube terminated by a grid.

The laser is a Photon Technology International (Ontario, Canada) model PL2300 ™ 1.2-mJ, 600-psec, 337-nm pulsed nitrogen laser. The laser beam is collimated by a 100-cm focal length lens, reflected 45°, and further focused by two additional lenses (with focal lengths of 20 cm and 6.5 cm, respectively), the second situated inside the vacuum chamber. A portion of the unreflected beam is detected by an Electro-Optics Technology (Freemont, CA) model ET2010 ™ fast photodiode, whose output provides the trigger pulse for a LeCroy (Spring Valley, NY) model 9400 ™, 100-MHz digital oscilloscope. The laser beam is attenuated by an iris and a variable optical density wheel.

Ions are detected by a Becton-Dickinson (Baltimore, MD) model MM1 ™ multiplier, whose output is connected directly to the vertical input of the digital oscilloscope or first amplified by a Stanford Research Instruments (Stanford, CA) model SR440 ™ 300-MHz, wide-band analog amplifier. The output of the digital oscilloscope is downloaded to a 386-based personal computer (PC) for data acquisition and processing, via a GPIB interface.

In general, peptide samples were dissolved in aqueous 0.1% trifluoroacetic acid to varying concentrations. 1 µl of peptide solution was then mixed with 4 µl of aqueous, concentrated caffeic acid solution, deposited by micropipette on one of the sample disks, and air dried at room temperature. The nitrogen laser was triggered manually to produce single-shot spectra, each of which was downloaded to the 386 PC. While mass spectra can often be obtained from a single laser shot, in general, spectra from several laser shots were added/averaged to produce good signal/noise.

2.3. Basic Strategies for Protein/Peptide Analysis

Ordinarily, molecular weights of proteins are determined by SDS-PAGE. If sufficient protein is available, the amino acid sequence can be determined directly by tryptic (or other enzymatic) digestion, followed by reversed phase HPLC separation and automated Edman sequencing of each of the tryptic fragments. Generally, two or more chemical (e.g., CNBr) or enzymatic digests are carried out to sequence overlapping regions and establish the correct order of the fragments. Alternatively, the amino acid sequence is deduced from the DNA sequence of the gene that expresses the protein.

There are a number of difficulties with such approaches: Molecular weight measurements by SDS-PAGE are quite approximate; Edman degradation approaches sequencing only from the N-terminus, is limited in the length of residues that can be sequenced, and cannot be carried out if the amino terminus is blocked. This is particularly critical if the N-terminal amino acid sequences is required to identify the gene that codes the protein; in addition, the DNA sequence for the gene-coding for the protein does not reflect post-translational modifications.

Plasma desorption, MALD, and ESI (within their respective mass limits) produce far more accurate measurements of peptide/protein molecular weights than SDS-PAGE. Fast atom bombardment mass spectrometry of purified tryptic peptides can reveal both N-terminal and C-terminal sequence ions, and it is not adversely effected by N-terminal blocking. When FAB is used with tandem mass spectrometers, sequence fragmentation can be enhanced by CID and (perhaps more importantly) peptides coeluting from off-line or on-line HPLC separations can be mass selected to ensure that fragment (sequence) ions originate from a single coeluting peptide.

Tandem mass spectrometers, particularly four-sector instruments employing high-mass-range electrostatic and magnetic analyzers, are expensive instruments not likely to be widely available to protein chemists or molecular biologists. At the same time, there are many structural questions that arise about peptides/proteins for which full or partial amino acid sequences are known. In the simplest case, a researcher may wish to determine the cleavage site of a precursor protein (expressed by a gene whose DNA sequence is known) by identifying the C-terminus of the fully processed protein. Two such examples are described later, and in one case, this information led to the identification of the enzyme responsible for the cleavage. Other questions include posttranslational modifications (glycosylation, phosphorylation, *etc.*), location of disulfide bonds, carbohydrate heterogeneity, *etc.* Such questions can often be addressed by molecular weight measurements on relatively inexpensive TOF mass spectrometers combined with exo- and endoproteinases.

Figure 4 summarizes the kinds of experiments used for peptides and proteins whose amino acid sequences are known or partially known. All involve measurements of molecular weights using TOF mass spectrometers. For proteins, large peptides, and the larger peptide fragments often generated by chemical digestion with CNBr, MALD would be used; for carboxypeptidase and aminopeptidase digestion, PD offers the opportunity

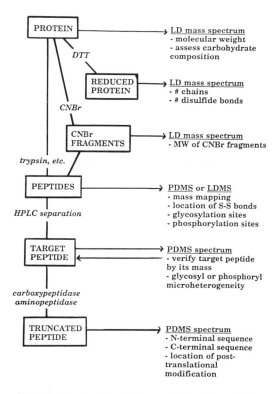

Figure 4. Basic strategies for peptide/protein analysis.

to carry out repeated *on-foil* digestions of the sample after it is removed from the instrument. Thus, the two techniques have some complementary advantages, and both are used in the examples that follow.

3. STRUCTURE OF PORPOISE RELAXIN

Bedekar *et al.*[49] have spent a number of years comparing the structures, biological activity, and insulin cross reactivity of relaxin from different mammalian species. Relaxin is a polypeptide hormone that is synthesized and stored in the corpus luteum during pregnancy; it is responsible for the dilation of the symphys pubis of most mammals during parturition. Relaxin has three-dimensional and disulfide homology (but no cross reactivity, biologically or immunologically) with insulin. Initial amino acid sequencing gave the following structure for relaxin obtained from porpoise:

$$\boxed{}$$
RMTLSEKCCQVGCIRKDIARLC

$$\mid \qquad\qquad\qquad\qquad\qquad \mid$$

Pca-RNTDFIKACGRELVIRLWVEICGSV ...   (3)

for which the C-terminal sequence of the B-chain could not be determined. The C-terminal sequence was obtained in our laboratory[50] using PDMS, and it is used to illustrate our methods.

3.1. Molecular Weight Measurements

Plasma desorption mass spectrometry of intact porpoise relaxin yielded a protonated molecular ion at m/z 6058.0 (data not shown), suggesting a molecular weight of 6057.0. Relaxin was then reduced during DTT (dithiothreitol), and the PDMS of the resultant unfractionated mixture (see Figure 5) revealed protonated molecular ions of m/z 2528.2

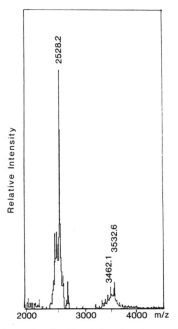

Figure 5. The PDMS spectrum of unfractionated, reduced porpoise relaxin. (Reprinted with permission from Reference 50.)

and 3532.6 for the A- and B-chains, respectively. The molecular weight of intact relaxin was then calculated from the measured masses of the A- and B-chains minus the mass of six hydrogen atoms for the fully oxidized peptide containing three disulfide bridges:

$$2527.2 + 3531.6 - 6(H) = 6052.8$$

which is 4.2 mass units lower than that obtained for intact relaxin. In the intact relaxin measurement, the presence of partially reduced relaxin or unresolved contributions to the molecular ion peak from $M + Na^+$ ions may account for this slight discrepancy.

Comparing the molecular weight determined for the B-chain (3531.6 Da) with the calculated mass for the known portion of the B-chain sequence (2875.2) gives the mass of the unknown C-terminal portion as 656.4 Da. In addition, the peak observed at m/z 3462.1 in Figure 5 is 70.5 Da less than the peak at 3532.6, suggesting the presence of a des-alanine structure. Thus, the C-terminal amino acid may be alanine.

3.2. Enzymatic Digestion and Peptide Mapping

Intact, unreduced porcine relaxin was digested with trypsin, and a plasma desorption *map* was obtained for the resultant unfractionated mixture of peptide fragments (see Figure 6). Cleavage at arginine and

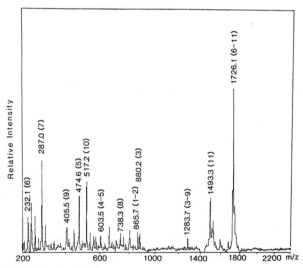

Figure 6. The PDMS map of the tryptic fragments of porpoise relaxin. (Reprinted with permission from Reference 50.)

Table 1. Molecular Weights of Tryptic Peptide Fragments of Porpoise Relaxin

Peptide fragments	Calculated molecular weight (M)	Molecular ion (MH$^+$) from tryptic map
A-chain fragments		
[1] *R*	174.0	
[2] *MTLSEK*	708.1	
[3] *CCQVGCIR*	881.2	880.2
[4] *K*	146.0	
[5] *DIAR*	473.6	474.6
[6] *LC*	234.3	232.1
B-chain fragments		
[7] *pGlu-R*	287.0	287.0
[8] *TNDFIK*	737.1	737.3
[9] *ACGR*	405.5	405.5
[10] *ELVR*	515.7	517.2
[11] *LWVEICGSV...*	1005.3	1493.3
Disulfide-linked fragments		
[3–9] *CCQVGCIR+ACGR*	1284.7	1283.7
[6–11] *LC+LWVEICGSV...*	1237.6	1726.1
Fragments from incomplete digestion		
[1–2] *R+MTLSEK*	864.1	865.7
[4–5] *K+DIAR*	601.8	603.5

Source: Adapted with permission from Reference 47.

lysine residues should result in six A-chain fragments, [1] to [6], and five B-chain fragments, [7] to [11], as shown in Table 1. All of these peptide fragments are accounted for, including peptides [1], [2], and [4], which are observed as fragments [1-2] and [4-5] resulting from incomplete digestion. The disulfide-linked fragment [3-9] is observed at m/z 1283.7, or 2 Da less than the sum of the molecular ions for peptides [3] and [9]:

$$880.2 + 405.5 - 2(H) = 1283.7$$

A protonated molecular ion for peptide [11], which contains the unknown C-terminal portion of the B-chain, is observed at m/z 1493.3. The molecular weight of this peptide (1492.3 Da) is 487.0 Da greater than the molecular weight (1005.3 Da) of the known portion of peptide [11]. The peak observed at m/z 1726.1 corresponds to the protonated molecular ion of the disulfide-linked fragment [6-11]. Its molecular weight (1725.1 Da) is 487.5 Da greater than the molecular weight (1237.6 Da) calculated for a disulfide-linked fragment containing peptide [6] and the known portion of peptide [11]. When these results are compared with the measured mass of the B-chain (3531.6 Da), there is an additional mass deficit of 172 Da, suggesting that there is an additional peptide [12] of mass $172 + 18 = 190$ Da; and that peptide [11] must terminate in either a lysine or an arginine.

Reduced relaxin was fractionated by reversed phase HPLC (see Figure 7) and PDMS obtained for fractions 72 and 76 (see Figure 8). In Figure 8a, the peak at m/z 3534.4 corresponds to the protonated molecular ion of the intact B-chain, while an additional peak at m/z 3462.8 corresponds to a des-alanine structure. While it is not clear whether reduction of the intact peptide with DTT results in truncated B-chain peptide or if this represents an additional form of the processed peptide, we note that this des-alanine species was observed in the mass spectrum of the unfractionated reduced peptide (see Figure 5) as well. If we then subtract the mass of an alanine residue (71 Da) from the mass deficit (172 Da) represented by peptide [12], this suggests that the next residue is threonine (101 Da).

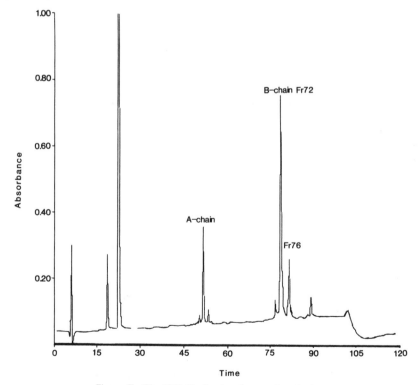

Figure 7. The HPLC of reduced porpoise relaxin.

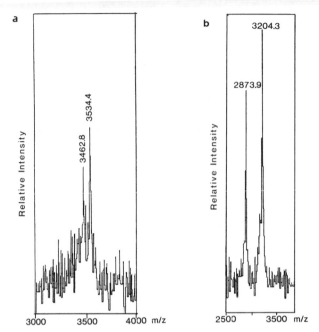

Figure 8. The PDMS spectra of (a) fraction 72 and (b) fraction 76 from the HPLC fractionation of reduced relaxin. (Reprinted with permission from Reference 50.)

The appearance of several minor peaks in the HPLC chromatogram shown in Figure 7 suggests that a number of truncated A- and B-chain peptides are formed during the reduction of intact peptide with DTT. If we substract the masses of alanine (71.1 Da) and threonine (101.1 Da) from the protonated molecular ion mass (3534.4) obtained from fraction 72, the resultant mass (3362.2) is 157.9 Da greater than the peak at m/z 3204.3 in Figure 8b. This suggests that the C-terminal amino acid for peptide [11] is arginine. The peak at m/z 2873.9 corresponds to the known portion of the B-chain.

Thus, considerable information about the sequence can be obtained from molecular weight measurements of the tryptic digest and reduced protein. We now know that the C-terminus of the B-chain is ...*RTA*. This information, and the remaining unknown sequence, can be obtained more directly using carboxypeptidase digestion of the intact B-chain.

3.3. Carboxypeptidase and Aminopeptidase Digestion

Chait *et al.*[51] described the use of carboxypeptidase-Y to reveal the amino acid sequence of bradykinin by PDMS. Such reactions can be carried out directly on the sample foil, followed by rinsing to remove excess enzyme; this enables repeated PDMS measurements of the same sample to follow the course of the reaction. Figure 9 shows the PDMS of fraction 72 (B-chain), following a 10-minute incubation with carboxy-peptidase-Y using a 1:1 molar ratio of enzyme:substrate. The peak at m/z 3533.2 corresponds to the intact B-chain, while the peaks at m/z 3462.6, 3361.0, 3206.6, 3149.2, and 2963.2 represent losses of 70.6, 101, 154.4, 57.4, and 186 Da, respectively, corresponding to the masses of alanine, threonine, arginine, glycine, and tryptophan, respectively. The peak at m/z 2963.2 is 85.4 larger than the mass calculated for the known portion of the sequence, suggesting that the final amino acid is a serine. Thus, the completed sequence for the B-chain is

Pca-RTNDFIKACGRELVRVWVEICGSVSWGRTA

This sequence was subsequently confirmed by automated Edman degrada-tion.[52]

In theory, molecular weight measurements combined with enzymatic digestion could be used to determine the entire sequence. Plasma desorp-tion mass spectra would be obtained for intact and reduced peptide, followed by mapping the unfractionated digest. Each of the peptides would

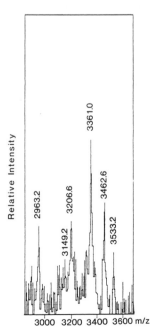

Figure 9. The PDMS spectrum of HPLC fraction 72 following carboxypeptidase digestion. (Reprinted with per-mission from Reference 50.)

then be isolated and subjected to carboxypeptidase and/or aminopeptidase digestion to determine the sequence from the molecular ions of the resulting series of nested, truncated peptides. Sequencing in this fashion is generally limited to only a few residues, so that initial digestions with different enzymes (trypsin, chymotrypsin, elastase, *etc.*) or chemicals (CNBr) could be used to provide overlapping fragments whose terminal sequences would then be determined by *on-foil* digestion and PDMS.

More often, all or part of the sequence is known, so that initial enzymatic digestion is used to identify the site of the interest, and carboxypeptidase (aminopeptidase) digestion can be targeted to that site. In such cases, this approach provides a real alternative to methods using CID on tandem instruments.

4. REACTIONS OF BrCCl₃ WITH HEMOGLOBIN

Davies *et al.*[53] have studied the inactivation of cytochrome P-450 by the hepatotoxic agent, carbon tetrachloride. Irreversible inactivation of cytochrome P-450 in the liver is thought to be mediated by the trichloromethyl radical that results from reductive dechlorination of CCl_4 by cytochrome P-450, leading to a number of altered heme products.[54] Previous studies of a model system, involving the reaction of $BrCCl_3$ with reduced myoglobin, resulted in the identification of several soluble heme

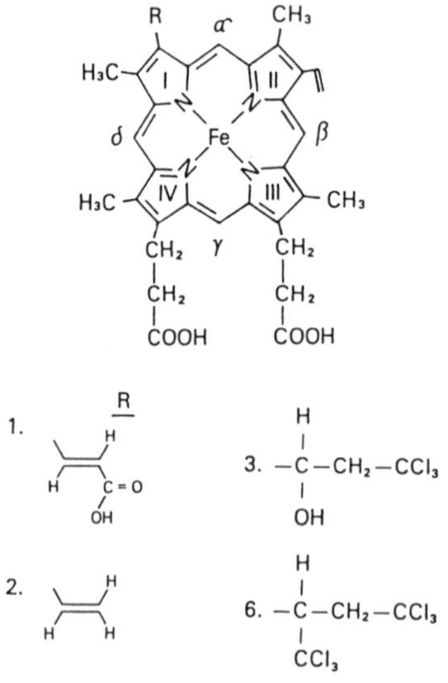

Figure 10. Structure of the soluble heme products resulting from the reaction of BrCCl₃ with reduced myoglobin. (Reprinted with permission from Reference 55.)

products[55] and a protein-bound product formed by covalent attachment of the prosthetic heme to the histidine residue *93* (normally, the axial ligand to the heme iron) via a CCl_2 moiety derived from $BrCCl_3$[56] Figure 10 shows the structures of the soluble heme products identified by using a combination of FAB MS and two-dimensional nuclear overhouser effect spectroscopy 2D (NOESY) and correlated spectroscopy (COSY) NMR.[55]

Initial studies of the reaction between $BrCCl_3$ and hemoglobin by Pohl and Osawa[57] revealed that the prosthetic heme was covalently attached to a pentapeptide *CDKLH*, corresponding to residues *93-97* of the β-subunit of the apoprotein. Unlike myoglobin, this peptide did not include the proximal histidine. Plasma desorption mass spectrometry was used to determine the exact location of covalent attachment of the prosthetic heme to this peptide as well as its structure.[58]

4.1. Digestion with Elastase

Reversed-phase HPLC separation of the products of the reaction between $BrCCl_3$ and hemoglobin revealed both soluble and a covalent heme product. Digestion of the covalent heme product with elastase (which cleaves at histidine) and further purification by HPLC resulted in the isolation of two heme peptides. The PDMS of the major product (not shown) revealed a protonated molecular ion (m/z 1349.3), corresponding to a structure containing the heme, pentapeptide, and an additional CCl_3 moiety (hemeCCl$_3$ + *CDKLH*). The presence of a CCl_3 moiety was confirmed by peaks corresponding to the mass of heme, hemeCCl, hemeCCl$_2$, and hemeCCl$_3$. In addition, a peak corresponding to the mass of hemeCCl$_3$ + *CDK* eliminated leucine and histidine as attachment sites and prompted our subsequent prolonged elastase digestion (3 days at 37 °C) of the covalent heme product. Plasma desorption mass spectra of the major and minor products (see Figures 11a and 11b) give molecular ions corresponding to hemeCCl$_3$ + *CDK* (m/z 1098) and hemeCCl$_2$ + *CDK* (m/z 1062), respectively.

4.2. On-Foil Digestion with Aminopeptidase

Each of these heme tripeptides was then subjected to digestion with aminopeptidase-M for 6 days. Plasma desorption mass spectra of the final products are shown in Figure 12. The peaks at m/z 855 and 818 correspond to attachment of the prosthetic heme to cysteine (hemeCCl$_3$ + *C* and hemeCCl$_2$ + *C*, respectively). It has been suggested[53, 58] that both the

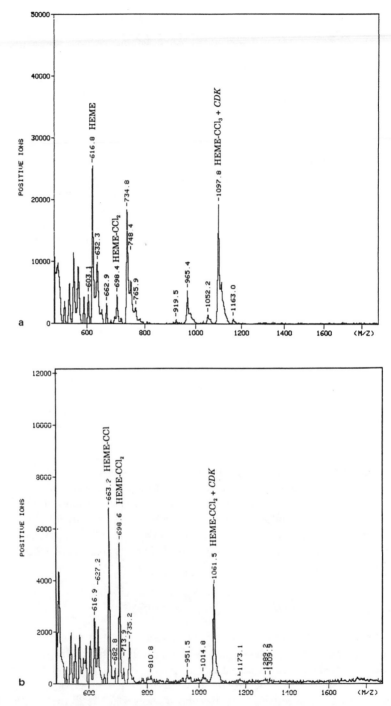

Figure 11. The PDMS mass spectra of the (a) major and (b) minor heme-peptide products following prolonged digestion with elastase. (Reprinted with permission from Reference 58.)

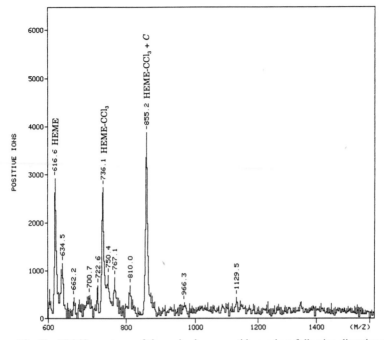

Figure 12. The PDMS spectrum of the major heme-peptide product following digestion with aminopeptidase. (Reprinted with permission from Reference 58.)

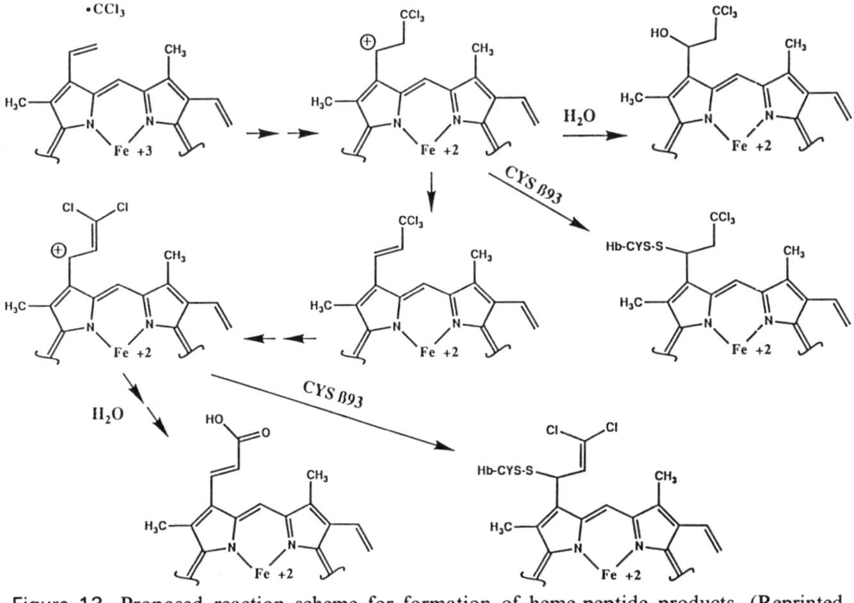

Figure 13. Proposed reaction scheme for formation of heme-peptide products. (Reprinted with permission from Reference 58.)

soluble products and the covalent attachment of the heme to the apoprotein result from an initial reaction of the trichloromethyl radical with the heme prosthetic group at the I-vinyl moiety, as shown in Figure 13.

Interestingly, FABMS and CID of the covalent heme-peptide products from both the myoglobin and hemoglobin studies did not produce amino acid sequence-specific fragmentation, which would have revealed the attachment site. Generally, the major fragment ion was the hemeCCl$_3$ moiety, accompanied by successive losses of chlorine atoms, suggesting that the positive charge was carried by the heme group. Thus, in this case, an enzymatic approach proved to be more effective in locating the attachment site.

5. CLEAVAGE OF THE AMYLOID PRECURSOR PROTEIN

The β-amyloid protein (β/$A4$) is a 4-kDa protein derived from a larger amyloid precursor protein (APP), the principal component of senile plaques in Alzheimer's disease. The APP is a transmembrane protein (see Figure 14) secreted as a carboxyl-terminal-truncated molecule, resulting from cleavage at a site above the membrane during normal processing. Because the β-amyloid protein also lies within the cell membrane and

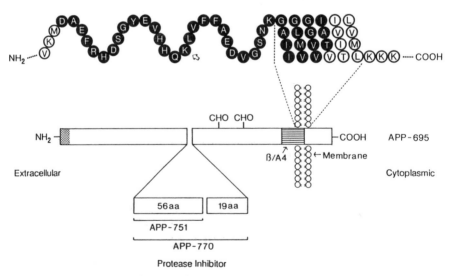

Figure 14. Structure of APP in the vicinity of the cell membrane. The region comprising the β-amyloid peptide is indicated in black.

cleavage occurs at a site within the β/*A4* region, the intact amyloidogenic β/*A4* fragment is not generated during normal catabolism. This would suggest that an early event in amyloid formation may involve altered APP processing that results in the release and subsequent deposition of intact β/*A4*.[59]

At the same time, the exact cleavage site of amyloid precursor protein during normal processing had not been determined. Thus, PDMS was used to determine the cleavage site.[60]

5.1. Determining the Cleavage Site of APP during Normal Processing

Chinese hamster ovary cells were transfected with the APP-770 gene; secreted proteins were purified by anion exchange HPLC and digested with cyanogen bromide (CNBr). The resulting peptide fragments were further purified by reverse-phase HPLC and their PDMS obtained. Because the sequence of APP is known, protonated molecular ions for peptide fragments resulting from cleavage at methionine correspond to their

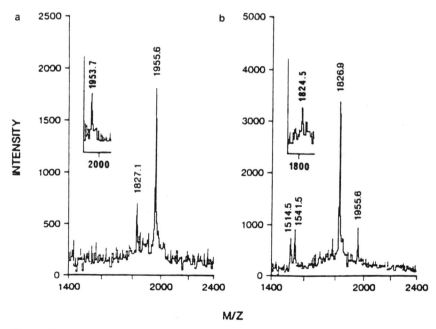

Figure 15. The PDMS spectra of the C-terminal peptides for the secreted product of APP. (a) The HPLC fraction 66, and (b) HPLC fraction 67. Insets are negative ion spectra. (Reprinted with permission from Reference 60.)

calculated masses, except for the C-terminal fragment, which is truncated. Two C-terminal fragments were identified, and their PDMS spectra are shown in Figure 15. The protonated molecular ion at m/z 1955.6 in Figure 15a is consistent with the sequence *DAEFRHDSGYEVHHQK* and suggests that the primary cleavage site during normal processing is at lysine, located within the β/*A4* region. The second fragment (see Figure 15b) resulted in a protonated molecular ion at m/z 1826.9, corresponding to the sequence *DAEFRHDSGYEVHHQ*. The presence of peaks at m/z 1827.1 and 1955.8 in Figures 15a and 15b, respectively, results from incomplete HPLC fractionation of these two peptides. Inserts in Figure 15 represent the negative ion PDMS spectra, for which the molecular ions $(M-H)^-$ are 2 mass units below those from the positive ion mass spectra. Negative ion mass spectra were obtained to verify that peaks in the positive ion spectra represented MH^+ ions, and not, for example, MNa^+ or other adduct species.

5.2. Carboxypeptidase Digestion

While a full accounting of all intact CNBr fragments and the correspondence in mass of the PDMS molecular ions with a truncated sequence produce a high level of confidence that we have correctly identified the C-terminal fragment, partial sequencing of that fragment provides additional specificity. Therefore, the peptide purported to result from cleavage at lysine was subjected to *on-foil* digestion with carboxypeptidase-Y (10:1 weight ratio of enzyme/substrate for 3 hours). The resulting PDMS spectrum is shown in Figure 16. The peak at m/z 1956 corresponds to the protonated molecular ion, while peaks at m/z 1827 and 1699 correspond to successive losses of lysine and glutamine, respectively.

During this study, a variety of enzyme:substrate concentrations were used, and these are shown in Table 2. Reactions were all carried out directly on the nitrocellulose-coated sample foil used in PD. Reasonable

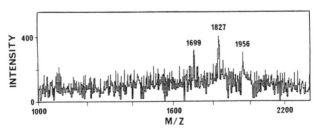

Figure 16. The PDMS spectrum of HPLC fraction 66 from secreted APP following digestion with carboxypeptidase-Y. The ratio of enzyme/substrate was 10:1.

Table 2. The Efficiency on *In Situ* Carboxypeptidase-Y Digestion

E/S^a	Ion ratio	20 m	2 h	3 h	14 h	24 h
1:1	$I_2/I_1{}^b$	0.00^c	0.37	0.42	0.91	1.08
	I_3/I_1	0.00	0.00	0.26	0.64	1.09
10:1	I_2/I_1	0.35	0.62	1.32		
	I_3/I_1	0.00	0.47	1.09		

[a] Weight ratio of enzyme to substrate.
[b] I_1, I_2, and I_3 are protonated molecular ions at m/z 1956, 1827, and 1699, respectively.
[c] Ratio of peak heights.

digestion periods were achieved when the molar ratio of enzyme to substrate was considerably higher than in solution. This may result from partial immobilization of enzyme molecules through hydrophobic attraction to the nitrocellulose surface. At the same time, there is considerable advantage in carrying out such reactions without having to transfer the peptide from the foil to solution or obtain spectra from several timed aliquots.

5.3. Synthetic C-Terminal Fragment

The C-terminal fragment *DAEFRHDSGYEVHHQK* was synthesized, and its PDMS spectrum recorded (see Figure 17). In contrast to the spectra shown in Figure 15, abundant sequence-specific fragments are observed. Such fragmentation is often observed in synthetic peptides for which considerably larger quanities are available.

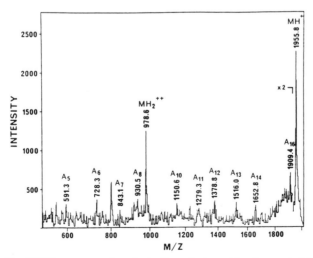

Figure 17. The PDMS spectrum of synthetic C-terminal peptide of a secreted APP product.

6. PROCESSING THE ASSEMBLY PROTEIN FROM CYTOMEGALOVIRUS

The B-capsids of simian (Colburn) strain cytomegalovirus (CMV) contain a phosphorylated protein that is absent from the mature virion: the assembly protein (AP), a 37-kDa (by SDS-PAGE) molecule believed to be involved in the capsid assembly.[61] The genomic region encoding the AP has been cloned, sequenced, and found to be organized as a nested set of four in-frame, 3′-coterminal genes (see Figure 18), suggesting that the viral genome could give rise to four independently transcribed 3′-coterminal RNAs coding for four overlapping in frame, carboxy-coterminal proteins. The four predicted proteins were identified by immunoassays in lysate of SCMV-infected cells using an antiserum specific for the carboxyl end of the AP precursor.[62] Their molecular weights, as determined by SDS-PAGE, were 85, 49, 40, and 33 kDa, corresponding to the 64, 46, 34, and 27-kDa proteins, respectively, shown in Figure 18.

Mature AP is processed from the 40-kDA (predicted 34-kDa) precursor—the preassembly protein by C-terminal cleavage.[63] The previously unknown cleavage site of the preassembly protein was determined in our laboratory using PDMS and enzymatic digestion to identify the carboxy terminus of the AP.[64] Determining the amino acid sequence at the cleavage site then enabled us to identify the protease acting at that site.

6.1. Determining the Cleavage Site

The amino acid sequences of the AP presursors have been deduced from the nucleotide sequence.[65] The amino terminus of the AP was determined by chemical digests, probing with antisera, and autoradiography;

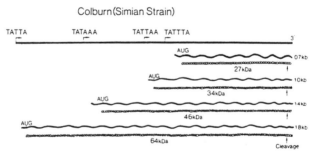

Figure 18. The model for the transcriptional and translational expression of the 3′-coterminal, nested assembly protein-related genes of CMV strain Colburn. (Adapted with permission from Reference 62.)

the carboxy terminus however has not been determined. From previous work, we know that the single cysteine in the deduced sequence (residue 279) is not part of the AP.[63] In our current approach, the AP was enzymatically digested, the resulting peptides fractionated by reversed-phase HPLC, and molecular weights of the purified peptides determined by PDMS and compared to molecular weights calculated from the sequence to reveal the carboxy terminal fragment.

Both Endoproteinase-Lys C and Endoproteinase-Glu C digestions were carried out to provide two independent confirmations of the cleavage site. These two enzymes cleave at lysine and glutamic acid residues, respectively; however, due to the long incubation times (24 hours) used in this study, additional cleavages at arginine and aspartic acid, respectively, were also observed. Plasma desorption mass spectra of the peptide fragments from the Endoproteinase-Lys C digest accounted for every residue from residue 1–277. The PDMS of fraction 24 gave a peak at m/z 903.3 (see Figure 19a), corresponding to a protonated molecular ion of a peptide with the sequence *SAERGVVNA* in the region

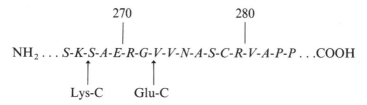

The Endoproteinase-Glu C digest accounted for a significant number of fragments. The PDMS of fraction 26 from this digest gave a peak at m/z 616.2 (see Figure 19b), corresponding to a molecular ion for the peptide *RGVVNA*, and an additional peak at m/z 744.5, corresponding to the molecular ion for the peptide *ERGVVNA*. These results suggest that the alanine at position 277 is the last residue in the amino acid sequence.

The identity of the C-terminus was verified by carboxypeptidase digestion of these two peptides, carried out directly on the sample foil. A 10 μl of an 0.4 μg/10 ul solution of Carboxypeptidase P in 25-mM sodium citrate buffer was added to each peptide sample and incubated for 5 minutes. Excess fluid was spun off and the foil washed with 30 ul of 0.1% TFA; the sample was then dried and reinserted into the mass spectrometer. The mass spectrum of fraction 24 from the Endo Lys-C digest containing peptide *SAERGVVNA* showed a significant reduction in the size of the peak at m/z 903.3, and a new peak appeared at m/z 462.2. This corresponds to the loss of the first five C-terminal residues (*GVVNA*) resulting in a peptide *SAER*, whose calculated protonated molecular ion mass is 462.5 amu. The mass

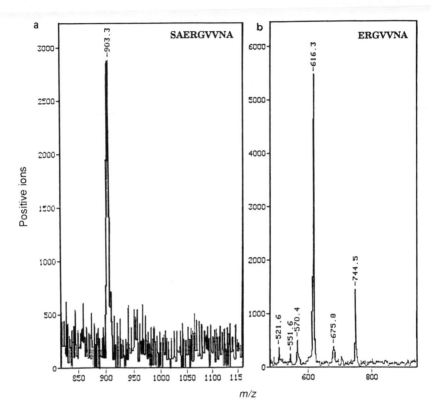

Figure 19. The PDMS spectra of the C-terminal peptides of the CMV assembly protein resulting from digestion with (a) Endoproteinase Lys-C and (b) Endoproteinase Glu-C. (Adapted with permission from Reference 64.)

spectrum of fraction 26 from the Endo Glu-C digest containing peptide *RGVVNA* shows an additional peak at m/z 547.6 resulting from the loss of the C-terminal alanine and corresponding to the protonated molecular ion of the peptide *RGVVN*, whose calculated mass is 544.7 amu. These results confirm that the last residue on the AP is an alanine.

The amino acid sequence *VNAS* surrounding the cleavage site of the preassembly protein corresponds to a four amino acid consensus sequence (*V*/L-X-A-*S*/V, where X is a polar amino acid, P_3 is usually Val, and P'_1 is usually Ser), highly conserved in the region near the carboxyl end of all herpesvirus assembly protein homologs previously analyzed. The proteinase responsible for cleavage of the preassembly protein was subsequently localized at the amino terminal 249 residues of the 85-kDa (64-kDa) protein.[64]

7. INSERTION PEPTIDES USED IN X-RAY CRYSTALLOGRAPHIC STUDIES

Recently, MALD was used in our laboratory to resolve a protein sequence ambiguity in an insertion mutant *Staph. nuclease* intended for X-ray crystallographic studies. Specifically, X-ray crystal structure data for a mutant peptide in which a putative alanine had been inserted in the C-terminal α-helix between residues Arg-126 and Lys-127[66] strongly suggested that in fact a glycine had been inserted.[67] The molecular weight of wild-type *Staph. nuclease*, calculated from its amino acid sequence, is 16,813.5 Da; calculated molecular weights for the glycine and alanine insertion peptides are 16,870.6 and 16,884.6 Da, respectively. For peptides or proteins in this mass range, we generally expect a mass accuracy of approximately 0.1% mass (in this case 17 Da). While this would certainly distinguish either of these insertion possibilities from the wild type, we note that the mass difference between the glycine and alanine insertion peptides is only 14 Da. Thus, we obtained and compared MALD mass spectra of

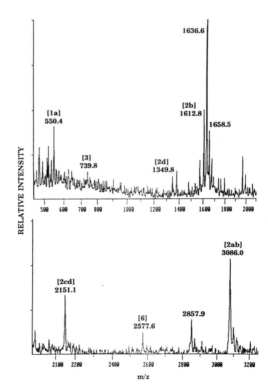

Figure 20. The MALD MS spectrum of the Endoproteinase Arg-C digest of wild-type *Staph. nuclease*. (Reprinted with permission from Reference 68.)

Table 3. Peptides Expected from Cleavage at Arginine Residues

Peptide fragments	Calculated molecular ion[a]	Observed wild type	Observed glycine insertion	Observed alanine insertion
[1] *ATSTKKLHKEPATLIKAIDGDTVKLMYKGQPMTFR*$_{1-35}$	3921.2	—[b]	—	—
[2] LLLVDPETKHPKKGVEKYGPEASAFTKKMVENAKKIEVEFDKGQR$_{36-81}$	5217.7	—	—	—
[3] TDKYGR$_{82-87}$	739.9	739.8	740.0	—
[4] GLAYIYADGKMVNEALVR$_{88-105}$	1984.6	—	—	—
[5] QGLAKVAYVYKPNNTHEQHLR$_{106-126}$	2468.1	—	—	—
[6] KSEAQAKKEKLNIWSEDNADSGQ$_{127-149}$ *(wild type)*	2578.1	2577.6	—	—
[6] GKSEAQAKKEKLNIWSEDNADSGQ$_{127-150}$ *(glycine insertion)*	2635.2	—	2635.4	—
[6] AKSEAQAKKEKLNIWSEDNADSGQ$_{127-150}$ *(alanine insertion)*	2649.2	—	—	2634.8

[a] Mass of the protonated molecule $(M+H)^+$.
[b] Not observed.
Source: Adapted with permission from Reference 67.

Table 4. Additional Peptides from Cleavages between Adjacent Lysine Residues

Peptide fragments	Calculated molecular ion[a]	Observed wild type	Observed glycine insertion	Observed alanine insertion
[1a] *ATSTK*$_{1-5}$	551.6[c]	550.4	—	—[b]
[1b] *KLHKEPATLIKAIDGDTVKLMYKGQPMTFR*$_{6-35}$	3432.6	—	—	—
[2a] *LLLVDTPETKHPK*$_{36-48}$	1491.9	—	—	—
[2b] *KGVEKYGPEASAFTK*$_{49-63}$	1613.0	1612.8	1613.3	1612.5
	1635.0[c]	1636.6	1635.9	1635.7
	1657.0[d]	1658.5	—	—
[2ab] *LLLVDTPETKHPKKGVEKYGPEASAFTK*$_{36-63}$	3086.0	3086.0	3085.9	3085.0
[2c] *KMVENAK*$_{64-70}$				
[2d] *KIEVEFDKGQR*$_{71-81}$	1349.7	1349.8	1348.9	1349.3
[2cd] *KMVENAKKIEVEFDKGQR*$_{64-81}$	2150.8	2151.1	2151.3	2150.6

[a] Mass of the pronated molecule (M+H)$^+$ unless otherwise indicated.
[b] Not observed.
[c] Calculated mass for the molecular ion species: (M+Na)$^+$.
[d] Calculated mass for the molecular ion species: (M+2Na−H)$^+$.
Source: Reprinted with permission from Reference 67.

enzymatic digests of the wild-type peptide, an authentic glycine insertion peptide and the putative alanine insertion peptide.[68] With this approach, a mass shift of 57.1 or 71.1 Da should be measurable within 1 Da for the peptide fragment containing the insertion, while all other peptide fragments should have the same mass as the wild type.

Endoproteinase Arg-C digestion was carried out on all three peptides. The peptide fragments (and their calculated molecular ion masses) expected from cleavage at arginine residues are shown in Table 3. As shown, the glycine or alanine insertion is residue 127 on the N-terminus of peptide fragment [6]. Figure 20 shows the LD mass spectrum of the Endo

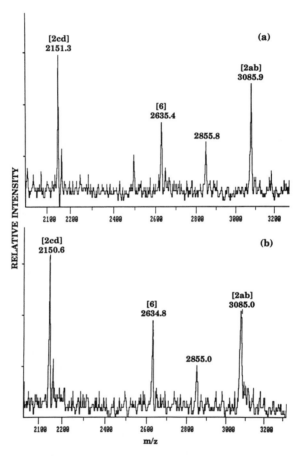

Figure 21. High-mass region of the MALD MS spectrum of the Endoproteinase Arg-C digests of (a) authentic glycine insertion peptide and (b) putative alanine insertion peptide. (Reprinted with permission from Reference 68.)

Arg-C digest of wild-type *Staph. nuclease.* Only peptides [3] and [6] are observed intact; other peaks in the mass spectrum can be rationalized as additional cleavages at locations where there are two adjacent lysine residues. There is one such site in peptide [1] and three sites in peptide [2]; sequences and calculated molecular ion masses for these additional peptide fragments are given in Table 4.

The protonated molecular ion for peptide [6] for the wild-type is observed at m/z 2577.6 and corresponds to the calculated molecular ion within 0.5 Da. Figures 21a and 21b show the upper mass regions of the LD mass spectra for the Endo Arg-C digests of the authentic glycine and putative alanine insertion peptides, respectively. Peptide fragment [6] from the authentic glycine insertion peptide gives a protonated molecular ion at m/z 2635.4 (0.2 Da greater than the calculated molecular ion for a glycine insertion). Peptide fragment [6] from the putative alanine insertion peptide gives a similar ion at m/z 2634.8, which is 0.4 Da less than the calculated value for a glycine insertion. Thus, the putative alanine insertion mutant is, indeed, a glycine mutant.[68]

8. DEVELOPING A TANDEM TIME-OF-FLIGHT MASS SPECTROMETER

Our approach has been to use low-resolution molecular ion mass measurements on TOF mass spectrometers, and appropriate enzymatic digestion, as an alternative to the fragmentation observed in FAB MS and FAB MS/MS for probing peptide/protein structure and processing. The primary advantage of this approach is that it is generally more sensitive, i.e., requires smaller amounts of sample. This results from the fact that the TOF mass analyzer records all of the ions over the entire mass range simultaneously in contrast to methods that scan a single mass at a time. At the same time, there are significant limitations to this approach. While it is possible to verify the identity of an enzymatic fragment using carboxy- or aminopeptidase digestion of the first few terminal residues, tandem techniques often reveal the entire amino acid sequence, which is an obvious advantage when the amino acid sequence is not known.

It has been one of our goals to develop a tandem *TOF/TOF* mass spectrometer combining the simultaneous recording advantages of the TOF mass analyzer with sequencing capability obtained from CID of mass-selected molecular ions. Tandem instruments using the TOF mass analyzer have been described previously; for example, Stults *et al.*[69] reported a combination magnetic-sector/TOF instrument (BTOF), while Strobel *et al.*[70] have more recently developed the double-focusing analog

(EBTOF). However, in both hybrid instruments, the first mass analyzer (MS1) is a scanning instrument, so that mass selection in the first analyzer precludes simultaneous observation of all molecular ions. In the instrument that we envision, MS1 would provide a map of the molecular ions of all fragments from an enzymatic digest, each of which would be gated through a collision chamber on successive TOF cycles to produce their sequence spectra in MS2. This would be accomplished using TOF analyzers for both MS1 and MS2.

Approaches to MS/MS using TOF analyzers alone have also been reported. Most notable is the *correlated reflex spectrum* method developed by Della-Negra and LeBeyec.[71] In their instrument, ions are produced by plasma desorption and mass analyzed on a reflectron TOF mass spectrometer. In each TOF cycle, the neutral species resulting from unimolecular dissociation of the molecular ion are recorded by a detector placed behind the reflectron. The arrival time of the neutral species is the same as that expected from the molecular ion. This time is then correlated with the arrival time of the reflected fragment ion; i.e., reflected ion spectra are obtained only when neutrals are detected within a preselected time window. There are two limitations to this approach: First, it is suitable for ionization techniques (such as plasma desorption) that produce only one or a few ions per TOF cycle; that is, correlation between a neutral and fragment ion arising from a single decomposition cannot be made using techniques that produce an entire mass spectrum as an analog signal (as for example, in LD) in each TOF cycle. The second limitation arises from the fact that when used in an MS/MS mode, a single reflecting TOF analyzer, is analogous to using a double-focusing sector instrument in a linked B/E mode; that is, the energy correction used to achieve high-mass resolution is compromised when the reflectron or electrostatic energy analyzer is (in effect) used as a second mass analyzer. Thus, to carry the analogy further, a high-performance TOF analog to the four-sector (*EBEB*) sector mass spectrometer should consist of two complete reflecting TOF analyzers.

At the same time, Standing *et al.*[72] have made some significant advances in the LeBeyec method, which they adapted for MALD using beam irradiances that promote the desorption of single ions.[73]

While the *correlated reflex spectrum* method relies primarily on unimolecular dissociations in the flight tube, Schey *et al.*[74] have reported a true tandem TOF flight instrument by using two linear mass analyzers in conjunction with surface-induced dissociation (SID). More recently, Jardine *et al.*[75] reported tandem TOF spectra using two linear (nonreflecting) mass analyzers.

However, in the instrument we envision, both mass analyzers would be of the reflectron type, so that high-mass resolution is achieved in both the molecular ion and product ion mass spectra. Instruments of this kind have been described by Bergmann *et al.*[76] using pulsed valves and laser-induced multiphoton ionization (MPI) of gaseous, high-molecular-weight, neutral clusters produced continuously, but these have not been described for the direct desorption of intact molecular ions from small quantities of peptides and proteins. As an introduction to our approach for unit mass resolution of peptides on a TOF mass spectrometer, we first review the basic issues effecting mass resolution.

8.1. Mass Resolution

In a TOF mass spectrometer, an ion of mass m is accelerated in the source region to a final kinetic energy

$$\frac{mv^2}{2} = eV \tag{4}$$

where V is the accelerating voltage. If the ionization/accelerating region is short, then TOF t is a simple inverse function of the velocities of ions of different mass as they traverse a flight tube of length D, leading to the TOF equation:

$$m = \frac{2eVt^2}{D^2} \tag{5}$$

Generally, ions are desorbed with a distribution of initial kinetic energies U_o prior to acceleration, so that the final energy of an ion of mass m is in fact

$$\frac{mv^2}{2} = eV + U_o \tag{6}$$

Deriving the TOF equation for ions with a distribution of initial kinetic energies[47] leads to an approximation of the mass resolution

$$\frac{m}{\Delta m} = \frac{eV}{U_o} \tag{7}$$

if the kinetic energy distribution is the major factor defining mass resolution. Thus, TOF mass spectrometers generally incorporate high-accelerating voltages (up to 30 kV) to improve mass resolution. Alternatively, reflectrons[48] are used to compensate for the initial kinetic energy distribution. Note that differentiating Equation 5 leads to the result

$$\Delta m = \left(\frac{2eV}{D^2}\right) 2\Delta t \qquad (8)$$

so that mass resolution for a TOF mass spectrometer is simply defined as

$$\frac{m}{\Delta m} = \frac{t}{2\Delta t} \qquad (9)$$

Thus, mass resolution can be improved if flight times are long. Since high-accelerating voltages (used to minimize the effects of kinetic energy distribution) decrease flight times, it is common to use long-drift regions, often several meters long. An extraordinary example of these principles was reported by Grix *et al.*[77] using a 4-m flight tube with a reflectron and pulsed ion extraction. They observed a molecular ion peak for triazine (m/z 866) at a flight time of 793 μsec and a peak width of 26 nsec, corresponding to a mass resolution in excess of 15,000!

On the other hand, if efforts can be made to decrease the time width Δt of ion packets extracted from the ion source, shorter drift tubes can be used to achieve reasonable mass resolutions. Furthermore, if Δt is constant for ions of all masses, then Equation 7 predicts that resolution increases with mass! Thus, our current efforts are aimed at producing TOF and TOF/TOF configurations that are compact, inexpensive, and capable of maintaining unit mass resolution at about 3–4 kDa. While such mass resolutions are considerably less than described by Grix *et al.*,[77] they are in fact comparable to mass resolutions normally used in far more expensive sector instrument when sample quantities are low and sensitivity is of major concern. This is a practical goal for the structural analysis of peptide fragments generated by enzymatic digestion of proteins, which generally fall in this mass range.

8.2. The DESKTOP Tandem Time-of-Flight Mass Spectrometer

A diagram of the compact tandem TOF mass spectrometer currently under development in our laboratory[78] is shown in Figure 22. The two reflecting flight tubes are 26-in. long, each electrically isolated from ground, and connected via a collision chamber. In the experiments carried out thus far, only MS1 has been used; flight tubes are at ground potential, with the sample placed on a probe surfase held at 2 kV. Samples are desorbed/ionized by the same 1.5-mJ, 600-psec pulsed nitrogen laser used in the

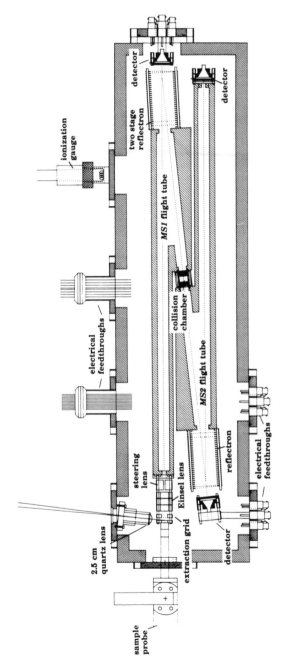

Figure 22. Diagram of the Desktop tandem (TOF/TOF) TOF mass spectrometer. (Reprinted with permission from Reference 78.)

instrument described in Section 2.2. In the nonreflecting mode, the first two-stage reflectron grids and lenses are also at ground potential, and ions are detected in the first detector located behind the reflectron. In the reflecting mode, the grid separating the two stages of the reflectron is at 1 kV, while the final grid is at 2.2 kV; ions are recorded on the detector located behind the second reflectron. In both cases, analog signals following each laser pulse are digitized on a LeCroy (Spring Valley, NY) model 9450, 400 Msamples/s digital oscilloscope.

8.3. Results of the Linear Portion of MS1

Figure 23 is a LD mass spectrum of tryptophan (MW 204.1), which is a strong UV absorber. Molecular ions are not observed; however, fragment ions and adducts that are observed up to m/z 295 illustrate mass resolution; in Figure 24, these same data are expanded for different regions of the mass spectrum. For the most part, peak widths are of the order of 5–6 nsec (considerably less than that reported by Grix *et al.*[77]), so that mass resolution increases with mass, from $m/\Delta m = 642$ for the sodium ion at m/z 23, to $m/\Delta m = 1528$ for the peak at m/z 295.3. The fact that peak widths are constant with mass suggests (in this experiment) that the initial kinetic energy distribution is not the limiting factor in mass resolution and results in excellent resolution without a reflectron.

8.4. Results of Reflected Ions in MS1

Figure 25 shows the LD mass spectrum of tryptophan after reflection; in Figure 25b, the region from m/z 130 to 146 has been expanded, reveal-

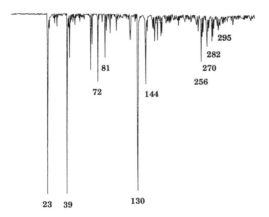

Figure 23. The LD mass spectrum of tryptophan, recorded for the linear portion of MS1. (Reprinted with permission from Reference 78.)

ing a mass resolution of $m/\Delta m = 2100$ for the peak at m/z 130. Because the initial energy distribution is not the limiting factor defining resolution, improved mass resolution for the reflected spectrum arises primarily from an increase in path length.

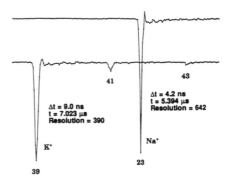

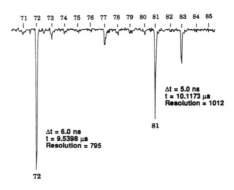

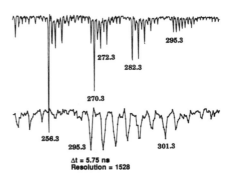

Figure 24. Expansion of the LDMS spectrum of tryptophan. (Reprinted with permission from Reference 78.)

a

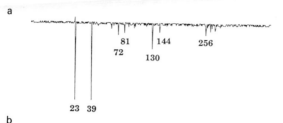

b

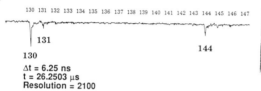

Δt = 6.25 ns
t = 26.2503 μs
Resolution = 2100

Figure 25. The LD mass spectrum of tryptophan (a) recorded after reflection in MS1 and (b) expansion of the mass region from m/z 130 to 146. (Reprinted with permission from Reference 78.)

9. CONCLUSIONS

It is likely that simple, inexpensive instruments will be developed for the structural analysis of proteins and peptides in much the same way that desktop gas chromatography mass spectrometer has been developed for carbohydrate and fatty acid analysis, drug testing, or identifying environmental pollutants. While neither TOF mass spectrometers or desktop GCMS possess the ultimate versatility of tandem, four-sector instruments, they provide a competitive and inexpensive approach to specific analytical problems to a larger number of research laboratories.

When combined with MALD, TOF mass spectrometers provide the highest molecular weight capability available; they also provide the highest possible sensitivity if we consider the ability to recover the sample from the mass spectrometer and carry out further enzymatic reactions on the same solid support, as previously described for PD. For very small quantities of proteins and peptides, sample transfer to and from solutions becomes a major problem. In addition, the absence of sequence-specific fragmentation in TOF mass spectrometers can be effectively overcome by the judicious use of exo- and endoproteinases and chemical cleavages.

The major impediment to universal acceptance of the TOF analyzer has been its low-mass resolution. While mass resolutions comparable to those from double-focusing sector instruments have (on rare occasions) been demonstrated,[77, 79] this has generally been accomplished on instruments whose size and cost are comparable to that of high-performance-sector instruments but lack the equivalent sensitivity, mass range, or versatility. Thus, we intend to solve the problem by developing (non extraordinary) compact and inexpensive unit mass resolution TOF instruments.

We note that peak widths for molecular ions of peptides and proteins reported in the literature for LD mass spectra are generally of the order of 30–40 nsec. These values are generally recorded (digitized) by transient recorders capable of digitizing analog signals at the rate of 100 Msamples/sec, i.e., a time resolution of 10 nsec. Our recording of 5–6 nsec peaks was carried out with a 400 Msamples/sec (2.5-nsec resolution) recorder, while Gsample/sec recorders are now commercially available with time resolutions of 1 nsec. At this point, both the length of the laser pulse and the detector response become major limitations in decreasing the peak width Δt. Thus, we have used a 600-psec laser and trimmed our detector response (using impedance-matched feedthroughs) to 0.89 nsec[78] Such considerations become increasingly important if time (and mass) resolution are to be achieved by decreasing peak widths rather than extending the size of the instrument.

At the same time, existing, commercially available TOF instruments (both PDMS and laser desorption mass spectrometry (LDMS) can be employed to solve analytical problems in protein structure and processing using methods discussed in Chapter 6. The approaches we have described use relatively simple, inexpensive instruments and enzymatic reactions familiar to the protein chemist.

ACKNOWLEDGEMENTS

Our analytical methods using TOF mass spectrometers were developed in collaboration with a number of other investigators. Thus, we acknowledge C. Schwabe and E. Büllesbach (Medical University of South Carolina); L. R. Pohl and Y. Osawa (NHBLI, NIH, Bethesda); S. Sisodia and J. Meschia (neuropathology laboratory, Johns Hopkins); W. Gibson, L. McNally, and A. Welch (pharmacology department, Johns Hopkins); and E. Lattman and L. Keefe (biophysics department, Johns Hopkins) for the opportunities they provided. Research was supported by grants (DIR 89-14549 and DIR 90-16567) from the National Science Foundation and (GM 33967) from the National Institutes of Health. Mass spectral analyses were carried out at the Middle Atlantic Mass Spectrometry Laboratory, an NSF-supported shared instrumentation facility.

REFERENCES

1. Barber, M.; Bordoli, R. S.; Sedgwick, R. D.; and Tyler, A. N., *J. Chem. Soc. Chem. Commun.* **1981**, 325.
2. Hill, J. A.; Martin, S. A.; Biller, J. E.; and Biemann, K., *Biomed. Environ. Mass Spectrom.* 1988, **17**, 147.

3. Macfarlane, R. D.; Skowronski, R. P.; and Torgerson, D. F., *Biochem. Biophys. Res. Commun.* 1974, **60**, 616.

4. Chait, B., *Int. J. Mass Spectrom. Ion Processes*, 1983, **53**, 227.

5. Demirev, P.; Olthoff, J. K.; Fenselau, C.; and Cotter, R. J., *Anal. Chem.* 1987, **59**, 1951.

6. Craig, A. G.; Engstrom, A.; Bennich, H.; and Kamensky, I., *Proc. Thirty-Fifth ASMS Conf. Mass Spectrom. Allied Topics,* (Denver, 1987), pp. 528–29.

7. Posthumus, M. A.; Kistemaker, P. G.; Meuselaar, H. L. C.; and Ten Noever de Brauw, M. C., *Anal. Chem.* 1978, **50**, 985.

8. Graham, S. W.; Dowd, P.; and Hercules, D. M., *Anal. Chem.* 1982, **54**, 649.

9. Van Breemen, R. B.; Snow, M.; and Cotter, R. J., *Int. J. Mass Spectrom. Ion and Phys.* 1883, **49**, 35.

10. McCreary, D. A.; Ledford, E. B., Jr.; and Gross, M. L., *Anal. Chem.* 1982, **54**, 1437.

11. Cotter, R. J.; Honovich, J. P.; Olthoff, J. K.; and Lattimer, R. D., *Macromolec.* 1986, **19**, 2996.

12. Brown, R. S.; Weil, D. A.; and Wilkins, C. L., *Macromolec.* 1986, **19**, 1255.

13. Coates, M. L.; Wilkins, C. L., *Biomed. Mass Spectrom.* 1985, **12**, 424.

14. Martin, W. B.; Silly, L.; Murphy, C. M.; Raley, T. J., Jr.; Cotter, R. J.; and Bean, M. F., *Int. J. Mass Spectrom. Ion Processes* 1989, **92**, 243.

15. Takayama, K.; Qureshi, N.; Hyver, K.; Honovich, J.; Cotter, R. J.; Mascagni, P.; and Schneider, H., *J. Biol. Chem.* 1986, **261**, 10624.

16. Cotter, R. J.; Honovich, J. P.; Qureshi, N.; and Takayama, K., *Biomed. Environ. Mass Spectrom* 1987, **14**, 591.

17. Karas, M.; and Hillenkamp, F., *Anal. Chem.* 1988, **60**, 2299.

18. Karas, M.; Ingendoh, A.; Bahr, U.; and Hillenkamp, F., *Biomed. Environ. Mass Spectrom.* 1989, **18**, 841.

19. Whitehouse, C. M.; Dreyer, R. N.; Yamashita, M.; and Fenn, J. B., *Anal. Chem.* 1985, **57**, 675.

20. Covey, T. R.; Bonner, R. F.; Shushan, B. I.; and Henion, J., *Rapid Comm. Mass Spectrom.* 1988, **2**, 249.

21. Smith, R. D.; Berinaga, C. J.; and Udseth, H. R., *Anal. Chem.* 1988, **60**, 1948.

22. VanBerkel, G. J.; McLuckey, S. A.; and Glish, G. L., *Anal. Chem.* 1991, **11**, 1098.

23. Larsen, B. S.; and McEwen, C. N., *J. Am. Soc. Mass Spectrom.* 1991, **3**, 205.

24. Sundqvist, B.; and Macfarlane, R. D., *Mass Spectrom. Rev.* 1985, **4**, 421.

25. McNeal, C. J.; Macfarlane, R. D.; and Thurston, E. L., *Anal. Chem.* 1979, **51**, 2036.

26. Alai, M.; Demirev, P.; Fenselau, C.; and Cotter, R. J., *Anal. Chem.* 1986, **58**, 1303.

27. Saxena, P.; and Wetlaufer, D. B., *Biochem.* 1970, **9**, 5015.

28. Jonsson, G.; Hedin, A.; Håkansson, P.; Sundqvist, B. U. R.; Sawe, G.; Nielsen, P. F.; Roepstorff, P.; Johansson, K. E.; Kamensky, I.; and Lindberg, M., *Anal. Chem.* 1986, **58**, 1084.

29. Chen, L.; Cotter, R. J.; and Stults, J. T., *Anal. Biochem.* 1989, **183**, 190.

30. Wang, R.; Chen, L.; and Cotter, R. J., *Anal. Chem.* 1990, **62**, 1700.

31. Macfarlane, R. D., *Accts. Chem. Res.* 1982, **15**, 268.

32. Tanaka, K.; Waki, H.; Ido, Y.; Akita, S.; Yoshida, Y.; and Yoshida, T., *Rapid Comm. Mass Spectrom.* 1988, **2**, 151.

33. Beavis, R. C.; and Chait, B. T., *Rapid Comm. Mass Spectrom.* 1989, **3**, 233.

34. Spengler, B.; and Cotter, R. J., *Anal. Chem.* 1990, **62**, 793.

35. Spengler, B.; Pan, Y.; Cotter, R. J.; and Kan, L.-S., *Rapid Comm. Mass Spectrom.* 1990, **4**, 99.

36. Salehpour, M.; Perera, I.; Kjellberg, J.; Hedin, A.; Islamian, M. A.; Håkansson, P.; and Sundqvist, B. U. R., *Rapid Comm. Mass Spectrom.* 1989, **3**, 259.

37. Perera, I. K.; Uzcategui, E.; Håkansson, P.; Brinkmalm, G.; Petterson, G.; Johansson, G.; and Sundqvist, B. U. R., *Rapid Comm. Mass Spectrom.* 1990, **4**, 285.
38. Beavis, R. C.; and Chait, B. T., *Rapid Comm. Mass Spectrom.* 1989, **3**, 436.
39. Beavis, R. C.; and Chait, B. T., *Rapid Comm. Mass Spectrom.* 1989, **3**, 432.
40. Nelson, R. W.; Thomas, R. M., and Williams, P., *Rapid Comm. Mass Spectrom.*, 1990, **4**, 348–51.
41. Overberg, A.; Karas, M.; Bahr, U.; Kaufmann, R.; and Hillenkamp, F., *Rapid Comm. Mass Spectrom.* 1990, **4**, 293.
42. Overberg, A.; Karas, M.; and Hillenkamp, F., *Rapid Comm. Mass Spectrom.* 1991, **5**, 128.
43. Chevrier, M. R.; and Cotter, R. J., *Rapid Comm. Mass Spectrom.*, in press.
44. Yergey, J. A., *Int. J. Mass Spectrom. and Phys.* 1983, **52**, 337.
45. Beavis, R. C.; and Chait, B. T., *Chem. Phys. Lett.* 1991, **5**, 479.
46. Pan, Y.; and Cotter, R. J., *Org. Mass Spectrom.*, in press.
47. Cotter, R. J., *Biomed. Environ. Mass Spectrom.* 1989, **18**, 513.
48. Mamyrin, B. A.; Karataev, V. I.; Shmikk, D. V.; and Zagulin, V. A., *Sovjet Phys. JETP* 1973, **37**, 45.
49. Bedekar, S.; Turnell, W. G.; Blundell, T. L.; and Schwabe, C., *NATRA* 1977, **270**, 449.
50. Woods, A. S.; Cotter, R. J.; Yoshioka, M.; Büllesbach, E.; and Schwabe, C., *Int. J. Mass Spectrom. Ion Processes*, in press.
51. Chait, B. T.; Chaudhary, T.; and Field, F. H., Mass Spectral Characterization of Microscale Enzyme Catalyzed Reactions of Surface-Bound Peptides and Proteins, in *Methods in Protein Sequence Analysis*, Walsh, K. A., ed. (Humana Press: Clifton NJ, 1987), pp. 483–92.
52. Schwabe, C.; and Büllesbach, E., private communication.
53. Davies, H. W.; Satoh, H.; Schulick, R. D.; and Pohl, L. R., *Biochem. Pharmacol.* 1985, **34**, 3203.
54. Levin, W.; Jacobson, M.; and Kuntzman, R., *Arch. Biochem. Biophys.* 1972, **148**, 262.
55. Osawa, Y.; Highet, R. J.; Murphy, C. M.; Cotter, R. J.; and Pohl, L. R., *J. Am. Chem. Soc.* 1989, **111**, 4462.
56. Osawa, Y.; Martin, B. M.; Griffin, P. R.; Yates III, J. R.; Shabanowitz, J.; Hunt, D. F.; Murphy, A.; Chen, L.; Cotter, R. J.; and Pohl, L. R., *J. Biol. Chem.* 1990, **265**, 10340.
57. Osawa, Y.; and Pohl, L. R., private communication.
58. Kindt, J. T.; Woods, A.; Martin, B. M.; Cotter, R. J.; and Osawa, Y., submitted.
59. Sisodia, S. S.; Koo, E. H.; Beyreuther, K.; Unterbeck, A.; and Price, D. L., *Science* 1990, **248**, 492.
60. Wang, R.; Meschia, J.; Cotter, R. J.; and Sisodia, S. S., *J. Biol. Chem.* 1991, **25**, 16960.
61. Gibson, W., *Virol.* 1980, **111**, 516.
62. Welch, A. R.; McNally, L. M.; and Gibson, W., *J. Virol.* 1991, **65**, 4091.
63. Gibson, W.; Marcy, A. I.; Comolli, J. C.; and Lee, J.., *J. Virol.* 1990, **64**, 1241.
64. Welch, A. R.; Woods, A. S.; McNally, L. M.; Cotter, R. J.; and Gibson, W., *Proc. Nat. Acad., Sci.*, in press.
65. McNally, L. M.; and Gibson, W., unpublished data.
66. Sondek, J.; and Shortle, D., *Proteins: Struct. Funct. Genet.* 1990, **7**, 299.
67. Keefe, L. J.; and Lattman, E., in preparation.
68. Keefe, L. J.; Lattman, E. E.; Wolkow, C.; Woods, A.; Chevrier, M.; and Cotter, R. J., *J. Appl. Crystallogr.*, in press.
69. Stults, J. T.; Enke, C. G.; and Holland, J. F., *Anal. Chem.* 1983, **55**, 1323.
70. Strobel, F. H.; Solouki, T.; White, M. A.; and Russell, D. H., *J. Am. Soc. Mass Spectrom.* 1991, **1**, 1.

71. Della-Negra, S.; and LeBeyec, Y., in *Ion Formation from Organic Solids IFOS III,* Benninghoven, A., ed. (Springer-Verlag: Berlin, 1986), pp. 42–45.
72. Standing, K. G.; Beavis, R.; Bollbach, G.; Ens, W.; Lafortune, F.; Main, D.; Schueler, B.; Tang, X.; and Westmore, J. B., *Anal. Instrum.* 1987, **16**, 1987.
73. Ens, W.; Mao, Y.; Mayer, F.; and Standing, K. G., *Rapid Comm. Mass Spectrom.* 1991, **5**, 117.
74. Schey, K.; Cooks, R. G.; Grix, R.; and Wollnick, H., *Int. J. Mass Spectrom. Ion Processes* 1987, **77**, 49.
75. Jardine, D. R.; Alderdic, D. S.; and Derrick, P. J., *Org. Mass Spectrom.* 1991, **26**, 215.
76. Bergmann, T.; Martin, T. P.; and Schaber, H.; *Rev. Sci. Instr.* 1989, **60**, 792.
77. Grix, R.; Kutscher, R.; Li, G.; Gruner, U.; and Wollnik, H., *Rapid Comm. Mass Spectrom.* 1988, **2**, 83.
78. Cornish, T.; and Cotter, R. J., submitted.
79. Grotemeyer, J.; and Schlag, E. W., *Org. Mass Spectrom.* 1987, **22**, 758.

7

Electrospray Ionization Mass Spectrometry and Tandem Mass Spectrometry of Large Biomolecules

Joseph A. Loo, Charles G. Edmonds, Rachel R. Ogorzalek Loo, Harold R. Udseth, and Richard D. Smith

1. INTRODUCTION

A significant research effort has been expended since the 1950s to develop mass spectrometry (MS) as a useful tool for applications in the biological sciences. Many biological materials are involatile and thermally unstable. As new ionization methods have been developed to propel and vaporize larger compounds into the gas phase without significant sample degradation, a parallel effort has gone into developing mass analyzers and detectors to accomodate the larger molecular species. From amino acids to peptides and finally to proteins, ionization methods such as field desorption (FD),[1] plasma desorption (PD),[2,3] and fast atom bombardment (FAB)[4,5] have enabled mass spectrometers to detect and determine molecular weight for biomolecules to ~40 kDa.

Joseph A. Loo, Charles G. Edmonds, Rachel R. Ogorzalek Loo, Harold R. Udseth, and Richard D. Smith • Chemical Methods and Separations Group, Chemical Sciences Department, Pacific Northwest Laboratory, Richland, Washington 99352. *Present address for J. A. L.:* Parke-Davis Pharmaceutical Research, Ann Arbor, Michigan 48105. *Present address for R. O. L.:* Department of Biological Chemistry, University of Michigan, Ann Arbor, Michigan 48109.

Experimental Mass Spectrometry, edited by David H. Russell. Plenum Press, New York, 1994.

The sensitive detection and accurate relative molecular mass (M_r) determination of contaminant proteins differing by as little as one amino acid residue is an analytical challenge that has necessitated the development of new approaches. Determining protein purity is important,[6] for example, in establishing product safety and effectiveness in the biotechnology industry, since trace impurities may have an important impact on product quality and regulatory certification. Recently, two new techniques for the desorption/ionization of even larger biomolecules have emerged. Infrared and ultraviolet (UV) laser desorption (LD), incorporating the analyte in an organic matrix, was pioneered by Karas et al.[7, 8] who have demonstrated the capability for ion formation for biomolecules greater than 200 kDa. Electrospray ionization (ESI),[9–16] which forms multiply charged gas-phase ions from highly charged liquid droplets at atmospheric pressure, was resurrected after a period of dormancy by Fenn et al.[11–14] as a viable analytical method for compounds of biochemical interest. The ESI-generated protein ions of over 100 kDa relative molecular mass include β-galactosidase (118 kDa),[17] bovine albumin dimer (133 kDa),[18, 19] IgG-class monoclonal antibodies (150 kDa), human α_2-macroglobulin subunit (186 kDa), and human complement C4 (197 kDa).[20]

The specificity and activity of proteins depend on whether they possess both the correct amino acid sequence (primary structure) along the polypeptide chain and whether the chain is folded to a precise three-dimensional structure.[21] Mass spectrometry offers the advantages of speed and sensitivity in protein primary structure determination and also in determining structures introduced into proteins, by posttranslational modifications, such as glycosylation, methylation, phosphorylation, and disulfide bonding, structural features that DNA sequences alone cannot reveal.[22, 23] More recent studies show that noncovalent associations, and the higher order structure of large molecules in solution, may also be amenable to study. The combination of tandem (MS/MS),[24, 25] in conjunction with an arsenal of selective proteolyses, is a powerful method for amino acid sequence elucidation for small polypeptides and proteins (for example, the elegant studies of thioredoxin, $M_r \sim 11,500$).[26, 27] The primary structure of proteins as large as bovine serum albumin ($M_r \sim 66,000$) can be deduced in this manner;[16, 28, 29] in fact, the 583 amino acid sequence for bovine albumin has recently been revised,[30] based on three residue corrections determined by MS and MS/MS.

Polypeptides with an M_r up to 3000 (and in some cases 4000, depending on the details of molecular structure and instrumentation) can be addressed by collisionally activated dissociation (CAD) with conventional ionization methods that produce primarily single charged species.[22, 23, 31–33] The principal limitations of these MS/MS methods arise from the combined effects of the decreasing efficiency with which large M_r molecular ions are formed and the lower efficiency of dissociating their singly charged ions.[34, 35] Photodissociation,[36, 37] surface-induced dissociation,[38] and the use of ion-trapping techniques,[39] which allow very large numbers of collisions, may offer promising alternatives.

It is conceivable that the CAD of large multiply charged molecules by ESI can be exploited to obtain protein sequence information from *intact* species more rapidly and sensitively than conventional techniques. In contrast to the conventional time-of-flight mass analyzers employed with LD, ESI used with quadrupole MS analyzers is well-suited to MS/MS experiments.[16, 28, 29, 40–42] In addition, CAD efficiencies of multiply charged ions can be very high due to the deposition of internal energy (i.e., "preheating") in the ESI atmosphere/vacuum interface region. These efficiencies may be enhanced by coulombic repulsive forces, allowing highly charged molecules to be more susceptible to fragmentation.

Chapter 7 discusses the physical aspects and the practice of ESI MS, including experimental techniques and their applications. Examples presented include polypeptides, intact protein, and oligonucleotides. Primary sequence information from CAD and MS/MS of ESI-generated ions are demonstrated for peptides and proteins with M_r 2000–66,000.

2. PRINCIPLES AND PRACTICE OF ELECTROSPRAY IONIZATION

2.1. Electrospray Ionization Process

The process of electrospray ionization occurs by nebulizing liquids in a large electrostatic field to yield highly charged droplets. These highly charged droplets, generally formed in a dry-bath gas at near atmospheric pressure, shrink in size by evaporating the neutral solvent until the charge repulsion overcomes the cohesive droplet forces, leading to a coulombic explosion. In the ion evaporation mechanism of Iribarne and Thomson,[43, 44] the evaporation process continues until the droplet surface curvature is sufficiently high to permit field-assisted evaporation of charged solutes. Alternatively, Röllgen et al.[45–47] argue that the actual field strength required for ion evaporation from even the smallest aqueous droplets almost always exceeds that for Rayleigh fission. Thus, an ion evaporation process yielding a single ion is unlikely because it is thought that such an event would almost certainly induce a Rayleigh explosion, causing a jet of very small charged droplets to be emitted. Additionally, droplet disintegration for ESI may be augmented by a nonuniform charge distribution (as likely for large multiply charged ions in a low-conductivity medium).[47] Such processes are well established for much larger droplets, but they have not been directly examined for the submicron diameter droplets of relevance to ESI due to the experimental challenges involved. The actual distinction between these two mechanisms may become blurred and a process yielding jets of very small droplets by Rayleigh fission may be mechanistically indistinguishable for smaller molecules from a process involving the field evaporation of highly solvated ions. Furthermore, molecular species exceeding 10^6 Da have recently been shown to be amenable to ESI by Nohmi and Fenn[48] and our laboratory. Thus, it becomes very difficult to imagine that an ion evaporation process where a molecule (whose extended length substantially exceeds the droplet diameter) carrying perhaps several thousand charges suddenly "lifts off" from a droplet surface. Indeed, in the large molecule limit, a single molecule could account for nearly all the droplet charge, and ion formation would simply involve evaporation of the residual solvent, as originally envisioned by Dole et al.[9] and Mack et al.[49]

Details of the ion formation mechanism await further experimental study. However, it is clear that multiply charged molecules (with an unknown initial extent of solvation) can be gently produced from solution by ESI. These ions generally arise from attachment of protons, alkali cations or ammonium ions for positive ion formation or, from polarity reversal of the nebulizing field, negative ions are formed by proton or other cation abstraction. The wide range of solutes already found to be amenable to ESIMS as well as the lack of substantial dependence of ionization efficiency on molecular weight suggest a highly nondiscriminating ionization process.

ESI was originally described by Dole et al.[9, 49, 50] in studies based on gaseous ion mobility measurements of intact ions from synthetic and natural polymers of molecular weights in excess of 100,000 Da. It is impressive to note that these researchers, employing a "plasma chromatograph," tentatively (and, for the most part, correctly) interpreted experiments on proteins as demonstrating a multiple-charging phenomenon. After a 10-year hiatus, these experiments were extended by Fenn et al.[11, 51, 52] employing atmospheric pressure sampling of ions and analysis with a quadrupole mass spectrometer, and essentially simultaneously, similar experiments were reported by Russian researchers[53, 54] using a magnetic sector instrument. These researchers outlined the fundamental aspects of ESI, demonstrating its ultimate utility in analyzing biomolecules and as a potential interface for the combination of liquid chromatography with MS. They demonstrated that the production of molecular ions bearing multiple charges provides access to higher molecular weights by extending the mass range for m/z limited mass spectrometers. The multiple-charging phenomenon has been demonstrated to apply to molecules of over 100 kDa in molecular weight measurement[17-20] employing quadrupole mass analysis of conventional mass-to-charge range, and it has permitted the measurement of relative mass with a precision of better than 0.002%.[10, 16, 55-58]

2.2. Atmospheric Pressure ESI Source

The design of the ESI source can be as simple as a metal capillary (e.g., hypodermic needle). Such an arrangement with a cylindrical counter electrode is the basis of source arrangements described by Whitehouse et al.[11] A 1–20 µl/min flow of analyte solution, typically as water-methanol mixtures and often with other additives such as acetic acid, is delivered to the capillary terminus from infusion syringes or liquid chromatographic columns. Alternatively, the ESI process can be accompanied by pneumatic nebulization accomplished by a high-velocity annular flow of gas at the liquid exit of the injection capillary, a process referred to as "ionspray".[10, 59] This method has the advantage of accomodating flow rates up to approximately 100 µL/min, making the method attractive for liquid chromatography (usually with eluent stream splitting). Ultrasonic nebulization as a means to assist the electrospray ionization process has been developed for high-flow LC/ESI MS applications.[60] Aerosol generation by an ultrasonic nebulizer allows the production of small diameter droplets where flow rates in excess of 200 µL/min are amenable for ESI MS experiments. The relative insensitivity of these methods to liquid solvent composition also allows compatibility with gradient elution methods.

A modified source developed at our laboratory for ESI MS and combined capillary electrophoresis MS[16, 61] is shown in Figure 1. An organic liquid sheath (typically pure methanol or acetonitrile, which may be augmented by small proportions of acetic acid or water) flows in the annular space between a 100-µm i.d.-fused silica capillary, that delivers the analyte solution to the ion source, and a cylindrical stainless steel electrode (400–500 µm i.d.). For positive ion ESI, a voltage of +4–6 kV is applied to this surrounding metal electrode. For experiments in which the sample solution is directly infused to the ESI source, syringe pumps deliver controlled flows of analyte and sheath liquids at rates of 0.1–1 µl/min and 2–4 µl/min, respectively. Stability of the ESI source depends critically on the stability of these flows. Our laboratory typically uses microliter

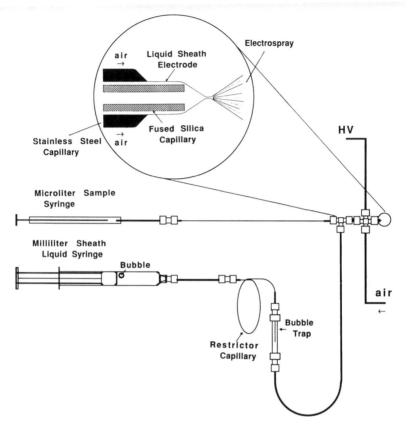

Figure 1. Schematic of the sheath flow interface for ESI MS showing the conventional arrangement for direct infusion of analyte solutions.

syringes installed in a syringe infusion pump to deliver analyte solution through 30–70 cm lengths of fused silica capillary. The electrical isolation provided by this length of fused silica is adequate for electrospray voltages over ± 6 kV. The sheath flow is further stabilized by using a fused silica flow restrictor (20 cm × 50 μm). Problems due to bubble formation in the connecting lines are minimized by including a trapping volume. Additional stability is obtained by degassing the organic solvents used in the sheath and avoiding unnecessarily high ambient temperatures. These steps are particularly useful for the combination of ESI mass spectrometry with capillary electrophoresis, as are (1) cooling the entire assembly to ~15 °C and (2) using a less volatile sheath solvent, such as 2-methoxyethanol.[62] Within this general scheme, many satisfactory arrangements for capillary tubing and ESI components are possible.

Positively or negatively charged droplets and ultimately analyte ions are produced, depending on the capillary bias voltage. In the negative ion mode, an electron scavenger, such as oxygen or SF_6,[52, 59, 63] is often helpful to inhibit electrical discharge at the capillary terminus. This is accomplished in our interface using a gas flow (250 mL/min) through a third annular passage in the electrospray ion source outside the sheath liquid. Its velocity is much lower than required for nebulization, but it serves to stabilize ESI operation and minimize the effects of air circulation around the ion source. Sample solutions of proteins and peptides are typically prepared in 1–5% v/v acetic acid for positive ion analysis and 1–3% NH_4OH for negative ion ESI.

Using electric fields for droplet nebulization (i.e., without a nebulizing gas flow and in the absence of heating) results in the optimum deposition of charge on the droplets and, at low flow rates, the most efficient production of analyte ions. However, this approach can cause practical restrictions on the range of solution conductivity and dielectric constant properties for stable operation at useful flow rates. Solution conductivities of $\leqslant 10^{-5}$ mho, corresponding to an aqueous electrolyte solution of $\leqslant 10^{-4}$ N, use the sample most efficiently at typical μL/min flow rates. Fluids with higher surface tensions require a higher threshold voltage for electrospray production (e.g., ~ 8 kV for water compared to ~ 4 kV for methanol given a 1-cm capillary-counter electrode distance), while higher dielectric liquids produce higher total ESI currents. For direct infusion experiments with typical 5% v/v aqueous acetic acid analyte solutions and a methanol sheath, total electrospray currents of 0.1–0.5 μA are typical. Solution electrolyte concentrations above $\sim 10^{-3}$ N can result in disabling instability in electrospray operations particularly at higher flow rates, unless assisted by other nebulization methods. This restriction can be reduced by using higher temperatures, a nebulizing gas, or indirectly adding a lower concentration sheath liquid as previously described.

2.3. Atmosphere/Vacuum Interface

The crucial features of an ionization source operated at atmospheric pressure involve efficiently sampling and transporting ions into the mass spectrometer at operating vacuum (e.g., $< 5 \times 10^{-5}$ Torr for a quadrupole mass filter). One method (available as a commercial instrument from Sciex, Thornhill, Ontario, CA) uses a single orifice (100–130 μm) entering directly into the high vacuum region of the mass spectrometer equipped with high-speed cryo-pumping capability. In this arrangement, positive ions produced at atmospheric pressure drift against a countercurrent flow of dry N_2 gas directed through an axial plenum toward the sampling aperture. The curtain of dry nitrogen serves to exclude large droplets and particles and aids desolvation of the ions. As ions pass through the orifice into the vacuum, further declustering is accomplished by collisions as ions are accelerated into the first radiofrequency- (rf-) only focusing quadrupole of a tandem quadrupole mass spectrometer.

Alternatively, instruments can be based on differentially pumped atmosphere/vacuum interfaces. Such a system employing a capillary inlet skimmer that has been adapted to several commercial instruments is described by Fenn *et al.*[11-14] The electrospray source, maintained at a few kilovolts with respect to the surrounding cylindrical electrode, produces an electrospray plume, which encounters a countercurrent flow of bath gas (typically N_2 at slightly above atmospheric pressure) that sweeps away uncharged material and solvent vapor from the mass spectrometer inlet. As droplets drift toward the sampling region, desorbed ions are entrained in the gas flow entering the glass capillary at the end of this chamber. Ions are swept by the gas flow through the sampling orifice, emerging as a charged component of a free-jet expansion in the first stage of differential pumping. Ions are then transmitted through a skimmer into the inlet optics of a (quadrupole) mass analyzer. The electrically insulating glass capillary inlet design provides a potential advantage by offering a wide choice of ion-accelerating voltages. This approach allows the outlet of a liquid chromatographic apparatus in a liquid chromatography (LC)/ESI MS experiment to be at ground potential. In this case, the electrode and cylinder end plate can be floated at high voltage, so that hydrodynamic flow through the glass capillary serves to deliver ions efficiently to the skimmer region. Ion mobilities at atmospheric pressure are sufficiently low so that the viscous drag can deliver ions from (or back to) ground potential across gradients of as much as 15 kV, allowing voltages appropriate for injecting ions into a magnetic sector instrument.[64]

An alternative approach to droplet desolvation relies solely on heating during droplet transport through a heated metal[56] or glass capillary.[65] A countercurrent gas flow is not essential. The electrospray source is typically more closely positioned with respect to the sampling capillary orifice (0.3–1.0 cm). Charged droplets are swept into the heated capillary, which provides effective ion desolvation, particularly when augmented by a capillary skimmer voltage gradient. The disadvantage of this approach and that of all sources not using a gas curtain or countercurrent flow is that much more solvent and residual material (i.e., solute particles) enter the mass spectrometer. Advantages of the heated capillary are however the ease with which it can be adapted to a variety of MS configurations, potentially offering higher transport efficiency due to the proximity to the electrospray tip, and its ability to heat ions for either desolvation or dissociation.[66]

The ESI MS instruments developed at our laboratory typically use a stage of differential (mechanical) pumping, as well as cryo-pumping, to increase the ion current sampled from atmospheric pressure. The ESI capillary tip is mounted 1–2 cm from the sampling orifice into the quadrupole mass spectrometer. Ions are sampled from atmospheric pressure through a 1-mm nozzle orifice to a 2-mm diameter beam skimmer in front of an rf-only quadrupole. The potential difference between the nozzle and skimmer elements ranges from +100–+400 V for positive ion studies. The nozzle/skimmer region is pumped using a 50-L/s single-stage Roots pump to maintain this region at approximately 2 Torr. The cryo-pumped quadrupole chamber typically reaches 10^{-6} Torr in normal operation. The ability to collisionally heat ions by applying electric fields in the differentially pumped regions is useful for both desolvation and at higher voltages, collisional activation and collisionally activated dissociation of analyte ions,[42] as discussed later.

2.4. ESI Mass Spectrometers

The widest use to date of ESI MS has been with quadrupole mass spectrometers of modest m/z range and resolution; however, several examples of ESI interfaces to instruments with special advantages over quadrupole mass spectrometers have been reported. The combination with magnetic instruments was accomplished several years ago by Russian researchers[53, 54] and more recently by others.[19, 64] The ability to exploit the capabilities of higher m/z resolution and higher energy CAD is promising. Quadrupole ion trap (i.e., ion trap mass spectrometer (ITMS)) results are particularly encouraging, although the mass measurement accuracy is currently much poorer than with quadrupole mass spectrometers.[67, 68] The ITMS sensitivity appears to be better than that of quadrupole mass spectrometers, suggesting very efficient ion injection and trapping. In addition, significant advantages in MS/MS sensitivity and especially resolution[68, 69] are obtainable. McLafferty and coworkers have pioneered the combination of ESI with Fourier-transform-ion cyclotron resonance (FTICR) MS.[58, 70, 71] Henry et al.[58] have demonstrated resolving powers greater than 60,000 with a 3 Tesla FTICR instrument, easily resolving the 18+ charge state for equine myoglobin (M_r 16,951) and the 16+ charge state for equine cytochrome c (M_r 12,360). Using a higher magnetic field strength (6.2 Tesla), a resolving power of nearly 10^6 for myoglobin and 5×10^5 for carbonic anhydrase (29 kDa) was achieved by the McLafferty group,[58b] and similar results have been achieved at our laboratory. Taking advantage of the ion-focusing properties of the magnetic field, Hofstadler and Laude[72] have demonstrated substantial gains in sensitivity by placing the ESI interface in the high magnetic field of the superconducting magnet. The potential for higher order MS, MS^n ($n \geqslant 3$), and efficient photodissociation is promising for both ITMS and FTICR experiments. Initial results with time-of-flight instrumentation suggest that improved sensitivity (by ion storage and accumulation) are necessary to yield results comparable to quadrupole instruments.[73]

3. ESI-MASS SPECTROMETRY

3.1. Positive-Ion ESI-MS of Proteins and M_r Determination

Large oligopeptides and small proteins so far examined by ESI MS typically show a distribution of multiply charged molecular ions arising (in positive ion mode) by either proton or alkali ion attachment, with no evidence of fragmentation observed unless dissociation is induced during

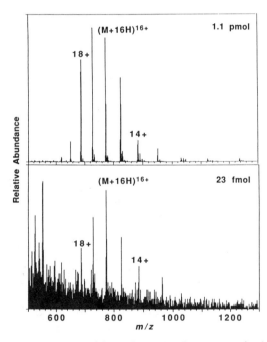

Figure 2. The ESI mass spectra of horse heart cytochrome c at the (*top*) 1.1-pmol and (*bottom*) 23-fmol level.

transport into the mass spectrometer vacuum by heating or collisional activation. The ESI mass spectra of horse heart cytochrome c (M_r 12,360) at the 1-pmol and 23-fmol (femtomole, 10^{-15} mol) levels are shown in Figure 2. A distinctive bell-shaped distribution of charge states is typically observed in which adjacent peaks differ by one charge. For the cytochrome c example illustrated, the highest m/z ion observed (with a quadrupole MS of m/z limit 1400) is at m/z 1374, corresponding to a net charge of 9+. The most abundant multiply protonated ion is observed at m/z 773.5 for the 16+ ion, and the highest charge state observed is 20+ (a small peak at m/z 619 in Figure 2). A striking feature of ESI mass spectra of most proteins is that the average charge state increases in an approximately linear fashion with molecular weight, extending the mass range of the MS

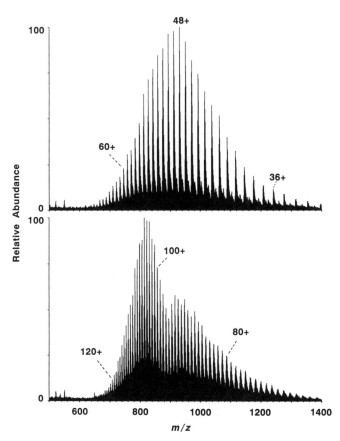

Figure 3. The ESI mass spectra of recombinant proteins (*top*) protein A (M_r ~44,650) and (*bottom*) α-galactosidase (M_r ~85,000).

experiment by the addition of compensating charge. This is demonstrated in the ESI mass spectra in Figure 3 for two recombinant proteins of more than 40 kDa.

The net number of possible protonation sites in solution is just one of the factors affecting the maximum extent of multiple charging observed in positive-ion ESI mass spectra. For most, but not all, proteins examined to date (prepared in aqueous solution, pH < 4), an approximately linear correlation is observed between the maximum number of charges and the number of basic amino acid residues (e.g., arginine, lysine, histidine, and the NH_2-terminus).[74] As an example, methionyl-human growth hormone (Met-hGH, M_r 22,256) is composed of 11 arginines, three histidines, nine lysines, and the NH_2-terminal methionine residue for a total of 24 basic and potentially favored protonation sites; the $(M+24H)^{24+}$ molecular ion at m/z 928 is the highest charged species observed. The acidic residues can be deprotonated in solution at pH greater than ~ 5, and such conditions would be expected to lower the observed ESI positive-charge state distribution. As discussed later, residual higher order structure can also result in significant changes in the observed charge state distribution; generally more "compact" structures yield lower charge state ions.

Many proteins contain disulfide bridges in their native structure that significantly affect higher order or three-dimensional structure. The native structure of bovine α-lactalbumin (M_r 14,175) is composed of 17 basic amino acid residues, and yet a maximum of only 13+ charges is observed in the ESI mass spectrum. It is composed of four cysteine-cysteine-disulfide bridges, and reduction of its disulfide bonds with 1,4-dithiothreitol (DTT, Cleland's reagent) results in higher charged molecules, up to 19+. The number of disulfide bonds can be estimated by the 8-Da increase in M_r on reduction. Likewise, carboxymethylation with iodoacetate of Cys residues after disulfide reduction results in the expected and more easily measured 472-Da increase for α-lactalbumin.

In some cases, the maximum number of charges has been observed to exceed the number of expected basic amino acid charge sites.[74] Bojesen's data[75] indicate that glutamine has a gas-phase proton affinity similar to lysine, and lactalbumin contains five such Gln residues as possible protonation sites. More recent results[76, 77] also support the speculation that glutamine residues may contribute to the more extensive protonation observed.

Another example of this characteristic is demonstrated in Figure 4. Bovine proinsulin, the precursor to insulin with an additional C-peptide joining the A- and B-chains, is composed of 81 residues (M_r 8680.8) and three disulfide bonds linking residues 7-67, 19-80, and 66-71. Thus, the majority of the polypeptide chain is enclosed by disulfide bonds, with the

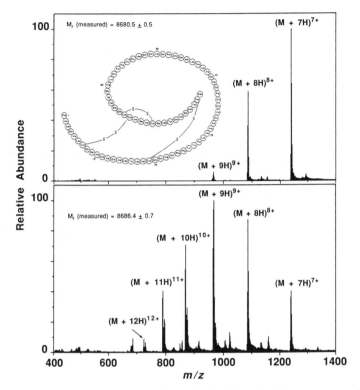

Figure 4. The ESI mass spectrum of bovine proinsulin (M_r 8680.8) (*top*) and after disulfide bond reduction (*bottom*) on adding dithiothreitol.

exception of the first six residues from the NH$_2$-terminus and the COOH-terminal asparagine residue. The measured molecular weight (8680.5 ± 0.46) from the m/z positions of the 7+ – 9+ molecular ions closely agrees with the expected value. Reduction of the three disulfide bonds with dithiothreitol results in the appearance of more highly charged molecular ions extending to the 12+ charge state. Again, although the primary structure of bovine proinsulin contains only nine basic amino acid sites (four Arg, two His, two Lys, and the NH$_2$-terminus) as potential protonation sites, "overcharging" on disulfide reduction is apparent. The experimentally measured M_r (8686.4 ± 0.67) is consistent for the reduced form of proinsulin (M_r 8686.9).

Since protein conformation in solution depends on intramolecular forces dictated by the amino acid sequence, we expect cleavage of disulfide bonds to affect greatly the higher order structure of the protein, allowing the molecule to "relax" into more extended conformations. Bovine serum albumin (M_r 66,430) is made up of 585 amino acids, 35 of which are cysteine residues that form 17 disulfide bridges. Its ESI mass spectrum shows the $(M+67H)^{67+}$ ion as the highest charged species; disulfide bond reduction with 1,4-dithiothreitol shows that up to 89 out of a possible (postulated) 100 sites are protonated.[16]

Increasing multiple charging for a given molecule is especially important for mass spectrometers with a limited m/z range, since it allows the instrument to detect molecular ions from large molecular species; performance is generally improved at lower m/z. Transferrin is an iron-binding glycoprotein of $M_r \sim 79,000$ with over 100 basic residues. The ESI mass spectrum of human apotransferrin (iron free) shows multiple charging to 55+ (m/z $\sim$1500),[57] beyond the m/z limit of our triple quadrupole MS. Reduction of its $\sim$13 disulfide linkages allows charging to 109+ (see Figure 5), bringing the entire charge distribution well within the instrumental range. (An additional polypeptide of M_r 6485 is observed, which is tentatively assigned as a peptide liberated from the COOH-terminus upon disulfide bond reduction.[78]) Fortunately, a majority of the proteins studied by electrospray ionization MS provide adequate charging to be detected by mass spectrometers (m/z limit 1400–2400), but some failures are evident, particularly among the glycoproteins.[78] Recent results by Allen *et al.*[79] and Mirza *et al.*[80] suggest that difficult proteins are often tractable using a heated capillary source and aqueous solutions. More recent results from our laboratory show that under certain ESI and interface conditions molecular ions at >m/z 10,000 are formed.

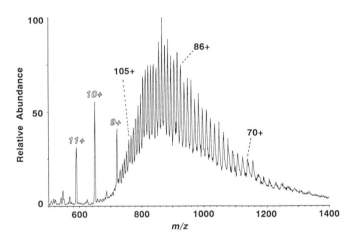

Figure 5. The ESI mass spectrum of human apotransferrin ($M_r \sim$ 79,500) after reduction of Cys-Cys bonds with dithiothreitol. Peaks labeled with open-face numbers are due to a polypeptide of M_r 6485.

The ESI mass spectra of proteins composed of noncovalently bound subunits show multiply charged ions characteristic of the subunit M_r. Human hemoglobin is a tetrameric protein consisting of two α- and two β-chains; its ESI mass spectrum from a solution pH of less than 6 show multiply charged ions for both the α (M_r 15,126) and the β- (M_r 15,865) chain units. Under typical solution and interface conditions, ESI mass spectra of heme-containing proteins do not usually exhibit multiply-charged ions of a heme-incorporated polypeptide when not covalently bound. An iron-porphyrin group is associated with hemoglobin; however, ions characteristic of the subunit minus the heme are found. The same is true for myoglobin, although its ESI mass spectrum may exhibit an ion series indicating heme association with the polypeptide backbone. This depends on the experimental parameters (e.g., nozzle skimmer potential, solution conditions), consistent with the weak chemical bonding.[81] At solution pH 6–7, conditions in which the heme-protein complex is stable in solution, ESI mass spectra show ions for the noncovalently-bound complex.[81] The heme-protein cytochrome *c* is covalently linked to the iron-porphyrin group through thioether bonds to two cysteine residues. Thus, the M_r calculated from its ESI mass spectrum from acidic pH solution (Figure 2) is consistent with a polypeptide plus the heme species.

Mass spectrometry with ESI has been used by Ganem, Li, and Henion to detect noncovalent enzyme substrate and receptor ligand complexes. A complex between the cytoplasmic receptor, human FK binding protein (FKBP, M_r 11,812), a member of the immunosuppressant binding protein family, and the recently discovered immunosuppressant FK 506 (M_r 804) were detected in the ESI mass spectrum.[82] Similarly, the enzymatic reaction of hen egg white lysozyme (HEWL) with N-acetylglucosamine (NAG) oligoaccharides was monitored by detecting the HEWL-NAG$_n$ complexes.[83] The hydrolysis of NAG$_6$ by HEWL produces predominantly NAG$_4$ and NAG$_2$. A solution containing HEWL and NAG$_6$ yields an ESI mass spectrum showing multiply charged ions for the HEWL-NAG$_6$ and HEWL-NAG$_4$ complexes. The fact that the HEWL-NAG$_2$ complex was not detected can be attributed to its known weak binding constant. During the reaction, the HEWL-NAG$_4$ complex grows in abundance as the HEWL-NAG$_6$ complex diminishes when monitored mass spectrometrically.

ESI MS should prove useful for probing other noncovalent binding interactions important in many biological systems; for example, protein–protein interactions (e.g., aggregation or quaternary structure)[84] and DNA-DNA (e.g., duplex formation)[85] interactions have been detected by ESI MS. Metal ion binding is an important function of many biological molecules. Metalloenzymes require the metal ion as an essential participant in the catalytic process, and, removing the metal ion may lead to loss of activity. Peptide-metal ion interactions in solution have been investigated by LD MS and ESI MS by Hutchens *et al.*[86] The 26-residue peptide, (Gly-His-His-Phe-His)₅ Gly was used as a model for the Cu(II)-binding histidine-rich glycoprotein and analyzed mass spectrometrically in the presence of copper ions; up to five Cu atoms were found to bind to the peptide by both LD and ESI experiments. A maximum binding capacity of five Cu atoms was determined by spectrophotomeric titration of the

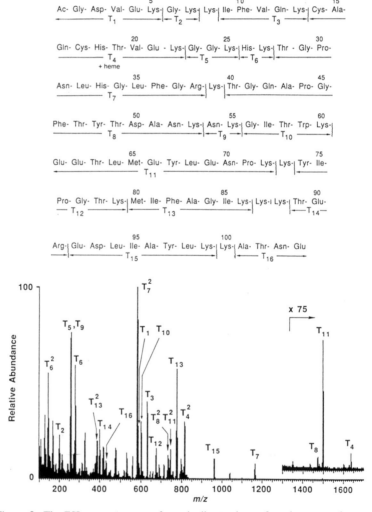

Figure 6. The ESI mass spectrum of tryptic digest mixture from horse cytochrome *c*.

peptide in aqueous solution. Multiply charged ions of the form $[M + (n - m) H + mCu]^{n+}$ were observed with $n = 3-7$. MS methods such as ESI may provide rapid evaluation of metal ion-binding stoichiometry.

Applying highly specific proteases is a common experimental method employed for protein structure elucidation, and analyzing the resulting digest mixture by FAB MS has been routinely used, often followed by MS/MS;[26, 27, 30, 31] however, complete identification of every component of protein digest mixtures is often not possible by FAB MS due to surface activity effects. This arises because samples are dissolved in a liquid matrix and desorbed from the surface layers. Components with high surface activity are localized at this interface and preferentially ionized by FAB; such a differential response seems substantially less important for electrospray ionization. The ESI MS analyses of tryptic digest mixtures of proteins (e.g., cytochrome *c* in Figure 6) permit detection of most components of both hydrophobic or hydrophilic character.[87] Synthetic equimolar mixtures of peptides with varying hydrophobicites do not show severe ionization bias with ESI MS; thus, surface activity does not appear to play so major a role in the ESI process as in FAB.

For positive-ion ESI MS, determining M_r is straightforward, assuming that adjacent peaks of a series differ by only one charge and charging is due to cation attachment (a proton or rarely, for small M_r oligopeptides, mixed H^+ and Na^+) to the molecular ion. With these assumptions, it follows that the observed m/z ratios for each member of the distribution of multiply protonated molecular ions (in the case of oligopeptides and proteins) are related by a series of simple simultaneous linear equations, where M_r, and charge are unknown. Any two peaks are sufficient to deter-

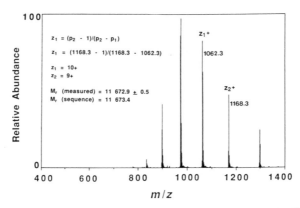

Figure 7. The ESI mass spectrum of thioredoxin (*E. Coli*). Peaks m/z 1298.0, 1168.27, 1062.27, 973.74, and 898.88 are centroid values for these multiply protonated ions (13^+–9^+). Calculating M_r is discussed in the text.

mine uniquely charge and M_r as shown by the protein thioredoxin example in Figure 7. These m/z values are obtained from the centroid of the peak, because isotopic contributions for these large multiply charged ions are not resolved by a quadrupole mass spectrometer. Equation 1 describes the relationship between a peak at m/z p_1 with charge z_1 and the M_r

$$p_1 z_1 = M_r + M_a z_1 = M_r + 1.0079 z_1 \tag{1}$$

where, for this case, we assume that the charge-carrying species (M_a) is a proton. A second peak at m/z p_2 (where $p_2 > p_1$) that is j peaks away from p_1 (e.g., $j = 1$ for two adjacent peaks) is given by

$$p_2(z_1 - j) = M_r + 1.0079(z_1 - j) \tag{2}$$

Solving Equation 1 and 2 simultaneously for z_1 yields

$$z_1 = j(p_2 - 1.0079)/(p_2 - p_1)$$

The charge for p_1 is obtained by taking z_1 as the nearest integer value, and thus the charge of each peak in the multiple-charge distribution is assigned.

Calculating of M_r for each of the observed m/z values provides enhanced precision. Thus, for the data shown in Figure 7, a standard deviation of ± 0.5 Da is observed for 5 m/z measurements from one spectrum. In this example, the error in the determination is $+0.004\%$. Typical M_r measurements results for peptides and proteins by ESI MS are summarized in Table 1. Mann et al. have developed an algorithm that deconvolutes the ESI mass spectrum and collapses the peaks into a single M_r for improved accuracy.[55] Such measurements are generally two to three orders of magnitude more accurate than those obtained using electrophoretic methods.

Relative molecular mass measurements for proteins of greater than 100 kDa have been demonstrated, and in favorable cases, measurement precision may be better than 0.005%. However, more accurate M_r measurements are required for >20 kDa proteins to differentiate, for example, single-residue mutations that produce 1-Da mass differences (i.e., Asp and Asn, Leu/Ile and Asn, Glu and Gln, etc.). Feng et al.[57] have demonstrated 0.001% (10-ppm) M_r precision (i.e., M_r 20,000 $\pm$ 0.2) employing a Gaussian curve-fitting method. The majority of the reported ESI MS experiments are done on relatively low-resolution quadrupole analyzers. With higher resolution instrumentation, such as FTICR mass spectrometers, charge state assignment in even complex mass spectra is possible by measuring the 1 dalton isotopic peak separation.[58, 71] This combination of high resolution and high accuracy are crucial for further advancement in identifying unknowns and complex mixture analysis.

Table 1. Calculated and Measured Relative Molecular Masses by ESI MS for a
Representative Group of Oligopeptide and Small Protein Standards

Polypeptide	Relative molecular mass (M_r)		Mass measurement error (%)
	Calculated	Measured	
A-chain bovine insulin (oxidized)	2531.6	2531.3	−0.01
B-chain bovine insulin (oxidized)	3495.9	3496.1	+0.01
Bovine insulin	5733.6	5733.9	+0.01
Porcine insulin	5777.6	5776.5	−0.02
Lactobionyl porcine insulin	6232.8	6234.5	+0.03
Dilactobionyl porcine insulin	6688.0	6688.4	+0.01
Bovine ubiquitin	8564.8	8564.7	−0.00
Bovine proinsulin	8680.8	8680.5	−0.00
Thioredoxin (*E. coli*)	11673.4	11672.9	−0.00
Horse heart cytochrome *c*	12360.1	12358.7	−0.01
Bovine ribonuclease A	13682.2	13681.3	−0.01
Bovine α-lactalbumin	14175.0	14173.3	−0.01
(carboxymethylated)	14647.4	14645.9	−0.01
Chicken egg white lysozyme	14306.2	14304.6	−0.01
Human lysozyme	14692.8	14695.2	+0.02
Methionyl-interleukin-2	15543.1	15547.1	+0.03
Human hemoglobin β-chain	15867.2	15867.4	+0.00
Chicken adenylate kinase	21593.8	21596.4	+0.02
Bovine carbonic anhydrase	29021.0	29017.0	−0.01
Bovine albumin	66430.3	66443.2	+0.02

3.2. Effects of Higher Order Structure on ESI Charge State Distribution

Protein function in biological systems generally involves interactions of specific structural conformations essential for activity,[21] so that information on higher order structure (secondary, tertiary, quaternary) is of fundamental importance to biological research. The three-dimensional protein structure in solution is generally governed by a large number of relatively weak noncovalent interactions. Changes in the protein environment, such as changes in pH, temperature, addition of reagents (guanidine hydrochloride or urea, for example), or other solvent conditions may cause the protein to unfold or denature and disrupt these cohesive forces. X-ray crystallography, nuclear magnetic resonance, and various spectroscopic methods have been used to probe conformational changes. Mass spectrometry potentially offers greater speed and sensitivity in qualitatively evaluating changes in higher order structure.

According to our present understanding, the charge distribution observed in ESI mass spectra of proteins is related to the solution phase structure. As shown for proinsulin in Figure 4 and Figure 5 for apotransferrin, protein charging depends dramatically on higher order protein structure. Reductive cleavage of disulfide bonds alters the higher order structure of the protein. The native disulfide-bridged conformation may have basic amino acid sites in the internal volume of the globular structure that are less readily protonated. Repulsive coulombic forces are also greater for the more compact native structure, and charge sites may be more labile in the gas phase to reactions (e.g., proton transfer). However, the precise details of the relationship between solution structure and ESI charging is unclear, because it is uncertain how ionization and subsequent gas-phase processes, such as proton transfer during desolvation, alter this initial charge distribution. Thus, it is uncertain to what extent higher order protein structure will be preserved through the electrospray process and be reflected in the observed ESI charge state distribution.

Chowdhury *et al.* have demonstrated a dramatic effect of pH on the ESI mass spectrum of cytochrome *c*.[89] A bimodal charge state distribution, reflecting a mixture of conformers, is also observed on adding acetic acid to an aqueous solution of hen egg lysozyme (M_r 14,306).[90] Lowering the pH from 3.3 to 1.6 denatures these proteins from a compact folded conformation to a random coil state. Thus, the higher charge state distribution (at lower m/z) may represent the contribution from the random coil state. Feng *et al.* have reported similar results for papain,[91] suggesting that the different charge state distributions observed for different pH conditions are consistent with the known solution structural changes. Recent results indicate that thermal denaturation of proteins may be studied by ESI MS (by heating the electrosprayed droplets), again causing the protein to unfold from its compact native form, indicated by a higher extent of charging.[92]

The use of organic solvents, such as those generally used for ESI, can also cause protein denaturation and lead to unfolding of the polypeptide chain.[93] A dramatic example of such behavior is found for bovine ubiquitin (M_r 8,565).[90] Figure 8a shows the ESI mass spectra of ubiquitin in a 18% acetonitrile solvent mixture (with 5% acetic acid) with an acetonitrile sheath. The ESI mass spectrum shows charging extending to the $(M+13H)^{13+}$ molecular ion, consistent with the 13 basic residues. Reducing the acetonitrile content to 12% produces a bimodal-charge distribution, indicating that a mixture of denatured (high-charge state) and native (low-charge state) forms are present (see Figure 8b). The native structure of ubiquitin is known to be globular in an aqueous environment.[94] The native globular form for ubiquitin dominates, as can be seen from the spectra dominated by the 7+ and 8+ species in aqueous solution

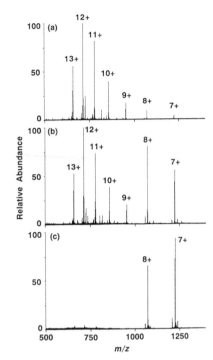

Figure 8. The ESI mass spectra of bovine ubiquitin in (a) 18% acetonitrile, (b) 12% acetonitrile with an acetonitrile sheath liquid, and (c) 0% acetonitrile with an aqueous sheath. Acetic acid to 5% was added to each analyte solution.

with an aqueous sheath (see Figure 8c). Recently, Katta and Chait have used the extent and rate of hydrogen/deuterium exchange in conjunction with ESI MS to probe changes in the conformation of proteins in solution.[95]

It is difficult to resolve all the factors contributing to the observed ESI charge state distribution. Changes in such physical parameters as pH, dielectric constant, and surface tension may have important effects on the desorption/ionization event(s) and ultimately affect the observed charge state distribution. Available results suggest that in some cases, multiple conformation states can be qualitatively monitored using the ESI mass spectrum. Although the charge state distribution from ESI is at least partially governed by solution-phase conformations, it is uncertain how liquid-phase structure is related to the relevant gas-phase structure(s). Even if the charge state distribution is locked in by factors that include solution-phase structure, the three-dimensional gas-phase structure may be altered by the heating required to complete desolvation and likely assisted by repulsive coulombic forces once the protein is stripped of mediating effects of the dielectric solvent.

3.3. Negative Ion ESI MS

The ESI MS of polypeptides and proteins has generally been performed in the positive ionization mode. The more basic sites of the amino acid residues (arginine, lysine, histidine, and the NH_2-terminus) have sufficiently large pK_a values to ensure that proteins will be multiply protonated in acidic pH solutions. Relatively few negative-ion ESI mass spectra of polypeptides have been reported. Most proteins in neutral and acidic aqueous solutions do not produce highly multiply charged anions because such acidic residues as glutamic acid (Glu) and aspartic acid (Asp) typically have pK_a's of around 5; i.e., the protein often retains a net positive charge in solution due to the protonated basic residues.

However, as shown in our laboratory,[96] multiply charged (deprotonated) molecules can be produced by ESI in basic pH aqueous solution (e.g., 1–3% ammonium hydroxide), where few basic residues could be charged. Horse heart myoglobin in aqueous 1% NH_4OH solution (pH ~ 11.3) yields an ESI mass spectrum with a bell-shaped multiple-charge state distribution extending up to approximately the $(M - 19H)^{19-}$ charge state. Horse myoglobin has 13 Glu and seven Asp residues, corresponding closely to the maximum negative-charge state observed. Similarly, A-chain bovine insulin (oxidized, M_r 2,531.6) yields multiple charging to the five-charge state in aqueous solution. However, in 1% NH_4OH, the observed multiple charging extends to the seven-charge state, presumably due to the deprotonation of the two glutamic acid residues.[96] Figure 9 shows the

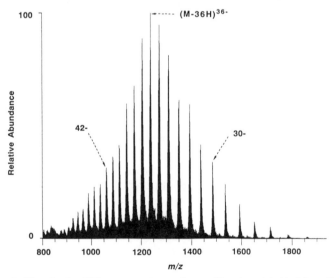

Figure 9. Negative-ion ESI mass spectrum of recombinant protein A (M_r ~44,650).

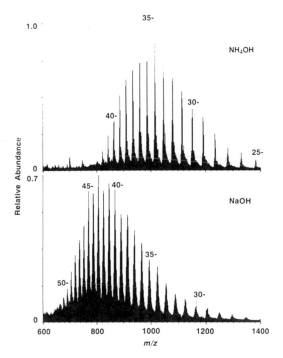

Figure 10. Negative-ion ESI mass spectra of porcine pepsin in (*top*) 1% NH₄OH and (*bottom*) 5×10^{-3} M NaOH.

negative ion ESI mass spectrum of recombinant protein A, desplaying charging to the 48-charge state (see Figure 3 for the corresponding positive-ion mass spectrum).

Figure 10a shows the negative-ion ESI mass spectrum for an aspartic proteinase found in gastric juice, pepsin (porcine, ~ 34.6 kDa). Pepsin does not afford a positive-ion ESI mass spectrum (with limited m/z range quadrupole instrumentation) because it has only two Arg, one Lys, and one His residues, apparently insufficient for producing multiply protonated molecules below m/z 2000. However, the 29 Asp and 13 Glu acidic amino acid residues permit multiple charging in the negative ionization mode up to $(M-42H)^{42-}$ from an aqueous 1% NH₄OH solution.

The effects of pH and cationic substitution are illustrated in Figure 10b for porcine pepsin. The 1% (0.15 M) NH₄OH solution used in Figure 10a contrasts significantly with results obtained for porcine pepsin in 5×10^{-3} M NaOH (pH 11.7). Slightly more charging, to the 51-charge state, is observed for pepsin in a 5×10^{-3} M NaOH aqueous solution. Cysteine and the phenolic hydroxy group of tyrosine residues may also deprotonate in high-pH solutions because they have expected pK_a ranges in representative proteins of 8.5–8.9 and 9.6–10.0, respectively. We have observed formation of triply charged anions from basic solutions of the hexapeptide Tyr-Tyr-Tyr-Tyr-Tyr-Tyr. It appears reasonable that some of the 16 tyrosine amino acids of pepsin may be deprotonated under the present ESI experimental solution conditions;[96] a large extent of sodium adduction is also observed. Higher charge states (>45-) show only small amounts of Na substitution, and the extent of Na substitution increase approximately linearly with decreasing charge. An unresolved distribution of sodium adducts is found for each charge state, so that lower charge states (higher m/z) show more sodium substitution than ions of higher charge. The m/z of each charge state shifts substantially because the most probable species contain numerous Na substitutions, i.e., $[M-(a+z)H+aNa]^{z-}$. The M_r calculations for conditions where cation adduction is not resolved are generally limited to ~0.5% precision and accuracy, because each charge state contains a different (and unknown) average number of sodium counterions. Clearly, the ability to obtain both positive- and negative-ion ESI spectra extends the utility of these methods and may provide new qualitative insights related to protein structure in solution.

Preliminary experiments have successfully applied ESI MS to oligonucleotides.[10, 16, 85, 97] The negative-ion ESI mass spectra of small oligodeoxyribonucleotides are characterized by multiply charged molecular

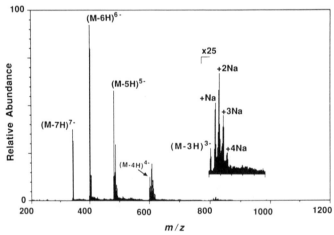

Figure 11. Negative-ion ESI mass spectrum of the Hae III linker, d(GGCCGGCC), M_r 2412. Peak multiplicity due to sodium association is particularly noticeable in the lower charge states of the molecular anion, e.g., ions grouped around m/z 481 and m/z 602.

anions of the form $(M - nH)^{n-}$, with the charge residing at the acidic-bridging phosphodiester and/or terminal phosphate groups. Each nucleotide residue can potentially accommodate one negative charge. Small oligonucleotides ($m = 3$–8, where m is the number of nucleotide units) give rise to molecular ions in the gas phase to near the maximum possible charge state. As the M_r increases, the extent of such multiple charging increases, maintaining the m/z of members of the distribution of molecular anions in the range of a quadrupole mass analyzer (between m/z 500–1500). Again, with increasing levels of polymerization and a decreasing charge state for a given molecular anion, an increased contribution of species arising from the substitution of sodium for the proton on the unionized phosphates is observed.[16] This substitution of Na$^+$ for H$^+$ among the increasing number of possible phosphate groups (which accompanies increasing polymerization and/or decreasing charge state) reduces the measurable current for any given ion and thus the sensitivity of the analysis. Where resolution permits, the separation of suites of members of molecular anions appear as shown in Figure 11 for a synthetic deoxyribonucleotide 8-mer. In these circumstances, the M_r of the parent molecule may be determined with comparable precision and accuracy to that observed for polypeptides under positive ionization conditions. For larger

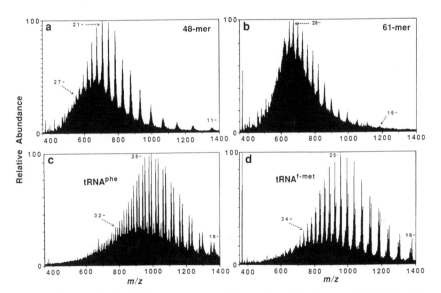

Figure 12. Negative-ion ESI mass spectra of large oligonucleotides: (a) d($A_9T_{17}G_{12}C_{10}$), M_r 14,771; (b) d($A_{18}T_{22}G_9C_{12}$), M_r 18,701; (c) tRNA^{Phe} (brewers' yeast), M_r 24,927; (d) tRNA$^{f\text{-met}}$ (*E. coli*) M_r 24,926.

synthetic deoxyribonucleotides, we observe multiply charged molecular anions with broadened peaks due to alkali attachment (see Figure 12). The M_r measurement for oligonucleotides is significantly affected by alkali metal association when resolution is insufficient for distinction among the sodium-associated peaks. In such cases, determinations based on the maximum or centroid of a broadened sodium-associated peak overestimates the M_r, with an error of the order of 1% for such measurements. Presently, the largest nucleotides successfully ionized are natural small oligoribonucleotide 76-mer transfer RNA's ($M_r \sim 25$ kDa).[16] Stults and Marsters[97] have demonstrated improved ESI sensitivity and M_r precision for large oligonucleotides after cold-ethanol precipitation and sodium displacement with ammonium acetate. For a 30-mer sample, signal strengths improve by a factor of more than 10 with M_r measurements of $\pm\ 0.03\%$ or better.

4. DISSOCIATION OF ESI-GENERATED MULTIPLY CHARGED IONS

A major characteristic of mass spectra using soft ionization techniques, particularly the matrix-assisted LD and ESI methods, is the predominance of a molecular ion, typically an even-electron-protonated or cationized species, and the lack of significant ions indicative of fragmentation. While the presence of one or several molecular ion species allows accurate measurement of molecular weight values, no additional direct data regarding molecular structures are obtained. Thus, a combination of these ionization methods with MS/MS[24, 25] is the procedure of choice for obtaining additional molecular structure information.

Tandem mass spectrometry of singly charged ions has not been fruitful beyond about m/z 4000 (to date) due in part to the decreased primary-ion signal and the decreasing efficiency for CAD with increasing M_r. Hunt and Shabanowitz[98] and Barber *et al.*[99] have previously demonstrated CAD of a doubly charged molecular ion (by FAB) from peptides in the 1–2 kDa molecular weight range, and Neumann and Derrick[100] demonstrated MS/MS of doubly charged bradykinin (M_r 1060) molecular ions produced by FD. The first tandem MS studies of more highly charged polypeptides were reported by our laboratory using ESI.[40]

A CAD mass spectrum of the $(M+2H)^{2+}$ ion of gramicidin D (M_r 1881) produced by electrospray ionization is shown in Figure 13. Gramicidin D is a mixture of four linear peptide components, with the Val¹-gramicidin A variety the most abundant species. MS/MS of the Val¹-gramicidin A doubly charged ion (see Figure 13) yields singly charged

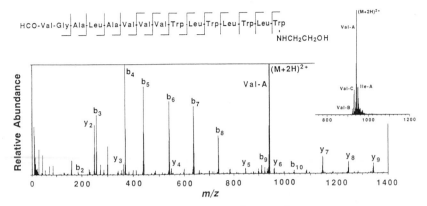

Figure 13. Tandem mass spectrum of the $(M+2H)^{2+}$ of Val^1-gramicidin A with laboratory frame collision energy approximately $+310$ eV. The inset shows the ESI mass spectrum region of the doubly charged molecule.

series of b_n and y_n product ions that allow for nearly complete sequence analysis of the molecule.*

Most of the fragmentation observed for Val^1-gramicidin A by Johnson *et al.*[101] in higher energy CAD experiments (10-keV energy with a tandem double-focusing mass spectrometer) of the singly charged analog are observed in the ESI MS/MS data, with the exception of the larger mass y_{10}–y_{14} and b_{13}–b_{15} product ion series, due to the limited m/z of our quadrupole instrument. Similarly, CAD of the double charged molecule ions of the remaining three components yields very similar fragmentation patterns, with the expected b_n and y_n ions and the shifts in product ion m/z. For example, CAD of Ile^1-gramicidin A yields the same y_n ions at the same m/z positions as Val^1-gramicidin A, but the b_n ions show an increase of 14 Da due to the substitution of an isoleucine for a valine residue at position 1. The Val^1-gramicidin C has a tyrosine at position 11 instead of tryptophan; therefore, the b_n ions are unchanged, but the y_n product ions are reduced by 23 Da. Likewise, phenylalanine substitution for tryptophan (position 11) for Val^1-gramicidin B results in a -39-Da shift for y_n product ions. Interestingly, an ion for the $(M+H+Na)^{2+}$ species of Val^1-gramicidin A is also evident in the ESI mass spectrum. Dissociation of this ion results in the same b_n fragment ions shown in Figure 13. However, the

---

* Fragmentation labeling is based on conventional notation[23] in which lower case letters are used and subscripts denote the residue position, counting from the NH_2-terminus for a_n, b_n, and c_n ions, and from the COOH-terminal for x_n, y_n, and z_n product ions. A superscript is added to indicate the fragment-ion charge state. Absence of a superscript denotes a singly charged fragment.

y_n series shows a $+23$ Da shift, which can be attributed to sodium adducted product ions. The sodium adduct appears to be bound exclusively in the vicinity of the COOH-terminal portion of the molecule.

The ESI MS/MS of tryptic peptides combined with on-line separations methods, such as liquid chromatography and capillary electrophoresis, have been demonstrated to be a powerful method for protein sequence analysis.[102–104] Doubly charged peptides result due to trypsin's activity on the COOH-terminus of lysine and arginine residues, leaving a positive charge on the COOH-terminal amino acid and the primary amine on the NH_2-terminus. The CAD of such doubly charged molecules results primarily in singly charged product ions. (Additional variations occur when a histidine residue is present, often increasing by one the maximum charge state of the molecule).

For larger molecules, the more highly charged ions are more susceptible to dissociation. We have demonstrated the production of sequence-specific fragment ions from multiply charged molecular ions of larger oligopeptides than might be efficiently dissociated as singly charged species.[40–42] Since collision energy is proportional to the number of charges at a given m/z, it is reasonable to assume that the greater translational energies of ions with a greater number of charges is the primary reason for increased CAD efficiency. In addition, efficiency of the MS/MS of multiply charged ions may be enhanced by the strained configuration from electrostatic repulsive forces.

The CAD of the $(M+3H)^{3+}$ to $(M+6H)^{6+}$ molecular ions of melittin (M_r 2845) from ESI yields multiply charged product ions (to $4+$) that can readily be ascribed to the known sequence of the polypeptide.[40] Dramatic differences among spectra of various charge states and the large number of fragment ions with various charge were observed. Although the CAD of multiply charged ions from larger peptides yield product ions that can readily be correlated to the sequence, *complete* sequence assignments are generally lacking. Studies to date indicate that a progressively smaller portion of the molecule is susceptible to lower energy CAD with increasing analyte size. For example, human parathyroid hormone (1-44, M_r 5064) yields multiply charged molecular ions from $4+-9+$. Collisional dissociation analysis of these parent ions, producing primarily singly charged b_n and y_n series sequence ions from both COOH- and NH_2-termini, affords sequence information for approximately one-third of the molecule.[42]

Larger multiply charged oligopeptide and protein ions can efficiently be dissociated by collisions with a neutral gas target, but interpreting the resulting product ion spectrum, even in cases where the primary structure is known, presents special difficulties due to the fact that MS does not directly provide information on charge state, especially with relatively low-resolution quadrupole instrumentation. (This complication is largely avoidable for doubly charged ions). With higher resolution methods, such as double-focusing mass spectrometers[88] and ion-trapping instruments, such as the ion trap mass spectrometer[68] and the Fourier-transform ion cyclotron resonance (ICR) mass spectrometer,[58, 71] measurement of the 1–dalton isotopic separation reveals the charge state. However, even from low-resolution instrumentation, interpreting multiply charged product ion mass spectra is feasible in many cases. For each peak in the product ion mass spectrum, all possible M_rs are calculated (m/z times the possible charge states [parent charge]) and compared to all possible b- and y-mode fragment ions from the known sequence. More probable assignments are initially selected by searching for features that appear to be common for most of the MS/MS spectra we have examined. Particularly useful is a progression of product ion charge states (e.g., y_{25}^{2+}, y_{25}^{3+}, y_{25}^{4+}, *etc.*) and product ion series (e.g., y_{25}^{2+}, y_{26}^{2+}, y_{27}^{2+}, *etc.*). In addition, by activation in the atmospheric pressure/vacuum interface, we are able to produce a fragment ion, select it in first quadrupole, and further dissociate it in the

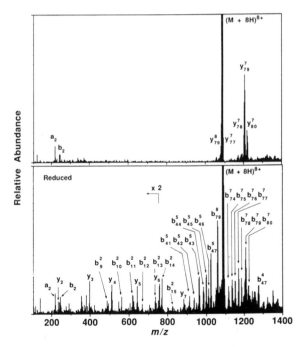

Figure 14. The MS/MS product ion spectra of the $(M+8H)^{8+}$ molecular ion of bovine proinsulin (*top*) and after reduction with DTT (*bottom*).

collision quadrupole, a quasi MS/MS/MS experiment.[16, 29, 41] This method can be used to confirm product ion assignments, and it often provides additional sequence information by probing molecular regions not previously examined in initial MS/MS experiments. Moreover, comparing CAD mass spectra from structurally related molecules, such as molecules with only a few amino acid variations (e.g., variants, proteins from different species with high homology or amino acid sequence alignment) or those with modified residues (e.g., carboxymethylation of free cysteines, acetylation of the NH_2-terminus, oxidation of a methionine residue, *etc.*), highlights the approximate location of dissociation by expected m/z shifts of the products.

Bovine proinsulin (see Section 3.1 and Figure 4) is an example of a larger molecule that we have studied. The MS/MS of the $(M+8H)^{8+}$ state for the native polypeptide (see Figure 14, *top*) with a laboratory frame collision energy of 1700 eV produces only highly charged y_n sequence ions from cleavage near the NH_2-terminus. The likelihood of observing fragmentation from regions enclosed by disulfide bonds in molecules of this size is low because (at least) two bonds must be ruptured. On the other hand, the CAD of multiply charged molecules from the disulfide-reduced form produces highly charged b_n product ions from cleavage near the COOH-terminus. The MS/MS spectrum of the $(M+8H)^{8+}$ ion for the reduced state (see Figure 14, *bottom*) contains a series of b^{7+}_{74-79} ions and a b^{8+}_{79} ion, as well as singly charged y_2-y_7 fragments from the COOH-terminal end, most of which were formerly enclosed by disulfide linkages in the native form. From regions closer to the NH_2-terminus, series of b^{2+}_{9-15} and b^{3+}_{13-15} sequence ions are found, but fragmentation also appears to originate from the middle region of the molecule, as suggested by the relatively intense ion series from m/z 1021 (b^{4+}_{47}) to m/z 913 (b^{5+}_{41}). This last observation is surprising based on theoretical expactation[105] and our observation of other large molecules studied to date show fragmentation occurring primarily from the ends of the molecule. A b_{47} sequence ion is created from the cleavage of the NH_2-terminal amide bond to Pro[48]. In CAD studies of multiply charged ions with M_r 2000-22,000,[40–42, 106] we have previously observed unusually enhanced dissociation directly from the NH_2-terminal amide bond to a proline residue. This phenomenon has been observed before in both high-energy tandem double-focusing CAD studies[107,108] and with low-energy quadrupole instrumentation[31, 108, 109] for small peptides. Proline is unique among the common amido acids in that the end of the side chain is covalently bound to the preceding amide nitrogen. The five-member ring imposes rigid constraints on N-C rotation and thus may have significant effects on the conformation of the polypeptide backbone. These effects are significant for CAD MS, since such constraints may make the adjacent peptide bond more susceptible to cleavage. However, proline residues are also present at positions 28, 37, 56, and 57 of the primary structure of proinsulin. It is unclear why fragmentation processes should be directed toward Pro[48], and not the other proline sites.

We can speculate on the role of the gas-phase tertiary structure of a multiply charged molecule ion on such dissociation processes. We have previously observed a possible tertiary structure effect on the CAD of multiply charged protein molecular ions.[41] For ESI MS/MS of bovine pancreatic ribonuclease A (M_r 13,682), composed of 124 amino acids and four disulfide bridges, only relatively small terminal regions of the molecule are directly probed by CAD using the present quadrupole instrumentation.[41] Nonetheless, CAD can be efficient for such multiply charged species, and product ion spectra may be interpreted on the basis of a known sequence and fragmentation modes. Additionally, differences in the product ion spectra of the same charge state from both native and reduced forms of the cystinyl-bridged protein suggest that higher order (secondary and tertiary) structures of these ions in the gas phase may influence the activation and dissociation processes.

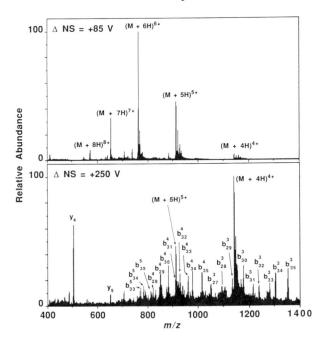

Figure 15. The ESI mass spectra of porcine ACTH (1-39) (M_r 4567) with $\Delta NS = +85$ V (*top*) and $+250$ V (*bottom*).

For certain proteins in which fragmentation is particularly diverse and abundant, and not presently fully interpretable, CAD spectra may provide qualitatively unique "fingerprinting". For example, the CAD mass spectra for the $(M+15H)^{15+}$ molecular ion of cytochrome c proteins from nine different species (bovine, chicken, dog, horse, pigeon, rabbit, rat, tuna, yeast) with a moderate M_r range have been obtained and compared.[42] Strong differences are observed in the spectra, even between species differing by as little as four amino acid residues out of over 100. Although presently required over 100 pmole of material, such results suggest future possibilities of using CAD mass spectral patterns to characterize unkown proteins rapidly or compare related molecular species (e.g., mutations and post-translational modifications).

While conventional CAD is normally performed in the collision quadrupole chamber (rf-only quadrupole, Q2), collisional processes in the differentially pumped atmosphere/vacuum interface can also occur prior to mass analysis. Our original observations showed that the voltage applied between the nozzle and skimmer (nozzle/skimmer voltage bias, ΔNS) in the atmosphere/vacuum interface of the mass spectrometer greatly influences the maximum number of charges observed for each polypeptide.[42, 110]

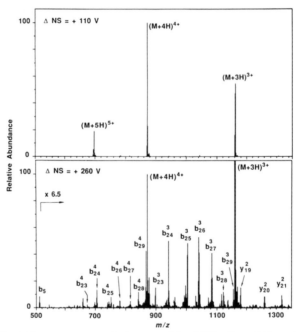

Figure 16. The ESI mass spectra of glucagon ($M_r = 3483$). *Top*: $\Delta NS = +110$ V; *bottom*: $\Delta NS = +260$ V.

Such electric fields, often applied for ion-focusing purposes, also result in collisions that can become increasingly energetic as pressure drops. For large molecules, more highly charged ions are more susceptible to dissociation. This observation, together with its converse, i.e., more highly charged species (at lower m/z) are most abundant at lower ΔNS (less energetic collisions), is consistent with a collisional activation process.

Internally excited or hot molecular ions produced by collisional excitation in the ESI atmosphere/vacuum interface undergo both metastable dissociation and CAD with greatly enhanced efficiency.[111] Cold molecular ions show no detectable metastable dissociation products, consistent with the low internal excitation expected from ESI; the relatively long drift time (~5 msec) at atmospheric pressure, which should aid thermalization; and an uncertain degree of cooling during expansion into the vacuum. The cold molecular ions require numerous collisions before CAD can occur, whereas hot ions (collisionally excited in the interface) require fewer collisions.

Collisional product ions are produced at elevated ΔNS, and, these are available for MS/MS analysis (i.e., CAD/MS/CAD/MS or MS/MS/MS), permitting confirmation of their structure or probing of regions not

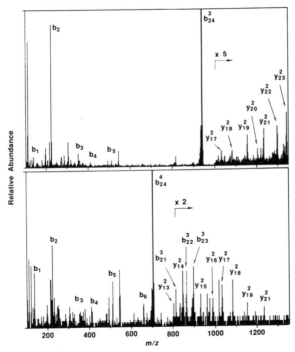

Figure 17. The further dissociation by CAD/MS/CAD/MS of two glucagon dissociation products "b_{24}^n." *Top*: b_{24}^{3+}; *bottom*: b_{24}^{4+}.

susceptible to fragmentation by CAD of the intact molecular ion.[29, 41, 42] A CAD spectrum generated in this manner is illustrated in Figure 15 for adrenocorticotropic hormone (ACTH 1-39, M_r 4567). The ACTH peptide has eight basic amino acid residues (plus the NH_2-terminus), many of which are located near the middle region of the molecule. With $\Delta NS =$ 85 V, multiple charging to the 8+ state is observed. At an elevated ΔNS of +250 V, most of the multiply charged molecular ions have dissociated to produce series of 3+, 4+, and 5+ b_{27}–b_{35} sequence ions, arising from cleavage near the COOH-terminus, in addition to an abundant singly charged y_4 ion. As previously mentioned, dissociation directed near a proline residue is favored, producing the y_4 product. All fragmentation assigned in Figure 15 has also been observed in the collective MS/MS experiments of the individual 5+, 6+ and 7+ molecular ions.[16] Further CAD of fragment ions generated in the interface can be used to confirm product ion assignments, as we have previously demonstrated for melittin,[42] ribonuclease A,[41] and serum albumins.[29] The MS/MS of

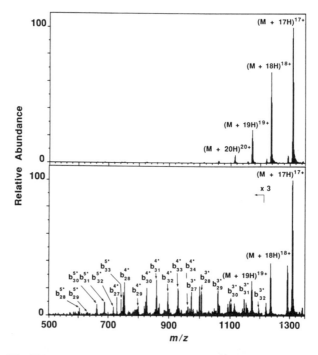

Figure 18. The ESI mass spectra of methionyl-human growth hormone ($M_r = 22,256$) with $\Delta NS = +110$ V (*top*) and +285 V (*bottom*). All fragment ions labeled with an asterisk are internal product ions in which the first two NH_2-terminal residues, Met-Phe, have been lost [e.g., $b_{30}^4{}^* = (b_{30}\text{-MetPhe})^4$]. See text for additional information.

the tentatively assigned y_4 ion at m/z 506 in Figure 15 produces singly charged a_2, b_2, b_3, and y_1–y_3 sequences ions consistent with the suspected Pro-Leu-Glu-Phe tetrapeptide.

A similar example is shown in Figure 16 for the polypeptide, glucagon (M_r 3483). At low ΔNS, multiply charged molecules to the 5+ charge state are observed (see Figure 16, *top*). Increasing ΔNS to +260 V produces a series of product ions from b_{23}–b_{29} with 3+ and 4+ charge states in addition to a doubly charged y series (see Figure 16, *bottom*). Further dissociation of the b_{24}^{3+} and b_{24}^{4+} sequence ions (see Figure 17) yields fragments that confirm its assignment and provide additional sequence information.

Even larger proteins can be effectively dissociated at sufficient collision energy in the interface region. Figure 18 shows ESI mass spectra of methionyl-human growth hormone at ΔNS = +110 V and +285 V. The Phe^2-Pro^3 bond from the NH_2-terminus is extremely labile, as evident in the collision spectrum with an elevated ΔNS. Multiply charged 3+–5+ sequence ions of type b_{25}–b_{35} in which the NH_2-terminal MetPhe residues have been lost dominate the spectrum [e.g., "internal" sequence ions, such as $(b_{28}$-MetPhe$)^{3+}$] in addition to an ion at m/z 279 (b_2). Also present are minor contributions for the conventional b_{27-32}^{3+} sequence ions without the loss of MetPhe (unlabeled in spectrum).[106]

These methods, as well as direct MS/MS, can be extended to molecules in excess of 50 kDa. We have obtained limited primary-sequence information from serum albumin proteins (~580 residues, M_r ~66,000) from several different species.[16, 28, 29] Again, multiply charged b-type product ions originating from regions near the termini of the molecule are formed. Product ion spectra for serum albumin proteins can be interpreted

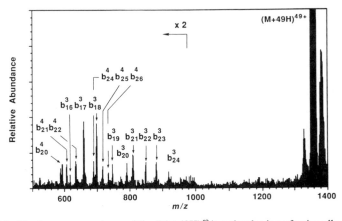

Figure 19. Tandem mass spectrum of the $(M+49H)^{49+}$ molecular ion of ovine albumin with laboratory frame collision energy approximately 6.9 keV.

based on primary structural similarities among serum albumins from various species. For example, a homology (primary-structure alignment) of 80% exists between bovine and human albumins (out of >580 residues) as well as between human and rat albumins, and 63% of the total residues are conserved between the three species. Assuming the majority of cleavage sites to be similar for all serum albumins studied, and excluding the possibility of CAD products originating from regions enclosed by disulfide linkages (since at least two bonds must be cleaved), nearly all the peaks in the ESI MS/MS spectra can be assigned as multiply charged b-fragment ions originating near residue 20 from the NH_2-terminus, as shown in Figure 19 for ovine albumin. Tandem mass spectrometry of a selected multiply charged ion yields fragment ions qualitatively similar to those produced in the atmospheric pressure/vacuum interface. This technique can be further extended by dissociation of CAD-produced ions to examine molecular regions unprobed by initial MS/MS experiments.

5. THE INTERFACE WITH CAPILLARY SEPARATION SCHEMES

A high-resolution separation prior to mass spectral analysis of biological materials is highly desirable because nearly all systems of interest are comprised of mixtures of varying complexity. Even proteins in a highly purified state are often mixtures due to naturally occurring microheterogeneity, or they are converted to mixtures before analysis by chemical or enzymatic processes. Biological sample sizes may be limited, and multiple purification stages are often precluded by other experimental considerations. Practical incentives exist to conduct biochemical research on the smallest scale possible; it is not surprising therefore that combined separations-MS analysis is of broad interest and greater sensitivity is almost always desired.

Liquid chromatography is a mainstay of biochemical analysis. The now commonplace method of "microbore" (1–2 mm i.d. columns with flow rates of 10–100 µl/min) liquid chromatography is particularly well-adapted to ESI practiced with pneumatically assisted nebulization (i.e., ionspray). Examining tryptic digests of proteins is a straightforward application of ESI methods with facile and immediate prospects. The practice of liquid chromatography in very narrow columns and capillary tubes (i.d. <250 µm and flow rates of 0.1–1 ml/min) is a variation that more laboratories are pursuing.

Relatively few examples of the application of LC/ESI MS to proteins have thus far been reported.[113] The use of LC/ESI MS for analysis of protein enzymatic digests has attracted great interest,[103, 104, 113, 114] particularly because sensitivity exists to allow on-line MS/MS of separated polypeptides (e.g., tryptic fragments). Conventional triple quadrupole instrumentation currently provides insufficient sensitivity for useful on-line MS/MS of proteins; however, the FTICR mass spectrometer[115] and the ion trap mass spectrometer[113] should provide superior performance for such applications. An advantage of LC separations currently is that relatively low-buffer or salt concentrations are generally employed, leading to a sensitivity advantage over capillary zone electrophoresis (CZE) (but not necessarily capillary isotachophoresis) at equivalent flow rates. The higher flow rate for LC separations often allow much better "trace" analysis than CZE separations, even though the latter would generally require much less sample and likely produces the lowest absolute mass (but not molar) detection limits.

Capillary electrophoresis (CE) in its various formats (free solution, isotachophoresis, isoelectric focusing, polyacrylamide gel, micellar electrokinetic chromatography) is attracting attention as a method for rapid high-resolution separations of very small sample volumes of complex mixtures. Combined with the inherent sensitivity and selectivity of MS, CE/MS is a powerful bioanalytical technique. The correspondence between CE and ESI flow rates and the fact that both are used primarily for ionic species in solution provide the basis for facile combination. Small peptides are easily amenable to capillary zone electrophoresis with ESI MS analysis with good reproducibility; for example, high-efficiency separations of biologically active peptides[116–118] and tryptic digests[117, 119] have been demonstrated. High-efficiency CZE with detection by ESI MS at the picomole level has been shown.[119, 120] The CZE/ESI MS method has been extended to the direct separation and detection of proteins.[115, 119–121] as shown in Figure 20 for a separation of three myoglobin species. In this example, approximately 100 fmol of each protein was injected. Horse and sheep myoglobins differ only slightly in molecular weight, and they are not resolved mass spectrometrically (see Figure 20c). The on-line CZE separation allows the two components to be distinguished, and more accurate molecular weight measurements can be obtained. Recent work by Hofstadler *et al.* with the combination of CZE/ESI and FTICR mass spectrometry have demonstrated impressive detection limits (6 fmol/component) combined with high resolution mass measurements.[115] However, the analysis of proteins by CZE/MS is challenging, due to the well-established difficulties associated with protein interactions with capillary surfaces. It is unlikely that one capillary or buffer system will be ideal for all protein separations, but rather (as for LC separations) procedures optimized for specific classes of proteins will have to be developed.

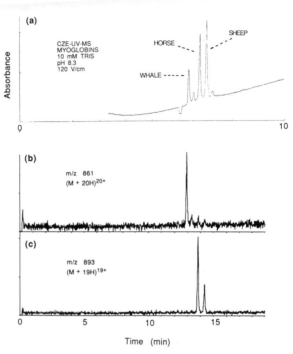

Figure 20. The CZE/MS single-ion electropherogram from an equimolar mixture (100 fmol per component) of myoglobin proteins (sperm whale (M_r 17,199), horse (M_r 16,951), and sheep (M_r 16,923), separated in a pH 8.3 Tris/HCl buffer in an untreated 50 μm i.d. capillary. The on-line UV response is shown on top.

Capillary isotachophoresis (CITP), or displacement electrophoresis, is an attractive complement to CZE and ideally suited for combination with MS.[122–124] The CITP method is amenable to low concentration samples where the amount of solution is relatively large, whereas CZE is ideal for analyzing minute solution quantities. Detection limits of approximately 10^{-11} M have been demonstrated by CITP/ESI MS, and substantial improvements appear feasible.[122] The CITP offers the potential for higher sample loading (and increased molar sensitivity), high-resolution separations, and actual concentration of separated sample bands. Analytes elute in CITP as flat-topped bands, where the length of the analyte band is proportional to the analyte concentration; most importantly, however, CITP provides a relatively pure analyte band to the ESI source without the large concentration of a supporting electrolyte demanded by CZE. This latter characteristic circumvents the principal challenge in the development of CZE/ESI MS, the disadvantageous effect of supporting electrolyte constituents on ESI. Thus, CITP/MS has the potential of allowing much greater sensitivities (and analyte ion currents) than are currently feasible with CZE/MS due to more efficient analyte ionization. The relatively wide and concentrated bands in CITP facilitate MS/MS experiments (which often require more concentrated samples than provided by CZE). These characteristics make CITP MS/MS potentially well-suited for characterizing enzymatic digests of proteins.[125] Significant research effort in all areas of separating biopolymers and their constituents by capillary electrophoresis is required to capitalize on the enormous potential that its combination with ESI MS has demonstrated.

6. CONCLUSIONS

With the emergence and rapid expansion of biochemical- and biotechnology-related fields, new and improved analytical methodologies have become vital for accessing purity and quality control. Electrospray ionization MS has already proven a sensitive detection method for a variety of biomolecular classes and molecular weight ranges. These developments may potentially often rapid and efficient sequencing methods when combined with capillary separation schemes and MS/MS. Work on interfacing ESI with other types of commercial mass spectrometers offering special advantages over quadrupole instruments has only just begun. Double-focusing sector instruments, popular in many research and support facilities, offer higher resolution analysis, a feature that may allow direct measurement of the charge state (by measuring the isotopic spacings). In addition, high-performance *tandem* double-focusing mass spectrometers allows higher keV energy CAD, which may lead to more efficient dissociation processes[126]. The production of multiply charged ions at lower m/z by ESI is particularly advantageous for FT ICR MS because resolution is inversely related to m/z. MS/MS and MS^m ($m > 2$) are easily implemented by ion-trapping devices, such as the ultra high resolution FTICR and the quadrupole ion trap. More complete primary-sequence information may be offered by extended MS/MS methods with ESI.

ACKNOWLEDGEMENT

This work was supported by the US Department of Energy, Office of Health and Environmental Research, (Contract DE-AC06-76RLO 1830), the National Institutes of Health, National Center for Human Genome Research (HG 00327), and the National Science Foundation, Instrumentation and Instrument Development Program (DIR 8908096). Pacific Northwest Laboratory is operated by Battelle Memorial Institute.

REFERENCES

1. Beckey, H. D., *Principles of Field Ionization and Field Desorption Mass Spectrometry* (Pergamon: Oxford, UK, 1977).
2. Sundqvist, B.; Macfarlane, R. D., *Mass Spectrom. Rev.* 1985, **4**, 421–60.
3. Jonsson, G.; Hedin, A.; Hákansson, P.; Sundqvist, B. U. R.; Bennich, H.; and Roepstorff, P., *Rapid. Comm. Mass Spectrom.* 1989, **3**, 190–91.
4. Barber, M.; Bordoli, R. S.; Elliott, G. J.; Sedgewick, R. D.; and Tyler, A. N., *Anal. Chem.* 1982, **54**, 645A–57A.
5. Barber, M.; and Green, B. N., *Rapid Comm. Mass Spectrom.* 1987, **1**, 80–83.
6. Briggs, J.; and Panfili, P. R., *Anal. Chem.* 1991, **63**, 850–59.
7. Karas, M.; and Hillenkamp, F., *Anal. Chem.* 1988, **60**, 2299–2301.
8. Karas, M.; Bahr, U.; Ingendoh, A.; and Hillenkamp, F., *Angew. Chem.* 1989, **101**, 805–6.
9. Dole, M.; Mack, L. L.; Hines, R. L.; Mobley, R. C.; Ferguson, L. D.; and Alice, M. B., *J. Chem. Phys.* 1968, **49**, 2240–49.
10. Covey, T. R.; Bonner, R. F.; Shushan, B. I.; and Henion, J., *Rapid Comm. Mass Spectrom.* 1988, **2**, 249–56.
11. Whitehouse, C. M.; Dreyer, R. N.; Yamashita, M.; and Fenn, J. B., *Anal. Chem.* 1985, **57**, 675–79.
12. Wong, S. F.; Meng, C. K.; and Fenn, J. B., *J. Phys. Chem.* 1988, **92**, 546–50.
13. Meng, C. K.; Mann, M.; and Fenn, J. B., *Z. Phys. D-Atoms, Molecules, and Clusters* 1988, **10**, 361–68.
14. Fenn, J. B.; Mann, M.; Meng, C. K.; Wong, S. F.; and Whitehouse, C. M., *Science* 1989, **246**, 64–71.
15. Loo, J. A.; Udseth, H. R.; and Smith, R. D., *Anal. Biochem.* 1989, **179**, 404–12.
16. Smith, R. D.; Loo, J. A.; Edmonds, C. G.; Barinaga, C. J.; and Udseth, H. R., *Anal. Chem.* 1990, **62**, 882–99.
17. Jardine, I.; Hail, M.; Lewis, S.; Zhou, J.; Schwartz, J.; and Whitehouse, C., in *Proc. Thirty-Eighth ASMS Conf. Mass Spectrom. Allied Topics* (Tuscon, 1990), pp. 16–17.
18. Loo, J. A.; Edmonds, C. G.; Udseth, H. R.; and Smith, R. D., *Anal. Chem.* 1990, **62**, 693–98.
19. Gallagher, R. T.; Chapman, J. R.; and Barton, E. C., in *Proc. Thirty-Ninth ASMS Conf. Mass Spectrom. Allied Topics* (Nashville, 1991), pp. 242–43.
20. Feng, R.; and Konishi, Y., *Anal. Chem.* 1992, **64**, 2090–95.
21. *Protein Folding: Deciphering the Second-Half of the Genetic Code*, Gierasch, L. M.; and King, J., eds. (Amercian Association for the Advancement of Science: Washington, DC, 1990).
22. Biemann, K.; and Scoble, H. A., *Science* 1987, **237**, 992–98.
23. Biemann, K., *Biomed. Mass Spectrom.* 1988, **16**, 99–111.
24. *Tandem Mass Spectrometry*, Mc Lafferty. F. W., ed. (Wiley: New York, 1983).
25. Busch, K. L.; Glish, G. L.; and McLuckey, S. A., *Mass Spectrometry/Mass Spectrometry: Techniques and Applications of Tandem Mass Spectrometry* (VCH: New York, 1988).
26. Johnson, R. S.; and Biemann, K., *Biochem.* 1987, **26**, 1209–14.
27. Johnson, R. S.; Matthews, W. R.; Biemann, K.; and Hopper, S., *J. Biol. Chem.* 1988, **263**, 9589–97.
28. Loo, J. A.; Edmonds, C. G.; Udseth, H. R.; and Smith, R. D., *Anal. Chim. Acta* 1990, **241**, 167–73.
29. Loo, J. A.; Edmonds, C. G.; and Smith, R. D., *Anal. Chem.* 1991, **63**, 2488–99.

30. Hirayama, K.; Akashi, S.; Furuya, M.; and Fukuhara, K., *Biochem. Biophys. Res. Commun.* 1990, **173**, 639–46.
31. Hunt, D. F.; Yates III, J. R.; Shabanowitz, J.; Winston, S.; and Hauer, C. R., *Proc. Nat. Acad. Sci. USA* 1986, **83**, 6233–37.
32. Carr, S. A.; Green, B. N.; Hemling, M. E.; Roberts, G. D.; Anderegg, R. J.; and Vickers, R., in *Proc. Thirty-Fifth ASMS Conf. Mass Spectrom. Allied Topics* (Denver, 1987), pp. 830–31.
33. Bean, M. F.; Carr, S. A.; Thorne, G. C.; Reilly, M. H.; and Gaskell, S. J., *Anal. Chem.* 1991, **63**, 1473–81.
34. Gross, M. L.; Tomer, K. B.; Cerny, R. L.; and Giblin, D. E., in *Mass Spectrometry in the Analyses of Large Molecules*; McNeal, C. J., ed. (Wiley, Chichester, UK, 1986), pp. 171–90.
35. Neumann, G. M.; Sheil, M. M.; and Derrick, P. J., *Z. Naturforsch.* 1984, **39a**, 584–92.
36. Hunt, D. F.; Shabanowitz, J.; Yates III, J. R., *J. Chem. Soc. Chem. Commun.* 1987, 548–50.
37. Lebrilla, C. B.; Wang, D. T.-S.; Mizoguchi, T. J.; and McIver, R. T., Jr., *J. Am. Chem. Soc.* 1989, **111**, 8593–98.
38. Mabud, M. A.; DeKrey, M. J.; and Cooks, R. G., *Int. J. Mass Spectrom. Ion Processes* 1985, **67**, 285–94.
39. Cooks, R. G.; and Kaiser, R. E., Jr., *Acc. Chem. Res.* 1990, **23**, 213–19.
40. Barinaga, C. J.; Edmonds, C. G.; Udseth, H. R.; and Smith, R. D., *Rapid Comm. Mass Spectrom.* 1989, **3**, 160–64.
41. Loo, J. A.; Edmonds, C. G.; and Smith, R. D., *Science* 1990, **248**, 201–4.
42. Smith, R. D.; Loo, J. A.; Barinaga, C. J.; Edmonds, C. G.; and Udseth, H. R., *J. Am. Soc. Mass Spectrom.* 1990, **1**, 53–65.
43. Iribarne, J. V.; and Thomson, B. A., *J. Chem. Phys.* 1976, **64**, 2287–94.
44. Thomson, B. A.; and Iribarne, J. V., *J. Chem. Phys.* 1979, **71**, 4451–63.
45. Röllgen, F. W.; Bramer-Wegen, E.; and Büffering, L., *J. Phys. Colloq.* 1987, **48**, 253–56.
46. Schmeizeisen-Redeker, G.; Büffering, L.; and Röllgen, F. W., *Int. J. Mass Spectrom. Ion Processes* 1989, **90**, 139–50.
47. Lüttgens, U.; Röllgen, F. W.; and Cook, K. D., in *Methods and Mechanisms for Producing Ions from Large Molecules*, Standing, K. G. and Ens, W., eds. (Plenum: New York, 1991), pp. 185–93.
48. Nohmi, T.; and Fenn, J. B., *J. Am. Chem. Soc.* 1992, **114**, 3241–46.
49. Mack, L. L.; Kralik, P.; Rheude, A.; and Dole, M., *J. Chem. Phys.* 1970, **52**, 4977–86.
50. Gieniec, J.; Mack, L. L.; Nakamae, K.; Gupta, C.; Kuman, V.; and Dole, M., *Biomed. Mass Spectrom.* 1984, **11**, 259–68.
51. Yamashita, M.; and Fenn, J. B., *J. Phys. Chem.* 1984, **88**, 4451–59.
52. Yamashita, M.; and Fenn, J. B., *J. Phys. Chem.* 1984, **88**, 4671–75.
53. Alexsandrov, M. L.; Gall, L. M.; Krasnov, N. V.; Nikolaev, V. I.; Pavlenko, V. A.; and Shkurov, V. A., *Dokl. Akad. Nauk SSSR* 1984, **277**, 379–83.
54. Alexsandrov, M. L.; Gall, L. M.; Krasnov, N. V.; Nikolaev, V. I.; and Shkurov, V. A., *J. Anal. Chem. USSR* 1986, **40**, 1227–36.
55. Mann, M.; Meng, C. K.; and Fenn, J. B., *Anal. Chem.* 1989, **61**, 1702–08.
56. Chowdhury, S. K.; Katta, V.; and Chait, B. T., *Rapid Comm. Mass Spectrom.* 1990, **4**, 81–87.
57. Feng, R.; Konishi, Y.; and Bell, A. W., *J. Am. Soc. Mass Spectrom.* 1991, **2**, 387–401.
58. (a) Henry, K. D.; Quinn, J. P.; and McLafferty, F. W., *J. Am. Chem. Soc.* 1991, **113**, 5447–49. (b) Beu, S. C.; Senko, M. W.; Quinn, J. P.; Wampler, F. M., III; and McLafferty, F. W., *J. Am. Soc. Mass Spectrom.* 1993, **4**, 557–65.

59. Bruins, A. P.; Covey, T. R.; and Henion, J. D., *Anal. Chem.* 1987, **59**, 2642–46.
60. Finch, J. W.; Musselman, B. D.; Banks, J. F.; and Whitehouse, C. M., in *Proc. Forty-First ASMS Conf. Mass Spectrom. Allied Topics* (San Francisco, 1993), p. 287.
61. Smith, R. D.; Barinaga, C. J.; and Udseth, H. R.; *Anal. Chem.* 1988, **60**, 1948–52.
62. Mylchreest, I.; Hail, M., in *Proc. Thirty-Ninth ASMS Conf. Mass Spectrom. Allied Topics* (Nashville, 1991), pp. 316–17.
63. Ikonomou, M. G.; Blades, A. T.; and Kebarle, P., *J. Am. Soc. Mass Spectrom.* 1991, **2**, 497–505.
64. Larsen, B. S.; and McEwen, C. N., *J. Am. Soc. Mass Spectrom.* 1991, **2**, 205–11.
65. Ogorzalek Loo, R. R.; Udseth, H. R.; and Smith, R. D., *J. Phys. Chem.* 1991, **95**, 6412–15.
66. Rockwood, A. L.; Busman, M.; Udseth, H. R.; and Smith, R. D., *Rapid Commun. Mass Spectrom.* 1991, **5**, 582–85.
67. Van Berkel, G. J.; Glish, G. L.; and McLuckey, S. A., *Anal. Chem.* 1990, **62**, 1284–95.
68. Schwartz, J. C.; Syka, J. E. P.; and Jardine, I., *J. Am. Soc. Mass Spectrom.* 1991, **2**, 198–204.
69. Williams, J. D.; Cox, K. A.; Cooks, R. G.; Kaiser, R. E., Jr.; and Schartz, J. C., *Rapid Comm. Mass Spectrom.* 1991, **5**, 327–29.
70. Henry, K. D.; Williams, E. R.; Wang, B. H.; McLafferty, F. W.; Shabanowitz, J.; and Hunt, D. F., *Proc. Nat. Acad. Sci. USA* 1989, **86**, 9075–78.
71. Loo, J. A.; Quinn, J. P.; Ryu, S. I.; Henry, K. D.; Senko, M. W.; and McLafferty, F. W., *Proc. Nat. Acad. Sci. USA* 1992, **89**, 286–89.
72. Hofstadler, S. A.; and Laude, D. A., Jr., *Anal. Chem.* 1992, **64**, 569–72.
73. Boyle, J. G.; Whitehouse, C. M.; and Fenn, J. B., *Rapid Comm. Mass Spectrom.* 1991, **5**, 400–5.
74. Loo, J. A.; Edmonds, C. G.; Udseth, H. R.; and Smith, R. D., *Anal. Chem.* 1990, **62**, 693–98.
75. Bojesen, G., *J. Am. Chem. Soc.* 1987, **109**, 5557–58.
76. Chait, B. T.; Chowdhurry, S. K.; and Katta, V., in *Proc. Thirty-Ninth ASMS Conf. Mass Spectrom. Allied Topics* (Nashville, 1991), pp. 447–48.
77. Moore, W. T.; Suter, M.-J. F.; Farmer, T. B.; and Caprioli, R. M., in *Proc. Thirty-Ninth ASMS Conf. Mass Spectrom. Allied Topics* (Nashville, 1991), pp. 256–57.
78. Loo, J. A.; and Smith, R. D., in *Proc. Fortieth ASMS Conf. on Mass Spectrom. and Allied Topics* (Washington, DC, 1992), pp. 627–28.
79. Allen, M. H.; Field, F. H., and Vestal, M. L., in *Proc. Thirty-Eight ASMS Conf. Mass Spectrom. Allied Topics* (Tuscon, 1990), pp. 431–32.
80. Mirza, U. A.; Cohen, S. L.; and Chait, B. T., *Anal. Chem.* 1993, **65**, 1–6.
81. Katta, V.; and Chait, B. T., *J. Am. Chem. Soc.* 1991, **113**, 8534–35.
82. Ganem, B.; Li, Y.-T.; and Henion, J. D., *J. Am. Chem. Soc.* 1991, **113**, 6294–95.
83. Ganem, B.; Li, Y.-T.; and Henion, J. D., *J. Am. Chem. Soc.* 1991, **113**, 7818–19.
84. (a) Baca, M.; and Kent, S. B. H., *J. Am. Chem. Soc.* 1992, **114**, 3992–93. (b) Ganguly, A. K.; Pramanik, B. N.; Tsarbopoulos, A.; Covey, T. R.; Huang, E.; and Fuhrman, S. A., *J. Am. Chem. Soc.* 1992, **114**, 6559–60. (c) Ogorzalek Loo, R. R.; Goodlett, D. R.; Smith, R. D.; and Loo, J. A., *J. Am. Chem. Soc.* 1993, **115**, 4391–92. (d) Light-Wahl, K. J.; Winger, B. E.; and Smith, R. D., *J. Am. Chem. Soc.* 1993, **115**, 5869–70.
85. (a) Ganem, B.; Li, Y. T.; and Henion, J. D., *Tetrahedron Lett.* 1993, **34**, 1445–48. (b) Light-Wahl, K. J.; Springer, D. L.; Winger, B. E.; Edmonds, C. G.; Camp, D. G., III; Thrall, B. D.; and Smith, R. D., *J. Am. Chem. Soc.* 1993, **115**, 803–4.
86. Hutchens, T. W.; Nelson, R. W.; Allen, M. H.; Li, C. M.; and Yip, T.-T., *Biol. Mass Spectrom.* 1992, **21**, 151–9.

87. Edmonds, C. G.; Loo, J. A.; Ogorzalek Loo, R. R.; and Smith, R. D., in *Techniques in Protein Chemistry II*, Villafranca, J. J., ed. (Academic Press: San Diego, 1991), pp. 487–95.

88. Cody, R. B.; Tamura, J.; and Musselman, B. D., *Anal. Chem.* 1992, **64**, 1561–70.

89. Chowdhury, S. K.; Katta, V.; and Chait, B. T., *J. Am. Chem. Soc.* 1990, **112**, 9012–13.

90. Loo, J. A.; Ogorzalek Loo, R. R.; Udseth, H. R.; Edmonds, C. G.; and Smith R. D., *Rapid Comm. Mass Spectrom.* 1991, **5**, 101–5.

91. Feng, R.; and Konishi, Y., in *Proc. Thirty-Ninth ASMS Conf. Mass Spectrom. Allied Topics* (Nashville, 1991), pp. 1432–33.

92. LeBlanc, J. C. Y.; Beuchemin, D.; Siu, K. W. M.; Guevremont, R.; and Berman, S. S., *Org. Mass Spectrom.* 1991, **26**, 831–39.

93. Franks, R.; and Eagland, D., *CRC Crit. Rev. Biochem.* 1975, **3**, 165–219.

94. *Ubiquitin*, Rechsteiner, M., ed. (Plenum: New York, 1988).

95. Katta, V.; and Chait, B. T., *J. Am. Chem. Soc.* 1993, **115**, 6317–21.

96. Loo, J. A.; Ogorzalek Loo, R. R.; Light, K. J.; Edmonds, C. G.; and Smith, R. D., *Anal. Chem.* 1992, **64**, 81–88.

97. Stults, J. T.; and Marsters, J. C., *Rapid Comm. Mass Spectrom.* 1991, **5**, 359–63.

98. Hunt, D. F.; Zhu, N.-Z.; and Shabanowitz, J., *Rapid Comm. Mass Spectrom.* 1989, **3**, 122–24.

99. Barber, M.; Bell, D. J.; Morris, M.; Tetler, L. W.; Woods, M. D.; Monaghan, J. J.; and Morden, W. E., *Org. Mass Spectrom.* 1989, **24**, 504–10.

100. Neumann, G. M.; and Derrick, P. J., *Aust. J. Chem.* 1984, **37**, 2262–77.

101. Johnson, R. S.; Martin, S. A.; and Biemann, K., *Int. J. Mass Spectrom. Ion Processes* 1988, **86**, 137–54.

102. Hail, M.; Lewis, S.; Jardine, I.; Liu, J.; and Novotny, M., *J. Microcolumn.* 1990, **2**, 285–92.

103. Huang, E. C.; and Henion, J. D., *J. Am. Soc. Mass Spectrom.* 1990, **1**, 158–65.

104. Covey, T. R.; Huang, E. C.; and Henion, J. D., *Anal. Chem.* 1991, **63**, 1193–1200.

105. Bunker, D. L.; and Wang, F.-M., *J. Am. Chem. Soc.* 1977, **99**, 7457–59.

106. Loo, J. A.; Edmonds, C. G.; and Smith, R. D., *Anal. Chem.* 1993, **65**, 425–38.

107. Martin, S. A.; and Biemann, K., *Int. J. Mass Spectrom. Ion Processes* 1987, **78**, 213–28.

108. Vestling, M.; Hua, S.; Murphy, C.; Orlando, R.; Wu, Z.; and Fenselau, C., in *Proc. Thirty-Ninth ASMS Conf. Mass Spectrom. Allied Topics* (Nashville, 1991), pp. 1473–74.

109. Gaskell, S. J.; and Reilly, M. H., *Rapid Comm. Mass Spectrom.* 1988, **2**, 188–91.

110. Loo, J. A.; Udseth, H. R.; and Smith, R. D., *Rapid Comm. Mass Spectrom.* 1988, **2**, 207–10.

111. Smith, R. D.; and Barinaga, C. J., *Rapid Comm. Mass Spectrom.* 1990, **4**, 54–57.

112. Feng, R.; and Konishi, Y., *Anal. Chem.* 1993, **65**, 645–49.

113. McLuckey, S. A.; Van Berkel, G. J.; Glish, G. L.; Huang, E. C.; and Henion, J. D., *Anal. Chem.* 1991, **63**, 375–83.

114. (a) Lee, E. D.; Henion, J. D.; and Covey, T. R., *J. Microcolumn.* 1989, **1**, 14–18. (b) Huang, E. C.; and Henion, J. D., *Anal. Chem.* 1991, **63**, 732–39. (c) Huang, E. C.; Wachs, T.; Conboy, J. J.; and Henion, J. D., *Anal. Chem.* 1990, **62**, 713A–25A.

115. Hofstadler, S. A.; Wahl, J. H.; Bruce, J. E.; and Smith, R. D., *J. Am. Chem. Soc.* 1993, **115**, 6983–84.

116. Lee, E. D.; Mück, W.; Henion, J. D.; and Covey, T. R., *J. Chromatogr.* 1988, **458**, 313–21.

117. Lee, E. D.; Mück, W.; Henion, J. D.; and Covey, T. R., *Biomed. Environ. Mass Spectrom.* 1989, **18**, 313–21.

118. Johansson, I. M.; Huang, E. C.; Henion, J. D.; and Zweigenbaum, J., *J. Chromatogr.* 1991, **554**, 311–27.
119. Smith, R. D.; Udseth, H. R.; Barinaga, C. J.; and Edmonds C. G., *J. Chromatogr.* 1991, **559**, 197–207.
120. Loo, J. A.; Jones, H. K.; Udseth, H. R.; and Smith, R. D., *J. Microcolumn Separations* 1989, **1**, 223–29.
121. Thibault, P.; Paris, C.; and Pleasance, S., *Rapid Comm. Mass Spectrom.* 1991, **5**, 484–90.
122. Udseth, H. R.; Loo, J. A.; and Smith, R. D., *Anal. Chem.* 1989, **61**, 228–32.
123. Smith, R. D.; Loo, J. A.; Barinaga, C. J.; Edmonds, C. G.; and Udseth, H. R., *J. Chromatogr.* 1989, **480**, 211–32.
124. Smith, R. D.; Fields, S. M.; Loo, J. A.; Barinaga, C. J.; Udseth, H. R.; and Edmonds, C. G., *Electrophoresis* 1990, **11**, 709–17.
125. Edmonds, C. G.; Loo, J. A.; Fields, S. M.; Barinaga, C. J.; Udseth, H. R.; and Smith, R. D., in *Biological Mass Spectrometry*, Burlingame, A. L.; and McCloskey, J. A., eds. (Elsevier: Amsterdam, 1990), pp. 77–100.
126. Fabris, D.; Kelly, M.; Murphy, C.; Wu, Z.; and Fenselau, C., *J. Am. Soc. Mass Spectrom.* 1993, **4**, 652–61.

8

Liquid Chromatography/Fast Atom Bombardment Mass Spectrometry

Justin G. Stroh and Kenneth L. Rinehart

1. INTRODUCTION

The fast atom bombardment (FAB) ionization technique,[1] introduced in 1981, opened new vistas in the analysis of compounds previously intractable to mass spectrometry (MS). Before its introduction, compounds that were polar or thermally unstable required laborious and difficult derivatization prior to mass spectral analysis. Although derivatization often produced compounds that were observable by electron ionization (EI) or chemical ionization (CI), the spectra were complex. FABMS, with no derivatization required, yields simpler and more easily interpretable spectra; since its introduction, FABMS has become the method of choice for polar compounds. Probe FABMS is fundamentally different from such ionization techniques as EI and CI in that FAB is a desorption technique. With FABMS, the analyte is dissolved or dispersed in a polar matrix [e.g., glycerol, thioglycerol, m-nitrobenzyl alcohol, diethanolamine, or a liquid mixture of dithiothreitol and dithioerythritol ("magic bullet")],[2] and the mixture is applied to a probe target. The target is inserted into the ion source of the mass spectrometer where its surface is bombarded with high-energy (8–10 keV) atoms [or in matrix secondary ion (SI)MS, with ions]. Cation attachment to, or proton removal from, the analyte occurs, producing ions that can then be analyzed in the mass spectrometer. Not only are molecular ion species produced, but often

Justin G. Stroh • Pfizer Central Research, Groton, Connecticut 06340. *Kenneth L. Rinehart* • School of Chemical Sciences, University of Illinois, Urbana, Illinois 61801.

Experimental Mass Spectrometry, edited by David H. Russell. Plenum Press, New York, 1994.

structurally useful fragment ions as well. Although FABMS is simple in design, it works best if the analyte is pure, unless tandem mass spectrometry (MS/MS) is used to study mixtures' individual components. The surface activity of impurities or other components can be a problem with FABMS, and their removal or separation is often carried out by using high-performance liquid chromatography (HPLC).

During the last two decades, the considerable interest in coupling HPLC with MS (LC/MS) has been driven largely by the desire to analyze compounds of biological and environmental importance. Many compounds in these two broad areas of research are polar, thermally unstable, and occur in small amounts in a vast matrix of impurities. The numerous LC/MS interfaces designed to analyze these types of compounds include direct liquid introduction (used with CI),[3] moving belt FAB,[4] continuous flow FAB,[5–7] particle beam/FAB,[8] and a number of interfaces that provide their own ionizations,[9] thermospray,[10] magic,[11] electrospray,[12,13] and ion spray.[14] Chapter 8 describes moving belt, continuous flow, and particle beam FABMS interfaces and some applications of the technique.

2. MOVING BELT INTERFACE

The moving belt interface[4] was the first LC/FABMS technique, initially introduced for use with EI and CI[15] and later adapted for FAB. Benninghoven et al.[16] developed a moving wire interface for SIMS of amino acids. Further refinement by Smith et al.[17] included a jet spray depositor between the LC and the moving belt interface for use with SIMS. Dobberstein et al.[4] were the first to use the moving belt interface directly with FABMS, while Lewis and Brooks[18] demonstrated the power of the technique by analyzing a synthetic mixture of uridine and adenosine. Subsequently, Stroh et al.[19–21] replaced the spray depositor with a contact frit depositor and employed microbore (1-mm i.d.) columns.

2.1. Design and Operation of the Moving Belt Interface

A schematic diagram of a moving belt interface is shown in Figure 1. Samples pass from the HPLC column through a heated jet spray depositor and onto a moving belt. The heated jet spray separates the analyte from most of the HPLC solvent by nebulization and evaporation, and the gaseous solvent is then removed under vacuum. The analyte deposited on the moving belt is transported through three vacuum locks, employing two additional pumps, and into the high-vacuum ion chamber of the mass

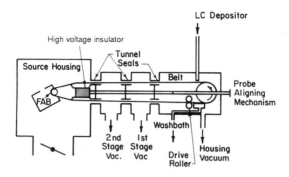

Figure 1. Schematic diagram of a moving belt interface.

spectrometer. Samples on the moving belt are then bombarded directly with a stream of atoms. Unlike probe FABMS, no matrix is employed with this technique; belt movement provides a continually replenished surface for the fast atoms to bombard. Excess sample is removed from the belt by a methanol clean-up washer located outside the high-vacuum region, and the belt is transported to collect new sample from the jet spray depositor. Difficulties were observed with the jet spray depositor when the heat required to vaporize the solvent degraded thermally labile compounds; moreover, puddling was a problem. These difficulties were overcome by the development of a contact frit depositor that allowed solvent leaving the HPLC to be deposited onto the moving belt in the liquid phase.[21] Once on the belt, the solvent was removed by vacuum alone, requiring the use of microbore columns at flow rates no larger than 50–100 μL/min.

A major advantage of the moving belt interface for LC/FABMS is its abundant production of fragment ions. Probe FABMS and moving belt LC/FABMS were compared in a study of paulomycins A (**1**) and B (**2**).[21]

1: R = CH₃
2: R = H

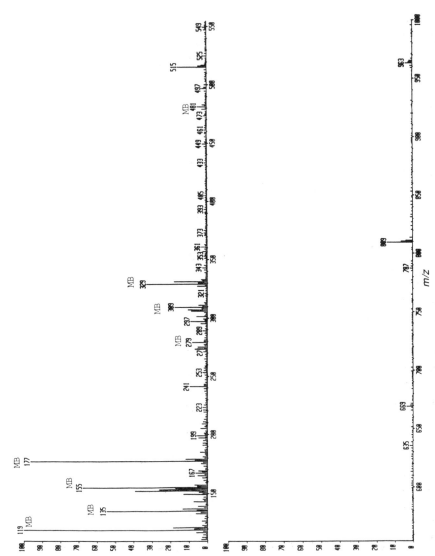

Figure 2. Paulomycin A by probe FABMS. *Top*: low mass range; *bottom*: high mass range. MB: magic bullet.

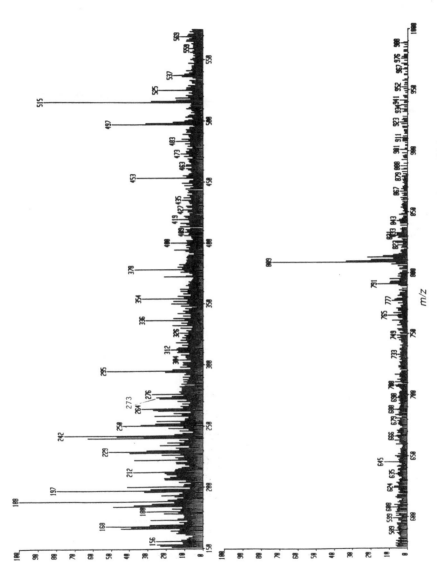

Figure 3. Paulomycin A by LC/FAB ms. *Top*: low mass range; *bottom*: high mass range.

Of the three matrices used for the probe experiment—glycerol, thio-glycerol, and magic bullet—magic bullet gave the best results, as shown for 1 in Figure 2. An $[M + Na]^+$ ion appears at m/z 809 with two fragment ions at m/z 515 and 241. Other ions in this spectrum arise from the matrix. In contrast to the probe FAB mass spectrum of 1, the moving belt LC/FAB mass spectrum shows not only the $[M + Na]^+$ ion at m/z 809, but numerous fragment ions as well (see Figure 3), suggesting that with its lack of a matrix the moving belt technique involves a higher energy process than probe FABMS. In the probe FABMS experiment, the matrix not only provides a continually replenished sample surface for ionization by the fast atom beam, but it also absorbs some of the excess internal energy created by such a process. When no matrix is available (as with the moving belt experiment), excess internal energy is translated into increased fragmenta-tion, which can be very useful in the structural analysis of unknown com-pounds. It has also been noted that the amount of fragmentation can be varied by varying belt speed,[22] thereby changing the energy imparted to the analyte molecules and producing different amounts of internal energy. If an organic compound is placed on the moving belt and the sample is rotated into the path of the atom beam and stopped, a sudden burst of ions is observed followed by a rapid decay in signal, decreasing to zero signal in less than 1 second. This phenomenon, reminiscent of SIMS, results from ruptured organic covalent bonds that generate a charred surface impenetrable by atom beams.

Major disadvantages of moving belt LC/FABMS are due to the nature of the moving belt interface, which is (1) costly, (2) available from only a few manufacturers, (3) difficult to use if there is a high percentage of water in the mobile phase, and (4) difficult to manage. Of these four drawbacks, difficulties with high water content in the mobile phase and in management are by far the major concerns, since many applications of LC/MS require an HPLC mobile phase that contains 50–100% water, and the moving belt interface requires three vacuum pumps, meticulously cleaned and conditioned belts; also, the ion beam is often unstable.

2.2. Applications of the Moving Belt Interface

Several applications of LC/FABMS using the moving belt interface have been reported. FABMS is quite useful in analyzing peptides, and moving belt LC/FABMS has been applied to antiamoebin I (3, a hexadeca-peptide antibiotic), a mixture of peptides obtained from the partial hydrolysis of antiamoebin I,[20] and to the zervamicins (4, zervamicin IC).[21] The moving belt technique revealed the presence of several new components of the peptide antibiotic leucinostatin.[23] The known leucino-

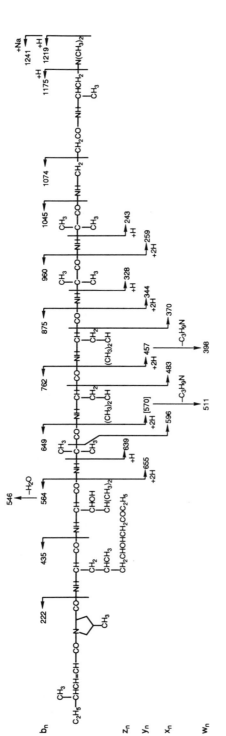

Scheme 1. The LC/FAB fragmentations for leucinostatin A.

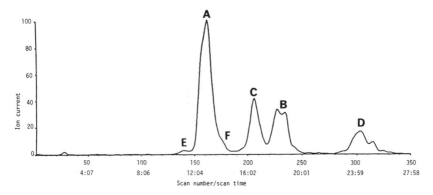

Figure 4. Reconstructed ion chromatogram of a leucinostatin mixture.

statins A ($\equiv$CC-1014) and B (see Figure 4) were identified by their
LC/FAB mass spectra, and the structures of two new compounds, leucinostatins C and D, were assigned based on comparison of their LC/FAB
mass spectra to that of leucinostatin A (see Scheme 1). Two other components observed in very low abundance in this peptides mixture have
molecular weights corresponding to leucinostatins A and C with an extra
oxygen. No fragment ions were observed for these trace components, and
therefore their structures could not be completed.

3: Ac-Phe-Aib-Aib-Aib-Iva-Gly-Leu-Aib-Aib-Hyp-Gln-Iva-Hyp-Aib-Pro-Phol
4: Ac-Trp-Ile-Glu-Iva-Ile-Thr-Aib-Leu-Aib-Hyp-Gln-Aib-Hyp-Aib-Pro-Phol

The antitumor, antiviral, and immunosuppressive cyclic depsipeptide
didemnins[24] were also studied by this technique.[21] The LC/FABMS
chromatogram (Figure 5) indicated the presence of four compounds; closer
examination, however, and the use of single-ion plots, showed that the first
peak in the chromatogram contained didemnins D (**7**) and E (**8**) and the
third peak contained didemnins A (**5**) and B (**6**). The observation of a new
didemnin, labeled didemnin X in the chromatogram, led to the identification of didemnins X (**9**) and Y (**10**).[25]

Another application of this technique is in the analysis of carbohydrates, including stachyose, maltopentaose, maltohexose, β-cyclodextrin,
$Man_5GlcNAc_2$, viridopentaose-B, and numerous oligosaccharides isolated
from sheep urine.[22] Moving belt LC/FABMS proved to be a key technique
in locating and identifying component ecteinascidins (Et's) in an HPLC
separation of a mixture of these antitumor heterocycles.[26–28] As seen in
Figure 6, ecteinascidins 743 (**11**), 745 (**12**), and two isomeric 759s (**13, 14**)

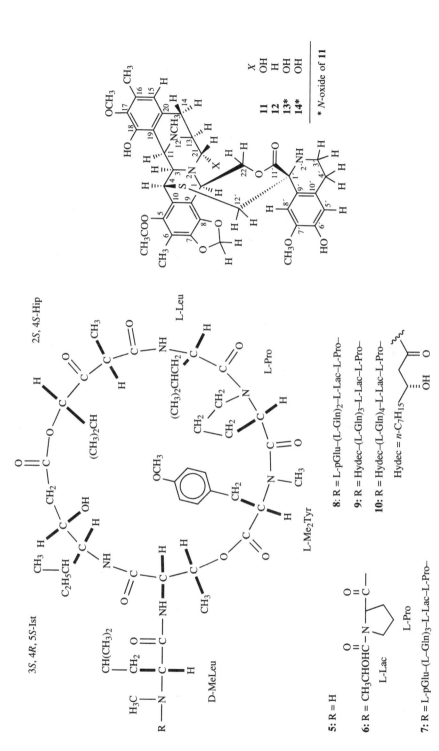

X

11   OH

12   H

13*   OH

14*   OH

* *N*-oxide of **11**

3*S*, 4*R*, 5*S*-Ist

2*S*, 4*S*-Hip

L-Leu

L-Pro

D-MeLeu

L-Me₂Tyr

5: R = H

6: R = CH₃CHOHC—N   —C—

    L-Lac   L-Pro

7: R = L-pGlu–(L-Gln)₃–L-Lac–L-Pro–

8: R = L-pGlu–(L-Gln)₂–L-Lac–L-Pro–

9: R = Hydec–(L-Gln)₃–L-Lac–L-Pro–

10: R = Hydec–(L-Gln)₄–L-Lac–L-Pro–

Hydec = *n*-C₇H₁₅

      OH

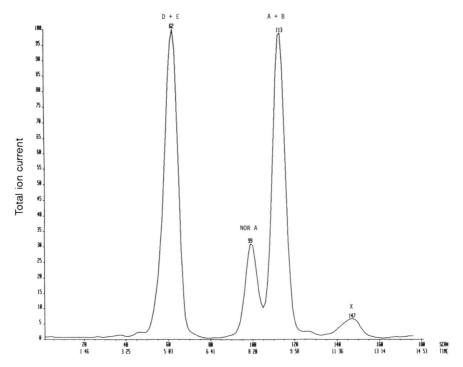

Figure 5. Reconstructed ion chromatogram of a didemnin mixture (77% methanol, 23% water, 0.23% triethylamine, 0.10% acetic acid, flow rate 75 μL/min, 5 μg injected).

were identified by single-ion plots. Their structures were subsequently assigned as shown. (It should be noted that the molecular ions observed at m/z 744, and 760 were actually $M + H - H_2O$ ions; $M - H$ ions were subsequently observed by negative-ion FABMS).[27]

The moving belt interface has been employed to monitor enzymatic reactions of peptides in real-time. Initially Caprioli *et al.*[29] employed FABMS with a standard probe to observe the time course of enzymatic reactions. Samples containing the substrate, enzyme, glycerol, and aqueous buffer were spotted on the FAB probe and inserted into the ion source for analysis. It was determined that enzymatic reactions could be monitored in real-time using this technique when up to 70% glycerol and 30% aqueous buffer were used; however, the activity of trypsin decreased over time due to evaporation of the water.[30] Murawski[31] demonstrated that these two disadvantages (use of a matrix and loss of enzyme activity due to loss of water) could be overcome by using the moving belt technique. The absence of a glycerol matrix is important for three reasons: (1) The matrix can

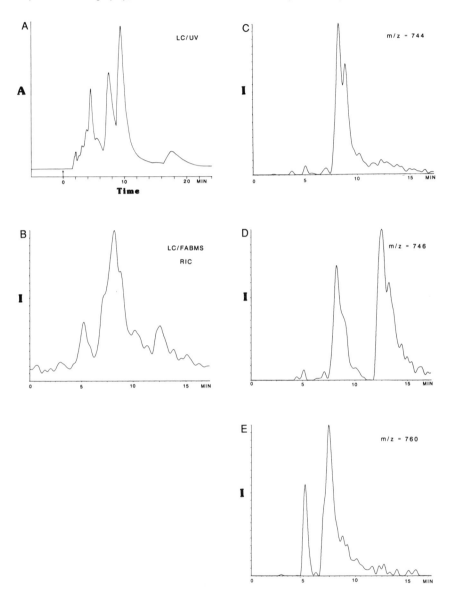

Figure 6. (A) The UV chromatogram for a mixture of ecteinascidins 743, 745, 759A, and 759B [Whatman Partisil 5 octadecylsilyl (ODS) column (5 mm, 4.6 × 250 mm), 70:30:0.1 : MeOH:H₂O:NH₄HCO₂, 1.0 mL/min]. (B) Reconstructed ion chromatogram (RIC) for the same mixture [Alltech C18 microbore column (10 mm, 1 × 250 mm), 70:30:0.1 : MeOH:H₂O:NH₄HCO₂, 100 mL/min]. (C) Ion chromatogram for m/z 744. (D) Ion chromatogram for m/z 746. (E) Ion chromatogram for m/z 760.

interfere with the activity of the enzyme; (2) the matrix produces intense ions at low mass, which can interfere with structurally useful ions in that mass range; and (3) the matrix can cause ion-suppression effects. It was found that by using the moving belt technique, transient products of enzymatic digestion could be observed, their kinetics determined, and 100% aqueous buffers could be used. The moving belt technique has been used to observe tryptic digestion of *N-p*-tosyl-L-arginine methyl ester (TAME), serum thymic factor, mastoparan, and carboxypeptidase Y digestion of bradykinin.[31]

3. CONTINUOUS FLOW FAB

Limits on the water content of the HPLC solvent system restrict the moving belt technique, since many compounds amenable to FAB are water-soluble. Development of the continuous flow FAB interface has alleviated this problem.

3.1. Design of the Continuous Flow FAB Interface

Continuous flow FAB (CF/FAB), also called frit FAB and dynamic FAB, was introduced by Ito *et al.* in 1985.[5] In the original design, samples were injected into a capillary HPLC column that ran the length of the FAB probe and terminated in the ion block of the mass spectrometer. A stainless steel frit was placed at the end of the column to disperse the eluent and act as a probe target for the ionization process. This design was improved by the addition of a position adjuster for the interface, and gradient elution chromatography was carried out in conjunction with this technique.

In 1986, Caprioli *et al.*[6] described an interface for the direct introduction of samples into a FAB ion source using a continuous flow of solvent (see Figure 7). With this interface, pure samples were injected into a solvent flow and delivered to the FAB ion source via a fused silica capillary just protruding from the end of the probe where the solvent covered a copper tip. No stainless steel frit was used.

It was immediately recognized that CF/FAB would be of great use for both LC/FABMS and analyzing single-component samples. While virtually any solvent system could be used to introduce the sample into the ion source of the mass spectrometer, a small amount of matrix had to be added to the solvent system to obtain a reasonable spectrum. Initially, approximately 10% glycerol was added to the solvent system for two

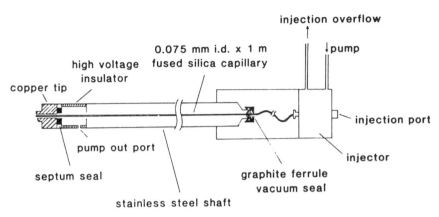

Figure 7. A continuous flow FAB interface for the direct introduction of sample. (Reprinted with permission from Reference 6, copyright 1986 American Chemical Society.)

reasons. As in probe FABMS, the matrix provides a dispersion medium for the analyte as well as a way of distributing excess internal energy. With CF/FAB, the matrix also reduces thermal cooling due to jet expansion of a stream flowing into a high-vacuum region. If no viscous matrix is used, water in the solvent system freezes at the tip of the interface, clogging the sample flow at the surface. Because 10% glycerol provided more matrix than required, 5% glycerol is now commonly used.

A major advantage of CF/FAB over probe FAB is an apparent increase in sensitivity. In general, approximately 1 nanomole of sample is required to obtain a FAB mass spectrum using a probe. On the other hand, a few picomoles are often sufficient for a FAB mass spectrum using a CF/FAB interface. Two basic reasons for the increased sensitivity of CF/FAB are reduced ion suppression effects and reduced background.[32] Ion suppression is a major difficulty with FAB; because it is a surface technique, whatever resides on the droplet surface of the target is observed in the FAB mass spectrum. If a 99% pure analyte with low surface activity contains 1% of an impurity with high surface activity, the probe FAB mass spectrum is representative of the 1% impurity. This problem is greatly reduced with CF/FAB because solvent and analyte are constantly mixed at the surface of the target. Reduction of the chemical background is significant because most of the substance on a probe FAB target is matrix, which gives rise to ion signals at almost every mass unit and contributes significantly to the total ion current. At low masses, the spectrum is dominated by signals arising from the matrix. These problems with the matrix are largely overcome with CF/FAB, because the analyte is admitted to the ion source with a low matrix concentration (only 5%).

One of the initial difficulties with the CF/FAB technique was that the required addition of glycerol (at 5–10% relative concentration) to the solvent system tended to interfere with chromatographic separations. For normal-bore HPLC, this could be easily remedied by post-column addition of the matrix. Post-column addition for capillary columns was introduced by Moseley et al.[33] with the development of a coaxial design in which the capillary HPLC column was surrounded by another column containing solvent and matrix. The capillary HPLC and matrix columns both terminate at the probe tip where mixing occurs.

In the operation of CF/FAB, there are four interrelated difficulties: temperature dependence of the probe tip, high-vacuum instability, ion beam instability, and memory effects. The heart of the problem lies in delivering approximately 5 µl of solvent (usually containing a large percentage of water) directly to a high-vacuum system ill-equipped to handle it. First, jet expansion causes substantial cooling of the solvent between the capillary tubing and the high-vacuum region. When the solvent system contains water, freezing is likely to occur at the probe tip. Because the addition of glycerol reduces, but does not eliminate, such freezing, the probe tip is also heated to 40–50 °C. If too much heat is applied, however, all the solvent evaporates inside the capillary, leaving analyte to plug the exit hole. Second, the solvent tends to form a bead on the capillary probe tip; when it becomes too large, the bead either falls off or "explodes" into the high vacuum, creating large pressure fluctuations in the ion source that translate directly into ion beam instability. Finally, memory effects arising from incomplete consumption of the analyte at the probe tip lead to tailing in chromatographic peaks. These problems can be minimized by choosing the solvent system, flow rate, and probe temperature to maintain a constant evaporation rate at the probe tip. This is sometimes quite difficult, especially if gradient elution chromatography is required. Another solution is to attach a wick made of cellulose absorbent strips to the probe tip.[34] The more even evaporation produced by the wick has been demonstrated to allow stable beam operation for more than 9 hours.

Another disadvantage of the CF/FAB technique is often the poor quality of the total ion chromatogram (TIC). Below 1000 mu, the matrix contributes significantly to the total ion current, making it difficult to analyze compounds with molecular weights in that range. Above 1000 mu, however, background derived from the matrix is much smaller, and the TICs more closely approximate chromatograms obtained with an ultraviolet (UV) detector.

Despite such disadvantages, CF/FAB has found numerous applications and been developed in conjunction with three types of HPLC, employing regular-bore columns (4.6-mm i.d.), microbore columns

(1.0-mm i.d.), and capillary columns. Normal-bore columns in common use for chromatography do not require column conditions to be redefined when transferred to an LC/MS system; however, a splitting tee must be used to reduce column flow to a rate acceptable to the high-vacuum requirements of the mass spectrometer (approximately 5 µL/min). Splitting tees reduce the overall sensitivity of the system by at least a factor equal to the ratio of the total flow to the flow into the ion source of the mass spectrometer. At an HPLC flow rate of 1 mL/min and a flow rate of 5 µL/min into the ion source, sensitivity is reduced by a factor of 200. Using microbore columns with flow rates of 50–100 µL/min reduces this disadvantage greatly. Though a splitter is still required, operating a microbore HPLC at a flow rate of 50 µL/min yields only a tenfold reduction in sensitivity. To obtain full sensitivity, capillary columns must be used at flow rates of 1–5 µL/min, so that the total solvent flow from the column can enter the ion source of the mass spectrometer. Capillary HPLC also gives increased separation efficiency and the ability to analyze very small samples. Unfortunately, capillary HPLC has many limitations. The columns are not generally available, because there are no manufacturers of capillary HPLC columns in the United States and only one in Europe, and in-house preparation of columns requires specialized knowledge not generally available to chromatographers. Capillary HPLC columns must be kept scrupulously clean since they become plugged much more easily than do normal or microbore columns. The HPLC connections are also more difficult to make with capillary columns than with normal bore columns. Finally, capillary HPLC columns are sensitive to overloading. In spite of these limitations, capillary columns may become the method of choice for CF/FABMS because of sensitivity considerations.

3.2. Applications of the Continuous Flow FAB Interface

In its short history, CF/FABMS has found many applications in LC/MS, direct analysis of single compound samples, in capillary zone electrophoresis (CZE), enzyme kinetics, and MS/MS.

FABMS has been widely used to study peptides, but some large peptides exceed the attainable mass range. One method for extending that mass range involves performing proteolytic digestions followed by direct FABMS analysis. Hydrophobic proteolytic peptide fragments, however, tend to be more easily ionized by FABMS than hydrophilic peptides, so that usually only the more hydrophobic peptides are observed. This can be partially remedied by the laborious and difficult procedure of collecting reversed-phase HPLC fractions and obtaining FAB spectra of the fractions.

As an alternative, CF/FABMS has been demonstrated to be quicker and more efficient in analyzing proteolytic peptide fragments than collecting fractions for analysis. Peptides are often digested with trypsin to give cleavage at arginine and lysine, both basic amino acids. This guarantees that there will be at least one basic residue in every peptide fragment (with the possible exception of the C-terminal fragment), an important consideration because peptides containing basic residues tend to give stronger molecular ions by FABMS. Tryptic digestion followed by CF/FABMS has been used analyzing human apolipoprotein A-1[35] (a 245-amino acid peptide), β-lactoglobulin A,[32] sperm whale myoglobin,[36] and horse heart cytochrome c.[37] Proteolytic peptide digestion using dipeptidyl aminopeptidase (DAP-1) followed by CF/FABMS has been used in analyzing angiotensin 1 (a decapeptide), as well as an octapeptide,[38] and enkephalins.[39] The CF/FABMS has also been used as an LC/MS interface in analyzing compounds other than peptides, including oligosaccharides,[40] permethylated oligosaccharides,[41] monobactams,[42] bile acids,[43] and the polymeric mixtures Triton-X and PEG-600.[44]

As previously stated, CF/FAB spectra contain reduced chemical background, and they are often quite useful for analyzing single-component samples, since they show relatively more intense molecular and fragment ions for peptides, such as substance P and the oxidized B chain of bovine insulin.[6] Reduced chemical background allows significantly better FAB mass spectra, as for vitamin B12 coenzyme, cholesterol, and glycerides.[45]

The CF/FABMS interface is used in conjunction with MS/MS to obtain increased fragmentation from molecular ion species. However, the intensity of a signal arising from fragmentation depends on the absolute number of molecular ions produced in the FAB experiment, and molecular ions can be easily suppressed with probe FAB. Using a CF/FAB interface reduces suppression effects and increases the relative molecular ion intensity, as demonstrated for numerous peptides[46] and drugs of addiction. [47]

Yet another application of CF/FAB is in measuring the kinetics of enzyme reactions. Previously, samples of an analyte undergoing enzymatic degradation had to be removed from the reaction vessel and placed on a probe target containing a relatively large amount of glycerol. Using CF/FAB as the introduction technique allows the reaction of the enzyme on the substrate to continue at the probe tip until the moment of analysis. Furthermore, solvent flowing through the capillary may be essentially aqueous, as required for the enzyme to react with the substrate. This technique has been demonstrated with the hydrolysis of ribonuclease S peptide with carboxypeptidases Y and P.[48] Thus, both CF/FAB and moving belt LC/FAB appear to be useful in studying enzyme kinetics.

Closely related to enzyme kinetics is the emerging field of on-line analysis of biological fluids *in vivo*. Caprioli and Lin[49] measured physiological concentrations of sodium penicillin G in a live rat by inserting a microdialysis probe into the rat using a syringe pump to push the dialysate into a CF/FAB ion source. In this case, the dialysate was analyzed by using a tandem mass spectrometer operating in the negative-ion FAB mode essentially as a separation device to isolate the drug from biological fluids. The major fragment ion (m/z 192) was monitored as a function of time, yielding a time course blood level of sodium penicillin G in the rat. This new technique holds great promise for the *in vivo* analysis of compounds in physiological fluids.

Finally, CF/FABMS has been coupled with capillary zone electrophoresis (CZE) to increase greatly the number of theoretical plates (relative to HPLC) obtainable with charged species. The major difficulties here are that CZE uses high voltage and the technique is restricted to compounds that can be separated according to their charge. This coupling has been applied to the analysis of small peptides,[50, 51] mixtures of enkephalins,[52] and fragments of β-endorphin.[6]

4. PARTICLE BEAM/FAB

As already stated, one of the major difficulties in using CF/FAB is the required low-flow rate into the ion source of the mass spectrometer. A recently developed technique that allows total flow rates of 0.5–2.0 mL/min into the vacuum region of the mass spectrometer is particle beam/FAB. The interface (see Figure 8) involves nebulization followed by solute enrichment by means of momentum separation. Originally designed for use with EI, this technique is useful in the FAB mode as well.[8] With this interface, the HPLC eluent containing solvent, analyte, and matrix is passed through fused silica capillary tubing where the eluent is heated and droplets that form at the exit of the tubing are desolvated by nebulization. Solvent molecules are then pumped away by a two-stage momentum separator. The analyte and matrix pass into the high-vacuum region of the FAB ion source to strike the FAB target and undergo normal FAB ionization. No major disadvantages of this very new technique have come to light. Its major advantage is that the entire contents of a 4.6-mm (i.d.) column may enter the vacuum region with no decrease in sensitivity due to splitting, and sensitivity in the picogram region (for erythromycin A)[53, 54] has been demonstrated. A few directly injected samples have been analyzed using particle beam FAB, including the oxidized B-chain of bovine insulin, somatostatin, gramicidin S, substance P and its fragment 2-11, and α-solanine.[8, 53, 54]

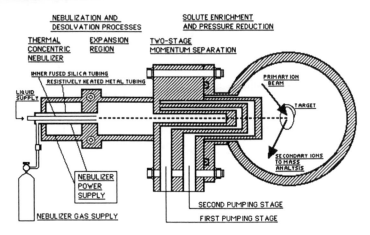

Figure 8. Schematic diagram of particle beam SIMS configuration. (Reprinted with permission from Reference 54.)

5. CONCLUSION

The many advances made since the introduction of the first LC/FABMS interface have allowed application to a wide range of solvents (e.g., 0–100% water) with improved sensitivity, more abundant fragment ion reduced chemical background, and less suppression due to the matrix and impurities. The interfaces remain difficult to handle (often due to stability problems), and the mass ranges of the three LC/FABMS techniques are limited inherently by the useful mass range of FABMS. In the future, FABMS may be extended to such compounds as oligosaccharides and peptides with molecular weights to *ca.* 10,000–20,000 mu, but electrospray ionization may become the method of choice beyond this range. Some classes of compounds are amenable to FAB and at least one other form of ionization (such as EI, CI, or thermospray); in these cases, ease of analysis will play a major role in the choice of technique. Nonetheless, online LC/FABMS techniques have become proven methods of analysis and will become increasingly useful with further development.

REFERENCES

1. Barber, M.; Bordoli, R. S.; Sedgwick, R. D.; and Tyler, A. N., *J. Chem. Soc. Chem. Commun.* 1981, 325–27.

2. Witten, J. L.; Schaffer, M. H.; O'Shea, M.; Cook, J. C.; Hemling, M. E.; and Rinehart, K. L., Jr., *Biochem. Biophys. Res. Commun.* 1984, **124**, 350–58.

3. Baldwin, M. A.; and McLafferty, F. W., *Org. Mass Spectrom.* 1973, **7**, 1111–12.

4. Dobberstein, P.; Korte, E.; Meyerhoff, G.; and Pesch, R., *Int. J. Mass Spectrom. Ion and Phys.* 1983, **46**, 185–88.

5. Ito, Y.; Takeuchi, T.; Ishii, D.; and Goto, M., *J. Chromatogr.* 1985, **346**, 161–66.

6. Caprioli, R. M.; Fan, T.; and Cottrell, J. S., *Anal. Chem.* 1986, **58**, 2949–54.

7. Caprioli, R. M., *Anal. Chem.* 1990, **62**, 477A–85A.

8. Sanders, P. E.; Willoughby, R. C.; and Mitrovich, S., *Proc. Thirty-Seventh ASMS Conf. Mass Spectrom. Allied Topics* (Miami Beach, 1989), pp. 110–11.

9. Yergey, A. L.; Edmonds, C. G.; Lewis, I. A. S.; and Vestal, M. L., *Liquid Chromatography/Mass Spectrometry. Techniques and Applications* (Plenum: New York, 1990).

10. Blakley, C. R.; Carmody, J. J.; and Vestal, M. L., *Anal. Chem.* 1980, **52**, 1636–41.

11. Willoughby, R. C.; and Browner, R. F., *Anal. Chem.* 1984, **56**, 2626–31.

12. Whitehouse, C. M.; Dreyer, R. N.; Yamashita, M.; and Fenn, J. B., *Anal. Chem.* 1985, **57**, 675–79.

13. Smith, R. D.; Loo, J. A.; Edmonds, C. G.; Barinaga, C. J.; and Udseth, H. R., *Anal. Chem.* 1990, **62**, 882–99.

14. Bruins, A. P.; Covey, T. R.; and Henion, J. D., *Anal. Chem.* 1987, **59**, 2642–46.

15. Hayes, M. J.; Lankmayer, E. P.; Vouros, P.; Karger, B. L.; and McGuire, J. M., *Anal. Chem.* 1983, **55**, 1745–52.

16. Benninghoven, A.; Eicke, A.; Junack, M.; Sichtermann, W.; Krizek, J.; and Peters, H., *Org. Mass Spectrom.* 1980, **15**, 459–62.

17. Smith, R. D.; Burger, J. E.; and Johnson, A. L., *Anal. Chem.* 1981, **53**, 1603–11.

18. Lewis, I. A. S.; and Brooks, P. W., *Proc. Thirty-First Ann. Conf. Mass Spectrom. Allied Topics* (Boston, 1983), pp. 850–51.

19. Stroh, J. G.; Cook, J. C.; and Rinehart, K. L., Jr., *Adv. Mass Spectrom. 1985* 1986, **10B**, 625–26.

20. Stroh, J. G.; Cook, J. C.; Milberg, R. M.; Brayton, L.; Kihara, T.; Huang, Z.; Rinehart, K. L., Jr.; and Lewis, I. A. S., *Anal. Chem.* 1985, **57**, 985–91.

21. Stroh, J. G., Ph.D. Diss., University of Illinois, Urbana, 1986; *Chem. Abstr.* 1987, **106**, 168128d.

22. Santikarn, S.; Her, G. R.; and Reinhold, V. N., *J. Carbohydr. Chem.* 1987, **6**, 141–54.

23. Stroh, J. G.; Rinehart, K. L., Jr.; Cook, J. C.; Kihara, T.; Suzuki, M.; and Arai, T., *J. Am. Chem. Soc.* 1986, **108**, 858–59.

24. Rinehart, K. L.; Kishore, V.; Nagarajan, S.; Lake, R. J.; Gloer, J. B.; Bozich, F. A.; Li, K-M.; Maleczka, R. E., Jr.; Todsen, W. L.; Munro, M. H. G.; Sullins, D. W.; and Sakai, R., *J. Am. Chem. Soc.* 1987, **109**, 6846–48.

25. Sakai, R.; and Rinehart, K. L., *197th ACS Nat. Mtg.* (Dallas, 1989); Abstract *ORGN* 171; Rinehart, K. L.; Sakai, R.; and Stroh, J. G., U.S. Patent Application Serial No. 335,903, 1989. '

26. Holt, T. G., Ph.D. Diss., University of Illinois, Urbana, 1986; *Chem. Abstr.* 1987, 193149u.

27. Rinehart, K. L.; Holt, T. G.; Fregeau, N. L.; Stroh, J. G.; Keifer, P. A.; Sun, F.; Li, L. H.; and Martin, D. G., *J. Org. Chem.* 1990, **55**, 4512-15.

28. Rinehart, K. L.; Holt, T. G.; Fregeau, N. L.; Keifer, P. A.; Wilson, G. R.; Perun, T. J., Jr.; Sakai, R.; Thompson, A. G.; Stroh, J. G.; Shield, L. S.; Seigler, D. S.; Li, L. H.; Martin, D. G.; Grimmelikhuijzen, C. J. P.; and Gäde, G., *J. Nat. Prod.* 1990, **53** (4), 771-792.

29. Smith, L. A.; and Caprioli, R. M., *Biomed. Mass Spectrom.* 1983, **10**, 98-102, Caprioli, R. M.; Smith, L. A.; and Beckner, C. F., *Int. J. Mass Spectrom. Ion and Phys.* 1983, **46**, 419-22.

30. Smith, L. A.; and Caprioli, R. M., *Biomed. Mass Spectrom.* 1984, **11**, 392-95.

31. Murawski, S. L., Ph.D. Diss., University of Illinois, Urbana, 1987; *Chem. Abstr.* 1987, **107**, 214022k.

32. Caprioli, R. M., *Trends Anal. Chem.* 1988, **7**, 328-33.

33. Moseley, M. A.; Deterding, L. J.; de Wit, J. S. M.; Tomer, K. B.; Kennedy, R. T.; Bragg, N.; and Jorgenson, J. W., *Anal. Chem.* 1989, **61**, 1577-84.

34. Shih, M-C.; Wang, T.-C. L.; and Markey, S. P., *Anal. Chem.* 1989, **61**, 2582-83.

35. Caprioli, R. M.; DaGue, B. B.; and Wilson, K., *J. Chromatogr. Sci.* 1988, **26**, 640-44.

36. Caprioli, R. M.; DaGue, B.; Fan, T.; and Moore, W. T., *Biochem. Biophys. Res. Commun.* 1987, **146**, 291-99.

37. Caprioli, R. M.; Moore, W. T.; DaGue, B.; and Martin, M., *J. Chromatogr.* 1988, **443**, 355-62.

38. Hutchinson, D. W.; Woolfitt, A. R.; and Ashcroft, A. E., *Org. Mass Spectrom.* 1987, **22**, 304-6.

39. Games, D. E.; Pleasance, S.; Ramsey, E. D.; and McDowall, M. A., *Biomed. Environ. Mass Spectrom.* 1988, **15**, 179-82.

40. Ito, Y.; Takeuchi, T.; Ishii, D.; Goto, M.; and Mizuno, T., *J. Chromatogr.* 1987, **391**, 296-02.

41. Boulenguer, P.; Leroy, Y.; Alonso, J. M.; Montreuil, J.; Ricart, G.; Colbert, C.; Duquet, D.; Dewaele, C.; and Fournet, B., *Anal. Biochem.* 1988, **168**, 164-70.

42. Ashcroft, A. E., *Org. Mass Spectrom.* 1987, **22**, 754-57.

43. Ito, Y.; Takeuchi, T.; Ishii, D.; Goto, M.; and Mizuno, T., *J. Chromatogr.* 1986, **358**, 201-7.

44. Takeuchi, T.; Watanabe, S.; Kondo, N.; Goto, M.; and Ishii, D., *Chromatogr.* 1988, **25**, 523-25.

45. Barber, M.; Tetler, L. W.; Bell, D.; Ashcroft, A. E.; Brown, R. S.; and Moore, C., *Org. Mass Spectrom.* 1987, **22**, 647-50.

46. Deterding, L. J.; Moseley, M. A.; Tomer, K. B.; and Jorgenson, J. W., *Anal. Chem.* 1989, **61**, 2504-11.

47. Seifert, W. E., Jr.; Ballatore, A.; and Caprioli, R. M., *Rapid Comm. Mass Spectrom.* 1989, **3**, 117-22.

48. Caprioli, R. M., *Biomed. Environ. Mass Spectrom.* 1988, **16**, 35-39.

49. Caprioli, R. M.; and Lin, S.-N., *Proc. Nat. Acad. Sci. USA* 1990, **87**, 240-43.

50. Moseley, M. A.; Deterding, L. J.; Tomer, K. B.; and Jorgenson, J. W., *Rapid Comm. Mass Spectrom.* 1989, **3**, 87-93.

51. Minard, R. D.; Chin-Fatt, D.; Curry, P., Jr.; and Ewing, A. G., *Proc. Thirty-Sixth ASMS Conf. Mass Spectrom. Allied Topics* (San Francisco, 1988), pp. 950-51.

52. Reinhoud, N. J.; Niessen, W. M. A.; Tjaden, U. R.; Gramberg, L. G.; Verheij, E. R.; and van der Greef, J., *Rapid Comm. Mass Spectrom.* 1989, **3**, 348-51.

53. Kirk, J. D.; and Browner, R. F., *Biomed. Environ. Mass Spectrom.* 1989, **18**, 355-57.

54. Sanders, P. E., *Rapid Comm. Mass Spectrom.* 1990, **4**, 123-24.

Index